T0327362

Fundamentals of Drug Development

Fundamentals of Drug Development

Jeffrey S. Barrett

Registered Office
John Wiley & Sons, Inc., 111 River Street, Hoboken, NJ 07030, USA

Editorial Office
111 River Street, Hoboken, NJ 07030, USA

For details of our global editorial offices, customer services, and more information about Wiley products visit us at www.wiley.com.

Wiley also publishes its books in a variety of electronic formats and by print-on-demand. Some content that appears in standard print versions of this book may not be available in other formats.

Library of Congress Cataloging-in-Publication Data

Names: Barrett, Jeffrey S., author.
Title: Fundamentals of drug development / Jeffrey S. Barrett.
Description: Hoboken, NJ : Wiley, 2022. | Includes bibliographical
    references and index.
Identifiers: LCCN 2022017493 (print) | LCCN 2022017494 (ebook) | ISBN
    9781119691693 (cloth) | ISBN 9781119691709 (adobe pdf) | ISBN
    9781119691730 (epub)
Subjects: MESH: Drug Development–methods | Technology,
    Pharmaceutical–methods | Pharmaceutical Research–methods
Classification: LCC RM301.25 (print) | LCC RM301.25 (ebook) | NLM QV 745
    | DDC 615.1072/4–dc23/eng/20220627
LC record available at https://lccn.loc.gov/2022017493
LC ebook record available at https://lccn.loc.gov/2022017494

Cover Design: Wiley
Cover Image: © pluie_r/Shutterstock

Set in 9.5/12.5pt STIXTwoText by Straive, Pondicherry, India

This book is dedicated to former students of the class who repeatedly called for a reference for the lectures. It is also dedicated to the many talented and rigorous scientists who have supported the pharmaceutical industry either directly or in a collaborative manner. It is great work that we have done and still continue to do. Ethical drug development happens every day on "the dark side." I continue to believe in the mission to advance science and bring new medicines to patients. I have been blessed with many enlightened friends and collaborators and am forever grateful to those who put pen to paper and help breathe life into this book. Finally, the work would not have been completed without the enduring support from my wife Ann, sons Kyle and Ryan and my parents Carmella and Frank.

Contents

Contributors Biographies

Jim Ottinger

Jim Ottinger is a lifelong pharmaceutical industry executive with more than 40 years of strategic regulatory experience, having supported the successful development, approval, and commercialization of a large number of new and generic drugs throughout his career. Mr. Ottinger is currently Executive Vice President of Regulatory and Quality at UroGen Pharma, a small biotech company. Prior to joining UroGen, Jim served as Senior Vice President of Global Regulatory Affairs at Teva Pharmaceutical Industries, Ltd., where he was responsible for global regulatory oversight of Teva's portfolio of branded, generic, and over-the-counter products.

Previously, Jim was Vice President of Worldwide Regulatory Affairs of Cephalon, Inc. (acquired by Teva). Earlier in his career, Jim held a variety of senior regulatory positions with Premier Research Group Limited and spent nearly 25 years in regulatory positions of increasing responsibility at Wyeth Research (acquired by Pfizer, Inc.). Jim holds a Bachelor of Science in Pharmacy from the Temple University School of Pharmacy and is a registered pharmacist in the State of Pennsylvania. He lives in North Wales, PA.

Robert Bell

Robert Bell, Ph.D. is President / Owner of Drug and Biotechnology Development LLC, a consultancy to the pharmaceutical industry and academia for biological, drug, and device development. Dr. Bell is a lifelong Gator with Orange and Blue running through his veins, receiving a BS in Chemistry, MS in Food Science and Human Nutrition, and PhD in Pharmaceutics from the University of Florida. His employment history includes Carter-Wallace, Inc., AL Pharma, UDL Laboratories, Inc., Somerset Pharmaceuticals, Inc., and Barr Laboratories, Inc. Dr. Bell is an Adjunct Professor of Pharmaceutics, National Advisory Board Member and recipient of the Distinguished Alumnus Award from the College of Pharmacy at the University of Florida and former Affiliate Faculty at the College of Pharmacy at

Virginia Commonwealth University. Dr. Bell has published and presented extensively and has been issued eleven patents with other patents pending. Research interests include novel formulation development and delivery technologies, pharmaceutical and biomedical analysis, development of vaccines, biosimilars, women's health products, oncology therapeutics, substance abuse, and addiction therapies, pharmaceutical quality, and green eco-friendly chemistry initiatives. Dr. Bell has served in various leadership capacities within the American Association of Pharmaceutical Scientists (AAPS) including Chair of the Analysis and Pharmaceutical Quality (APQ) section, Editorial Advisory Board for the Journal of Pharmaceutical and Biomedical Analysis, reviewer for AAPS PharmSciTech and AAPS Journal, 2006 National Biotechnology Conference Chair, AAPS Executive Council Member-At-Large (2009), and the 2010 Chair of the Americas for the joint Pharmaceutical Sciences World Congress / FIP / AAPS Annual Meeting and chaired the AAPS Blog Committee. Dr. Bell was a Member of the Council of Experts, General Chapters-Biological Analysis, the Joint Standards Subcommittee, and continues to Chair the Cell Banking Expert Panel for United States Pharmacopeia, participates with the FDA Pharmaceutical Science and Clinical Pharmacology Advisory Committee (Alt), Editorial Advisory Board Member for the *Journal of Chemical and Pharmaceutical Sciences* and *Enliven: Biosimilars and Bioavailability,* and Co-Editor of the book *Poorly Soluble Drugs: Dissolution and Drug Release.* Dr. Bell is a member of the American Society of Clinical Oncology, American Urology Association, American Chemical Society, American College of Clinical Pharmacology, Parenteral Drug Association, and the Visiting Scientist Program and serves on several corporate boards.

Zhaoling Meng

Dr. Zhaoling Meng is Associate Vice President, Global Head of Clinical Modeling and Evidence Integration in Data and Data Sciences, Sanofi R&D. Zhaoling has 20+ years of experience in drug development ranging from discovery/nonclinical, clinical development and statistical methodology, modeling, and simulation (M&S). She advocates and provides leadership to support and promote clinical trial modeling and simulation, quantitative decision-making, real-world evidence, AI/deep learning application, and other innovations across drug development stages and various disease areas. In recent two years, Zhaoling successfully adapts her love for travel to virtual travel as she promotes the use of virtual patients and trials at work.

Eileen (Doyle)Castranova

Eileen is a drug development scientist who applies clinical pharmacology and pharmacometrics experience to clinical program strategy and trial design in support of early to late phase programs for small and large molecules. She has contributed to numerous NDAs, BLAs, and other regulatory submissions. In addition to her work as a consultant with Certara, Inc., she is an adjunct professor at the University of Pennsylvania's Institute for Translational Medicine and Therapeutics, where she teaches an introduction to drug development course. When not at the computer, Eileen enjoys kayaking and sailing with her husband and stepdaughter.

John (Chengfeng) Zhuang

John Zhuang has more than 20 years of experience in the biopharma industry, working for a start-up company (Asieris Pharmaceuticals) as well as for multinational companies (Johnson & Johnson, Eli Lilly, and Procter & Gamble). His experience includes regulatory affairs and toxicology/preclinical safety assessment for new drugs and medical devices. He received his PhD in toxicology from the Massachusetts Institute of Technology (MIT) and BS in chemistry from the Zhongshan University of China. He enjoys ocean fishing and other outdoor activities.

Jenny Zhuang

Jenny Zhuang attended the University of Pittsburgh and received a BS in bioengineering. She currently works for Veeva Systems, a software company focusing on pharmaceutical and life science industry applications. Jenny has over 3 years of experience implementing R&D applications for life science companies, specifically focusing on regulatory information management systems for medical device companies, and also interned for several years as human factors engineer at ZOLL LifeVest. She currently resides in New York City with her beloved cat Bean.

Donna Humski

Donna has dedicated a 35-year career to continuous improvement in the supply chain from all of its functional disciplines. Over the course of her career, she has led Planning and Sourcing organizations for some of the most iconic brands in the consumer OTC health and beauty care businesses. Donna led Manufacturing and Deliver teams servicing global patients for OTC, pharmaceutical products, and medical devices. She has played a role in nearly two dozen SAP implementations, groundbreaking new product introductions and acquisitions, divestitures, and establishing strategic collaborations. A graduate of Drexel and The Pennsylvania State University, Donna focuses time to introduce students to the opportunities of supply chain careers. She is also a guest presenter on supply chain principles at various universities. When free from supply chain challenges, Donna enjoys hiking, photography, and constructing unique cakes for her family and friends.

Preface

This introductory text lays the foundation for explaining how pharmaceutical research is conducted and the context for how the pharmaceutical industry fits into the global healthcare environment. It begins with a brief review of the history of drug development and explains the phases of drug development in detail. The decision-making process, drug development milestones, and compound progression metrics are defined and explained with examples. Corporate structures are discussed, along with the role and function of the project team to facilitate development. The various contributors and sectors that participate in pharmaceutical research and development, in general, are identified as well as the emerging trends which are shaping the future of drug development. The various disciplines involved are highlighted along with an assessment of the complexity and risks associated across the various stages of development. Differences in the nature and scope of development programs due to the therapeutic area of interest as well as the associated costs and resources required are also explored via examples.

As the pharmaceutical industry is constantly evolving in response to advances in science and technology as well as in response to an overall global healthcare economy that seeks to provide treatments to patients faster and cheaper, this text also provides some forward-looking analyses on current industry trends and economic indicators. In addition, attrition rates are discussed from both a historical perspective and in response to regulatory guidance, which has sought to improve this situation. It is clear that the era of the blockbuster drug has ended, and the sustainability of that approach is unlikely. Out of necessity, paradigms for drug development must likewise change for the benefit of both the industry and the patients that are depending on the industry to advance cures for a myriad of diseases.

This text was born out of a course initially directed and taught by Dr. Jeff Barrett. The course, Introduction to Drug Development (REG 612), has been a prerequisite in the Regulatory Track curriculum of the Masters in Translational Research Degree program offered in the Perelman School of Medicine at the University of

Pennsylvania (https://www.itmat.upenn.edu/itmat/education-and-training/itmat-education-courses.html). Elements of the course materials have also been taught at the University of Michigan as part of their Pharmaceutical Sciences 101 course, "From Molecules to Drugs and Drug Products," taught in the College of Pharmacy. After the third offering of the course, it became apparent that a text would supplement the lectures and provide a cohesive backbone for the subject matter. Indeed, many of the chapters coincide with the various course lectures though additional topics are provided in the text. At the conclusion of this course, students are expected to have a working knowledge of the drug development process, understand the regulatory basis by which new chemical entities are evaluated and ultimately approved, and appreciate the time and expense of drug development.

Dr. Barrett has over 30 years of experience in pharmaceutical research and development experience, 13 years of which were spent in the pharmaceutical industry from 1990 to 2003, followed by over 10 years as a Professor of Pediatrics at the University of Pennsylvania and The Children's Hospital of Philadelphia (2001–2013) and then back to the pharmaceutical industry at Sanofi Pharmaceuticals from 2013 to 2017 while still serving as an adjunct faculty member at the University of Pennsylvania where he continued to direct and teach the course. Dr. Barrett has also worked in the non-profit, global health sector for the Bill & Melinda Gates Medical Research Institute (MRI), where he led Quantitative Sciences from 2018 to 2020. In this capacity, he implemented a model-based drug development strategy employing PK/PD modeling, statistics, and clinical trial simulations to advance the discovery and development of new medicines and vaccines to treat malaria, tuberculosis (TB), and enteric diarrheal disease (EDD). Since 2020, Dr. Barrett has been working at the Critical Path Institute (https://c-path.org/about/people), where he led the Rare Disease Cures Accelerator – Data Analytics Platform (RDCA-DAP; https://c-path.org/programs/rdca-dap) development as a Senior Vice-President of the Institute. At present, Dr. Barrett is the Chief Science Officer at Aridhia Development Environment (https://www.aridhia.com/) promoting life science partners to collaborate, access and share secure data to deliver better patient outcomes.

Dr. Barrett's lectures are filled with anecdotes from industry experience, and he shares numerous examples from personal interactions. In all, Dr. Barrett has authored 18 chapters in this textbook. Additionally, invited contributors with targeted experience in drug development have further enhanced the text. Dr. Robert Bell has authored chapters on Formulations (Chapter 15) and Generic Drugs and the Generic Industry (Chapter 22). Dr. Zhaoling Meng is an accomplished statistician with experience in all phases of development and has authored the chapter on Phase 3 drug development (Chapter 9). Mrs. Donna Humski has held many positions in commercial operations including being a plant manager at J&J, and has authored the chapter on Distribution and the Supply Chain (Chapter 20).

Dr. John (Chengfeng) Zhaung co-authored the chapter on medical devices together with his daughter Jenny Zhaung (Chapter 19). Both have experience in this arena, and Dr. Zhaung has submitted several medical device NDAs for approval as part of his industry tenure. Dr. Eileen (Doyle) Castronova is the current instructor for the University of Pennsylvania's Introduction to Drug Development course and has authored chapters on Regulatory Milestones and the Submission Process (Chapter 13) and Attrition Rates and Evolving Corporate Strategies (Chapter 18). Dr. James Ottinger has over 40 years of experience in regulatory science and has authored the chapter on the Generic Drug approval process (Chapter 23). The textbook provides a more comprehensive perspective for the individual topics and is extensively referenced so that the student /reader can easily identify additional resources for further reading. An extensive glossary is provided, along with self-assessments and quiz questions for each chapter.

About the Companion Website

This book is accompanied by a companion website:

www.wiley.com/go/Barrett/FundamentalsDrugDevelopment

The website includes data sets and computer code.

Introduction

Drug development is the process by which new chemical entities are discovered, studied in laboratory and preclinical experiments, and investigated clinically in healthy volunteers and patients to determine if they are safe and efficacious. Assuming the compound under investigation passes rigorously defined milestones, submission of documentation to regulatory authorities (e.g. US FDA) can ensue and, pending a favorable review, market access can be granted. The process is highly regulated, and there is significant risk and cost involved for pharmaceutical sponsors to research and develop drugs, with the entire process averaging around 12 years once a product is discovered. Current estimates of the cost of developing a single new chemical entity through all phases of development suggest an average cost of greater than $2.6 billion USD (Tufts CSDD 2019). Once market access is granted, the sponsor company proceeds with the manufacture, packing, labeling, distribution, sales, and marketing of the commercial drug product. The further expansion of the product development continues through life cycle management via the development of new formulations, the clinical evaluation of additional patient populations, and the pursuit of entirely new indications.

Current drug development paradigms span several broad phases, which are loosely tied to milestones that confirm the positive progression criteria for each phase. The phases are generally defined as discovery, Phase 1, Phase 2, Phase 3, Phase 4 (post-marketing), and commercial development (sales, marketing, life cycle management, etc.). These individual phases will be defined and explained in detail in subsequent chapters. While they appear to represent distinct stages, they are often somewhat vague with respect to duration and content.

There are many layers of uncertainty that contribute to the assessment of risk for continuing to invest in compounds of interest. Many scientists representing

Fundamentals of Drug Development, First Edition. Edited by Jeffrey S. Barrett.
© 2022 John Wiley & Sons, Inc. Published 2022 by John Wiley & Sons, Inc.
Companion website: www.wiley.com/go/Barrett/FundamentalsDrugDevelopment

a multi-disciplinary workforce contribute to the research and development of new molecular entities (NMEs), devices, and generic drugs. Together they attempt to remove uncertainty by performing targeted experiments, developing models that describe the biology of the drug target and the drug characteristics, and discussing and evaluating key assumptions upon which these models are developed. Experimentation, modeling, and analysis with periodic review and decision making define the basic cyclic sub-process within each phase. Despite the rigor, we sometimes fail in our attempt to bring new medicines to market. Failures occur at every phase and are not solely linked to R&D shortcomings. In many instances, there is a disconnect between R&D and commercial assessment and planning, leading to manufacturing inadequacies, poor market evaluation, and product launches, or an inability to appropriately promote the product or otherwise ensure clinical utilization and ultimately product performance. The cost of drug development is very much influenced by the failures of development compounds that never make it to market. The current landscape for the pharmaceutical industry provides little margin for error for pharmaceutical sponsors making the cliché "fail quick" a metric for abandoning the development of compound candidates which do not provide an adequate probability of success.

The global healthcare economy also represents a dynamic and highly volatile environment on which drug development must operate and is only one of the multiple sectors which influence this arena. Efforts to control prices are commonly scrutinized by the medical, insurance, and political stakeholders and are often the subject of much criticism given the historical profitability of the pharmaceutical industry. Related current issues such as the payer's influence on market access, the necessity of reducing confusion to patients and prescribers, and the overall impact of these factors on choice reduction contribute to the health economic complexity and the potential for politicizing the landscape.

It is incumbent upon those that participate in drug development in one capacity or another to maintain high standards in our scientific rigor and decision making and work with regulators to ensure that the medicines we help to discover, produce, and recommend to patients can be safely and effectively administered with confidence. When new information regarding these products emerges or is otherwise discovered, we must act quickly to resolve or remedy the situation and remove the product from the market if warranted. There is an implicit trust that must be expected and maintained between patients, caregivers, and drug manufacturers for this industry to survive and sustain itself. Given the risk and ever-increasing costs, drug manufacturers must continue to expand the science without sacrificing rigor and quality with the hope of addressing current unmet medical needs in a timely manner. More specifically, the industry must learn to work together in a broader ecosystem than in the past and collaborate outside the past boundaries of an individual company's intellectual property. Future success in drug development will require sharing and collaboration beyond what was the past and current practice.

Reference

Tufts Center for the Study of Drug Development (2019). Press Release: Tufts CSDD Impact Report July/August 2019, Vol. 21 No. 4, https://csdd.tufts.edu/impact-reports.

1

The History of Drug Development

Jeffrey S. Barrett

Aridhia Digital Research Environment

Introduction

The history of drug development spans more than 10 centuries and is essentially coincident with the history of pharmacy although the modern era is much shorter. When we speak of drug development, we are typically referring to the development of small molecules as that constitutes a historical baseline for the most part but our scope in this text will also examine generic drugs, biologics, and therapeutic proteins as well as vaccines.

Most drug development milestones have occurred over the last hundred years consistent with the formal establishment of the pharmaceutical industry and the global regulatory community. However, we cannot consider drug development to have occurred from a completely undefined origin and there is indeed a prehistory to be defined and explained in order to appreciate the setting for modern drug development. There is an undeniable human behavior associated with the recognition of illness and the interest in aiding our fellow human being to improve our survival. We are very comfortable with the common cliché, "that's what separates us from the animals," but this may not be an entirely legitimate claim as we will see. A more poignant and relevant point of view is the appreciation for ethical drug development as we attempt to bring new medicines to market in an ever-competitive and complex healthcare environment.

The history of drug development also coincides with the history of regulatory oversight. Quite often the recognition of the need for regulation and oversight came on the heels of a tragedy in which a lack of understanding, negligence, or simple greed instigated an event that was ultimately dangerous for patients. In today's world, we are inundated with details regarding a new medicine's defendable attributes derived from years of research. In many cases, the prehistory of regulatory oversight also coincided with the marketing of non-pharmaceutical entities in which medical benefit was associated (see Figure 1.1

Fundamentals of Drug Development, First Edition. Edited by Jeffrey S. Barrett.
© 2022 John Wiley & Sons, Inc. Published 2022 by John Wiley & Sons, Inc.
Companion website: www.wiley.com/go/Barrett/FundamentalsDrugDevelopment

Duffy's Pure Malt Whiskey
for Medicinal Use. Sold by all Druggists, and bears
the highest endorsement of the leading chemists and medical authorities of the Country.

Figure 1.1 Advertisement from Duffy's Malt Whisky (circa 1880s) promoting alcohol for medicinal use.

Duffy Malt Whiskey Co.
ROCHESTER, N. Y.

For Sale by

SMITH, KLINE & FRENCH CO.
PHILADELPHIA, PA.

touting the presumed but unsubstantiated health benefits of malt whiskey by Whiskey distiller and manufacturer noted pharmaceutical manufacturer and reseller, Smith Kline and French – circa 1880s). It was commonplace during this time for alcohol distillers such as Duffy to make false claims like a tonic that "Makes the Weak Strong." Likewise, products promoted as patent medicines often contained alcohol, codeine, cocaine, and other opiates. The rapid expansion of glass bottle manufacturing via mechanical means helped fuel this expansion and promotion of medicines containing alcohol, a sign of the times in an era without regulation.

While the modern pharmaceutical marketer tries to encourage potential patients that they might benefit from drug or therapy, there is also a staggering amount of detail shared regarding the safety of the drug and the potential for anticipated or unanticipated, though perhaps rare side effects. It should also be appreciated that given the nature of the data content, such information is not always properly interpreted or completely objective. We must again remind ourselves of the historical perspective, however, as this was certainly not the case prior to the late 1990s. Shown next are two ads from pharmaceutical products sold without evidence of safety or efficacy (see Figure 1.2).

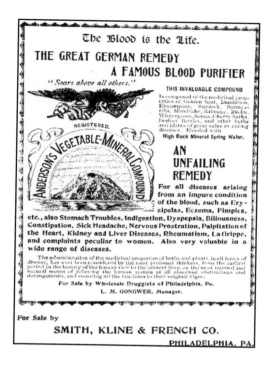

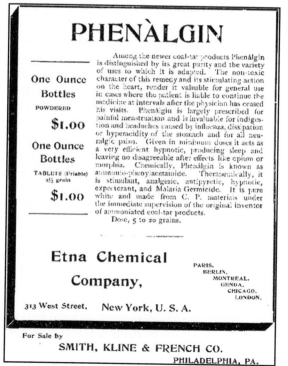

Figure 1.2 Early pharmaceutical advertisements making unsubstantiated medical claims before regulated drug development and FDA oversight.

The objectives of this chapter are to describe the origins of pharmacy and humankind's quest to medicate. This necessity has roots in the natural world as we will describe and was the impetus behind traditional medicines and the evolution of scientific practices that sowed the seeds of drug development. With technological advances came the ability and necessity of manufacturing and distributing goods and services to a global economy. Some discussion of early pharmacies that in many cases became early pharmaceutical companies is provided along with the timeline and necessity of regulatory oversight in response to the tragedy in many cases. Finally, we will touch on the modern era and how the age of mergers and acquisition and the evolution of healthcare has affected the industry.

Wild Health and Nature's Pharmacy

A few years ago, I stumbled upon Cindy Engle's book, "Wild Health" (Engle 2003). It is the first book dedicated to the field of zoopharmacognosy, or animal self-medication. The parallels to clinical pharmacology are many and, while the field is still in its infancy, it will likely borrow concepts well established from more mature science focused on human and domesticated animal health. Engle's fundamental premise is "when wild animals are free to range over undisturbed habitat, not exposed to high levels of pollution and not exposed to extremes of environmental change, they are generally in good health. They live within an ecosystem to which their physiology and behavior are, by virtue of their very survival, well adapted (Wild Health, p. 13)." Perhaps not surprisingly, many wild animals when brought into captivity experience poor and declining health and often die within a short time period. Likewise, it can easily be shown that the health of their immune system is intimately linked with their behavior in a particular environment. If we draw a parallel to the human condition, we may conclude that the various epidemics that have plagued our species over time were largely influenced by migration, exploration, climate extremes, and perhaps exposure to various environmental toxins (see Jared Diamond's Gun, Germs, and Steel). Not to be excluded is the dynamic nature of the human, particularly the Western diet as our lifestyles have advanced and we struggle to feed an ever-growing and dispersed population. The healthcare industry, as we will see, has evolved primarily to help us cope with disease conditions born out of this developing human environment that is far away from where we began.

One of the more fascinating aspects of zoopharmacognosy is the evidence of self-medicating behavior of wild animals. As Engle is quick to point out, we don't need to overinterpret the knowledge regarding the behavior (i.e. assume that we can learn something about medical treatments from animals). The salient point is that these behaviors are associated with improved survival and consistent with attributes favoring natural selection (Costa-Neto 2012). Fundamentally, this is what we try to achieve with the development of new drugs. Table 1.1 provides some examples of self-medicating behaviors observed in the wild. The diversity, purpose, and species represented in this table highlight the fact that these behaviors are indeed a part of successful strategies to combat seasonal or other external intrusions to the environment. For example, as food supply changes with the seasons, animals and insects change their diet to maintain their nutritional needs (e.g. fallow deer change their grazing habits from grass to fruits and/or nuts to brambles, ivy, and holly). There is a constant adaption of diet and other behaviors based on changing circumstances. These behaviors are taught or otherwise passed on to future generations as part of such strategies and those that do not employ them are less likely to thrive.

Likewise, these species have been able to inventory their surroundings and through empirical means determine the source of their medication and the specific location and perhaps season of the year when the timing or yield is appropriate for consumption. Nature's pharmacy is replete with antibacterial, antifungal, and antiviral compounds. While observing their effects on animals is difficult, it is clear that animals gain benefit from consuming natural antimicrobials. More impressive is that animals seemingly learn from each other and pass on these behaviors to subsequent generations.

While the list of example behaviors across species shown in Table 1.1 is impressive, higher order species definitely seem to embrace self-medication in a directed manner. Primates are particularly good at exploiting the medicinal properties of plants. Chimpanzees, bonobos, and gorillas have all figured out that swallowing rough leaves can purge their intestines of parasites. Chimps plagued by roundworm infections have been known to eat plants with anti-parasitic properties, despite their bitter flavor and lack of nutritional value. They also seem to empirically get the dose right. It is not known if there are common practices in dosing of such remedies that adjust for age and size differences, but individual animals do seem to "personalize" their treatment.

Humans too have long benefited from the compounds extracted from natural sources, and many of these have made their way into today's formularies and pharmacies. Table 1.2 lists some of our more well-known plant-derived compounds.

Table 1.2 is of course biased by the modern pharmacopeia, but there also exists a long history of natural remedies and holistic approaches used by ancient and modern human cultures across many geographic regions.

Table 1.1 Examples of behaviors associated with self-medicating and health maintenance in nature.

Species	Behavior	Benefit
Moose, deer, and caribou	Chew on cast antlers to replace calcium and phosphorous loss during antler growth (400 grams of antler tissue/day in the moose).	Prevents osteoporosis.
Free-ranging cattle in Venezuela and Gorillas in Rwanda	Based on seasonal changes in diet, cattle dig out and eat clay subsoils and gorillas mine and eat volcanic rock – both are forms of geophagy.	Acting as an antacid (clay) or bacterial adsorbent (volcanic rock), diarrhea is reduced, and fluid is retained.
Blue tits (*Parus caeruleus*)	Males bring empty snail shells to their nest for females to consume during egg-laying season.	Supplements mineral deficiency during pregnancy.
Worker honeybees[a]	Bees remove the cell cap and carry disease brood from the nest.	Reduces the spread of infection
Chimpanzees	Chimpanzees with upset stomachs seek and consume (roll in their mouth without chewing, then swallow) the whole leaves of the Aspilia plant.	The leaves contain thiarubrine-A, a chemical active against intestinal nematode parasites.
North American brown bears (*Ursos arctos*)	Bears make a paste of Osha roots (*Ligusticum porteri*) and saliva and rub it through their fur to repel insects or soothe bites.	This plant (known as "bear root") contains 105 active compounds (e.g. coumarins) that may repel insects when topically applied.

[a] There are honeybee workers that contain a "hygiene gene" that enables them to detect disease brood in larval or pupal stage of development.

Traditional Medicine, Traditional Remedies

Most of the world's cultures share a history of natural remedies that have either with the times and become embedded in current holistic medical paradigms, been determined to be unsafe or ineffective and abandoned, or been consumed into more modern clinical practices. Most are based on the belief that health and wellness depend on a delicate balance between the mind, body, and spirit. Table 1.3

Table 1.2 Examples of available resources from nature's pharmacy.

Natural product	Chemical constituents	Ailments/treatments
Willow bark	Salicin (origin of the analgesic aspirin)	Pain relief
Pine and juniper trees	Camphene and pinene	Antimicrobial
Pacific Yew (*Taxus brevifolia*)	Diterpenes (e.g. Taxol)	Anticancer (solid and leukemia-like cancers)
Chinchoa tree bark	Alkaloids (quinine)	Antipyretic, antimalarial, analgesic, and anti-inflammatory properties
Foxglove plant (*Digitalis lanata*)	Digoxin	Treatment of various heart conditions (atrial fibrillation, flutter, and heart failure)
Calabar bean	Physostigmine	Reversible cholinesterase inhibitor used to treat glaucoma, Alzheimer's disease, and delayed gastric emptying
Bark of the South American plant *Chondrodendron tomentosum*	d-Tubocurarine	Toxic alkaloid historically used as an arrow poison; evaluated clinically in conjunction with an anesthetic to provide skeletal muscle relaxation during surgery or mechanical ventilation
Leaves of tropical South American shrubs from the genus *Pilocarpus*	Pilocarpine	Non-selective muscarinic receptor agonist used to treat dry mouth (xerostomia), particularly in Sjögren's syndrome, but also as a side effect of radiation therapy for head and neck cancer.
Obtained from the plant *Ephedra sinica* and other members of the *Ephedra* genus	Ephedrine	Sympathomimetic amine and substituted amphetamine commonly used as a stimulant, concentration aid, decongestant, appetite suppressant, and to treat hypotension associated with anesthesia.

provides a comprehensive though certainly not complete list of various traditional medicines that have been used to support their local cultures prior to the adoption of more modern practices. As we will discuss, many of these are still in use. There is a great effort at present to consider the integration of these approaches with more modern medical treatments as healthcare considers a shift in emphasis from treatment to prevention strategies. This is especially true given the greater

Table 1.3 Examples of traditional medicine cultures.

Common names	Region/culture	Time of influence
Traditional Chinese medicine	China/Chinese and neighboring cultures	Origins predate the Shang Dynasty (1766–1122 BCE); the practice of TCM continues today though more commonly integrated with Western medical approaches.
Ayurvedic medicine	Indian/Hindu	Origins attributed to Atharva Veda (Vedic Sanskrit text ~ 20 books). From the sixth century BCE to the seventh century CE systematic development of science (Samhita period); classical works produced with evidence of organized medical care. Lingering practice.
Arabic Indigenous medicine	Eastern Europe and Middle East/Unani, ancient Greek, and Arabic contributions	Developed over the Islamic Golden Age (eighth to thirteenth century CE). Medieval Islam developed hospitals and expanded the practice of surgery.
Aboriginal bush medicine	Aboriginal Australia	Origin was several thousand years ago though not documented; declines today as knowledge is passed through singing and dancing rituals, less commonly practiced.
Traditional Celtic medicine	Central European tribes settling in Western Europe, Britain, and Ireland	Early tribes identified by 300 ADE. "Leeches" (Gaelic "lighiche" = physician) provided medical craft for clan elders. Wise woman or healers versed in herbal remedies and charms took care of the common folk. Some attempts to revive traditions today.
Native American medicine	Over 2000 tribes representing indigenous people in North America	Native medicine may be as old as 40000 years. No written language, so no documentation of Native American medicine until Europeans arrived 500 years ago. The medicine man, woman, or healer secures the help of the spirit world for the benefit of the community or an individual. Services included herbal medicine, bone-setting, midwifery, and counseling.

African herbal folk remedies, traditional African medicine	Throughout African continent	The Ebers papyrus (Egypt, 1500 BCE) is the earliest surviving record of medicinal plants though TAM predates. Traditional African practitioners a vital part of current healthcare system. About 60–80% of the African population relies on traditional remedies. Some attempts to combine traditional methods with homeopathy, iridology, and other Western healing methods, including some traditional Asian medicine.
Japanese Kampo medicine	Japan (via China originally)	Originally a Chinese tradition (221–210 BCE); first Chinese emperor sent emissaries to find the herb of immortality. In 701 CE the Taiho Code established a ministry of health. Empress Komyo (701–760) CE established a dispensary system to supply free medicine to the needy in 730 CE. Today in Japan, Kampo is integrated into the national healthcare system.

appreciation for holistic approaches on maintaining health homeostasis. What is eerily similar to the discussion on wild health is the focus of the historical traditional medicine approaches on utilizing knowledge from the local environment and the reliance on the various natural remedies from local resources common to the geographic region of origin. While this should not surprise us, it reinforces the notion of natural selection and the priority placed on strategies that favor survival. It is also fascinating to appreciate how early human species scavenged the landscape, evaluated empirically the available natural resources, and produced recipes for various treatment modalities. Comparing these early treatments and formulations across cultures, regions, and time periods gives us a sense of the early appreciation for both basic and clinical pharmacology. It also fundamentally provides a glimpse of how the medical profession evolved and how trust in the knowledge of early caregivers was established.

While a comprehensive review of traditional medicine is outside the scope of this chapter, we will look closer at traditional Chinese medicine (TCM) and Ayurvedic medicine as both are not only still being used to treat their countries/ cultures of origin but also the subject of recent interest in integrated (with modern medicine) approaches. Likewise, the continued funding of the National Center for Complementary and Alternative Medicine (NCCAM) and now the National Center for Complementary and Integrative Health (NCCIH, https://nccih.nih. gov/) through the National Institute of Health (NIH) suggests that there is more openness to the possibility of re-evaluating the science behind many of these approaches where clinical practice preceded more rigorous testing.

Traditional Chinese Medicine

TCM is a combination of traditional practices and beliefs developed over thousands of years in China. Common TCM practices include herbal medicine, acupuncture, massage (Tui na), exercise (qigong), and dietary therapy. The practices are based on belief in an energy source called *qi*, considerations of Chinese astrology and numerology, traditional use of herbs and other substances found in China, a belief that a map of the body is contained on the tongue that reflects changes in the body, and a model of the anatomy and physiology of internal organs.

Modern TCM is endorsed by both the industry itself and the government. The current version of the Pharmacopoeia of the People's Republic of China (10th edition, 2015) contains 4 volumes, one of which is entirely dedicated to TCM. It should be noted that the research and development of TCM in addition to its manufacturing and distribution is managed very differently than its more modern counterpart – the pharmaceutical industry. Academic researchers and their institutions have a much more intimate collaboration with the commercial TCM partners, and the separation between academic and commercial interests is not nearly as formalized.

During a teaching visit at several universities in China in 2009, I had the good fortune to observe the management of a TCM formulary and discuss the current climate for the practice of TCM in China (see Figure 1.3). It is quite clear that the strong tradition of TCM in China continues to be supported as part of a comprehensive healthcare solution, but also the evaluation of potential synergies of TCM with Western medical approaches is an exciting new frontier. One could argue that integrated, more holistic approaches are being more commonly considered globally, but indeed the point of reference is very different.

Despite the interest in integrated approaches, problems remain in order to reconcile TCM, which traditionally values empiricism and holistic philosophy, with the Western approach to disease treatment, such as a reproducible standardization of herbs using quantifiable lead compounds (biologically active ingredients), the frequent lack of rigorous stratification of patients or absence of a double-blind, randomized, placebo-controlled clinical trial design (Zhang and Schuppan 2014). Mechanistic preclinical validation of TCM drugs is still in its infancy but offers a potential bridge for target validation and dose selection. Some TCM drugs have been accused of negligent safety evaluation, based on case reports of toxicity, largely due to contamination with heavy metals or toxic

Figure 1.3 Storage of TCM ingredients at the Traditional Chinese Medicine Hospital pharmacy in Beijing, China. Compounding and additional preparations are also managed in the pharmacy, and the major route of administration for most prescriptions is via the oral route. Tea consumption remains the primary formulation vehicle.

alkaloids that erodes prescriber and patient confidence within and external to China. The Chinese government has initiated a national safety plan in 2011, investing in the modernization of TCM ($100 million from the National Natural Science Foundation in 2012), to promote research on lead compound identification and mechanisms of action, on a better standardization and well-controlled clinical trials that is essential if TCM is to move into the modern era.

The Chinese government strongly endorses TCM as part of the country's heritage as well as an integral part of its healthcare system. In the past, Communist Party Chairman Mao Zedong, in response to the lack of modern medical practitioners at that time, revived acupuncture and its theory was rewritten to adhere to the political, economic, and logistic necessities of providing for the medical needs of China's population. In the 1950s the "history" and theory of TCM was rewritten as communist propaganda, at Mao's insistence, to correct the supposed "bourgeois thought of Western doctors of medicine." More recently however a revamped Chinese FDA has made great strides in improving the efficiency of its regulatory reviews while still coping with the large volume of new TCM submissions illustrating that the TCM industry is alive and well. As part of the investment in the continued support for TCM, TCM teaching hospitals (Figure 1.4) are still a part of the landscape and the education of future TCM practitioners is seen as necessary to both maintain the practice and ensure that integrated approaches are not overly weighted toward Western medicine.

Ayurvedic Medicine

Indian Ayurvedic medicine includes a belief that the spiritual balance of mind influences disease. The Vedas (oldest Indian literature, 5000–1000 BCE) contain references regarding plants and natural resources for various treatments. There are three basic principles of Ayurveda: Roga vigyan, Vikritivigyan (the science of disease process), and Chikitsa vigyan (various therapeutic modalities). The pathological processes are described as Panchanidana (five etiological factors): these are Nidana (cause), Purvarupa (premonitory symptoms), Rupa (symptomatology), Upa-saya (therapeutic measures), and Samprapti (pathogenesis) (Mukherjee et al. 2017).

Ayurvedic medicine is a traditional medicine of India with a long history that has spread throughout the world based on empiric results and through globalization. Ayurveda believes in the existence of three elemental substances, the doshas (called Vata, Pitta, and Kapha), and states that a balance of the doshas results in health, while imbalance results in disease. Disease-inducing imbalances are believed to be able to be adjusted and balanced using traditional herbs, minerals, and heavy metals. Ayurveda stresses the use of plant-based medicines and

Figure 1.4 The necessity of creating teaching hospital settings is greatly appreciated by TCM practitioners. The First Clinical Medical College provides didactic and clinical training for young physicians trained in TCM approaches under the academic administration of the Beijing University of Chinese Medicine.

treatments, with some animal products, and added minerals, including sulfur, arsenic, lead, and copper sulfate. Ayurveda is focused on treating the patient and not the disease alone. This system emphasizes the uniqueness of each person regarding bio-identity, socioeconomic status, and biochemical and physiological conditions, which may promote illness.

Safety concerns have been raised about Ayurveda, with two US studies finding about 20% of Ayurvedic Indian-manufactured patent medicines contained toxic levels of heavy metals such as lead, mercury, and arsenic. Other concerns include the use of herbs containing toxic compounds and the lack of quality control in Ayurvedic facilities. Incidents of heavy metal poisoning have been attributed to the use of these compounds in the United States.

Obviously, these concerns are problematic for modern-day supporters of Ayurvedic medicine including researchers, practitioners, and their patients. The necessity of modernizing the R&D supporting Ayurveda, manufacturing processes, and the regulatory review of Ayurvedic products either alone or in combination with Western medicines is well appreciated as with TCM, and there are evolving strategies (Mukherjee et al. 2017) to address current gaps and limitations assuming Ayurveda will have a place in the evolving integrated medicine strategies of the future.

Alchemists, Chemists, and Pharmacists

As the practice of traditional medicine became more ingrained in the caregiver support provided by many early cultures, the practice of collecting, extracting, and formulating therapies to be administered to patients evolved and became more standardized. The training and expertise of early caregivers and medical practitioners varied dramatically and most of these individuals were engaged in some sort of apprenticeship prior to being entrusted with the regular duties. Various disciplines were borne out of this role and were critical to the development of more modern sciences that now support the pharmaceutical industry (e.g. pharmacists and physicians). Alchemy can be defined as a protoscience that contributed to the development of modern chemistry and medicine. While we tend to focus on some of the more ambitious goals of alchemy – the creation of the philosopher's stone, the ability to transmute base metals into the noble metals, and development of an elixir of life, early alchemists also developed basic laboratory techniques, theory, terminology, and experimental methods, much of which are still in use today. The alchemists, obsessed with secrecy, deliberately described their experiments in metaphorical terms laden with obscure references to mythology and history, which is why their contributions to the history of science and specifically drug development are often minimized.

The origins of alchemy are difficult to track down. In the East, in India and China, alchemy started sometime before the Common Era (CE) with meditation and medicine designed to purify the spirit and body and to thereby achieve immortality. In the West, alchemy probably evolved from Egyptian metallurgy as far back as the fourth millennium BCE. The ideas of Aristotle (384–322 BCE), who proposed that all matter was composed of the four "elements" – earth, air, fire, and water – began to influence alchemical practices when his student Alexander the Great (356–323 BCE) established Alexandria as a center of learning.

Not well appreciated is the time and geographic spans that alchemy presided over, essentially four millennia and three continents. With this broad familiarity came certain credibility eventually with high profile members of the religious community (Pope Innocent III, Martin Luther, etc.) being at least casual practitioners. Alchemy likewise is intimately linked with the origins of chemistry, which shares some overlap in time period. Disconnects, particularly around the emerging science of metallurgy, drove a wedge between alchemists and early chemists. Fundamentally, however, though alchemy played a significant role in the development of early modern science; its inclusion of Hermetic principles and practices related to mythology, magic, religion, and spirituality limited its utility and importance as more data-driven sciences evolved.

Chemistry is considered to have become an established science with the work of Antoine Lavoisier (1743–1794), who developed a law of conservation of mass that demanded careful measurement and quantitative observations of chemical phenomena. He is generally considered the driving force behind changing chemistry from a qualitative to a quantitative science and for discovering the role of oxygen in combustion. Likewise, with French chemists Louis-Bernard Guyton, Claude Louis Berthollet, and Antoine Francois, Lavoisier published a work titled "Méthode de nomenclature chimique" (Method of Chemical Nomenclature) in 1787. This was the first proper system of chemical nomenclature, i.e. a system of names describing the structure of chemical compounds, essential work for the developing medicinal chemistry that supports the pharmaceutical industry today, and patent law that protects pharmaceutical sponsors' intellectual property.

The discipline of pharmacy also overlaps with alchemy and chemistry to some extent with the first pharmaceutical text written on clay tablets by the Mesopotamians (approximately 2100 BCE). Some of the formulas and instructions written on the tablets include pulverization, infusion, boiling, filtering, and spreading – the basics of compounding and early formulations. In addition to herbs, ingredients such as beer, tree bark, and wine are also mentioned. Later, Galen (130–200 CE) formally introduced compounding, "a process of mixing two or more medicines to meet the individual needs of a patient" – a process practiced today for patients with special needs or for unique prescriptions.

In European countries exposed to Arabian influence, public pharmacies began to appear in the seventeenth century. About 1240 CE in Sicily and southern Italy, pharmacy was first separated from Medicine. Frederick II of Hohenstaufen (Emperor of Germany and King of Sicily) presented subject pharmacists with the first European edict completely separating their responsibilities from those of medicine, and prescribing regulations for their professional practice. The first official pharmacopeia, the "Nuovo Receptario" (originally written in Italian) became the legal standard for the city-state in 1498. It was the result of collaboration of the Guild of Apothecaries and the Medical Society – one of the earliest examples of interprofessional collaboration.

A more modern example can be observed in the early works of two French pharmacists, Messrs. Pierre-Joseph Pelletier and Joseph-Bienaimé Caventou. Together they isolated emetine from ipecacuanha in 1817 and strychnine and brucinefrom nux vomica in 1818. Then, in their laboratory in the back of a Parisian apothecary shop, they extracted ingredients from Peruvian barks that were effective against malaria. In 1820 Caventou and Pelletier announced the methods for separation of quinine and cinchonine from the cinchona barks, prepared pure salts, had them tested clinically, and set up manufacturing facilities. These early pharmacy laboratories provided the seeds for what the discipline would become and provided examples of the necessary separation sciences used in the extraction, isolation, and purification of active chemical entities from natural products – a key competence for what would become the early pharmaceutical industry.

The Birth of Pharmacy

The first record of an actual pharmacy dates to 774 CE in Baghdad. The Arabs separated the arts of apothecary and physician, establishing the first privately owned drug stores. They preserved much of the Greco-Roman wisdom, added to it, expanding services with the aid of their natural resources, syrups, confections, conserves, distilled waters, and alcoholic liquids.

Trade in drugs and spices was lucrative in the Middle Ages. In the British Isles, it was monopolized by the Guild of Grocers, which had jurisdiction over the apothecaries. After years of effort, the apothecaries found allies among court physicians. King James I with persuasion by the philosopher-politician, Francis Bacon, granted a charter in 1617, which formed a separate company known as the "Master, Wardens and Society of the Art and Mystery of the Apothecaries of the City of London" over vigorous protests of the grocers. This was the first organization of pharmacists in the Anglo-Saxon world.

Pharmacy Becomes an Industry

The modern pharmaceutical industry traces its origin to two sources: apothecaries that moved into wholesale production of drugs such as morphine, quinine, and strychnine in the middle of the nineteenth century and dye and chemical companies that established research labs and discovered medical applications for their products starting in the 1880s (Jones 2011). Merck, for example, began as a small apothecary shop in Darmstadt, Germany, in 1668, only beginning wholesale production of drugs in the 1840s. Likewise, Schering in Germany; Hoffmann-La Roche in Switzerland; Burroughs Wellcome in England; Etienne Poulenc in France; and Abbott, Smith Kline, Parke-Davis, Eli Lilly, Squibb, and Upjohn in the United States all started as apothecaries and drug suppliers between the early 1830s and late 1890s. Other firms whose names carry recognition today began with the production of organic chemicals (especially dyestuffs) before moving into pharmaceuticals. These include Agfa, Bayer, and Hoechst in Germany; Ciba, Geigy, and Sandoz in Switzerland; Imperial Chemical Industries in England; and Pfizer in the United States (Table 1.4).

Table 1.4 Early pharmaceutical businesses and their connection to the modern industry.

Date of origin	Milestone	Ties to the modern industry
1668	Rationalism and experimentation focused on improved production of goods	• Merck Pharmacy founded in Darmstadt, Germany (oldest pharmaceutical company)
1715	Apothecaries were the most common medical practitioners, offering medical advice and selling medicinal products	• Silvanus Bevan opened Plough Court pharmacy, an apothecary shop, in London, the predecessor of today's GlaxoSmithKline
1849	Targeted therapeutic areas of interest based on demand for goods during wartime	• Pfizer founded in the United States by two German immigrants, initially as a fine chemicals business; expanded rapidly during the American Civil War as demand for painkillers and antiseptics grew
1758	Switzerland's lack of patent laws led to it being accused of being a "pirate state" – framed the basis of patent protection later	• Materials, chemicals, dyes, and drugs sold by Johann Rudolf Geigy-Gemuseus in Switzerland became J. R. Geigy Ltd in 1914 and Novartis in 1996 (merger of Ciba-Geigy and Sandoz)
1858	Improved laboratory methods with focus on drug product quality	• Edward Robinson Squibb, a naval doctor during the Mexican American war of 1846–1848, sets up a laboratory in 1858, supplying Union armies in the civil war; the basis for today's Bristol Myers Squibb (BMS)
1863	The less strict delineation between "pharmaceutical" and "chemical" industries – consumer products	• Dye maker Bayer (Wuppertal, Germany) switches to pharmaceuticals and commercializes aspirin production
1876	First to focus on R&D as well as manufacturing	• Colonel Eli Lilly serving in the Union army during the Civil War, a trained pharmaceutical chemist, starts a pharmaceutical business that becomes Lily Pharmaceuticals

Drug Development in the Modern Era

Much of what we consider the modern era in drug development has occurred over the past 100 years (see Figure 1.5, bottom panel). The history of drug development and the pharmaceutical industry is very much associated with the necessity of manufacturing and distributing adequate quantities of drug products to the developed world (Barrett and Heaton 2019). Coincidentally, regulation of the processes underlying the R&D and manufacturing became a necessity often in response to tragedy (e.g. thalidomide in pregnant women in the 1950s) with an eventual global regulatory oversight in place for the developed world. This is not the same trajectory for all parts of the world, however, and global health considerations are managed in a somewhat protracted manner. But how did we get here?

Key discoveries of the 1920s and 1930s, such as insulin and penicillin, became mass manufactured and distributed, creating the infrastructure for R&D and manufacturing under one roof for a few companies. Switzerland, Germany, and Italy led the way with early strong pharmaceutical companies, with United Kingdom, the United States, Belgium, and the Netherlands following. Around the same time, legislation was enacted (Krantz 1966) to test and approve drugs and to require appropriate labeling. Prescription and non-prescription drugs became legally distinguished from one another at this time as well.

The growth of the pharmaceutical industry is spurred on by the development of systematic scientific approaches, understanding of human biology (including the discovery of DNA), and sophisticated manufacturing practices. Numerous new drugs were developed during the 1950s and mass produced and marketed through the 1960s. These included the first oral contraceptive, "The Pill," cortisone, blood pressure drugs, and other heart medications. Valium (diazepam), discovered in 1960, was marketed from 1963 and rapidly became the most prescribed drug in history, prior to controversy over dependency and habituation. Attempts were made to increase regulation and limit financial links between companies and prescribing physicians, including by the then relatively new US Food and Drug Administration (FDA). Calls for additional regulation increased in the 1960s after the thalidomide tragedy (severe birth defects).

In 1964, the World Medical Association issued its Declaration of Helsinki, which set standards for clinical research and demanded that subjects give their informed consent before enrolling in a clinical trial. Pharmaceutical companies became required to prove efficacy in clinical trials before marketing drugs around this time as well. From 1978, India took over as the primary center of pharmaceutical production without patent protection. Legislation allowing for strong patents, covering both the process of manufacture and the specific products, was eventually enacted in most countries. By the mid-80s, small biotech firms were struggling for survival and this led to the formation of partnerships with large

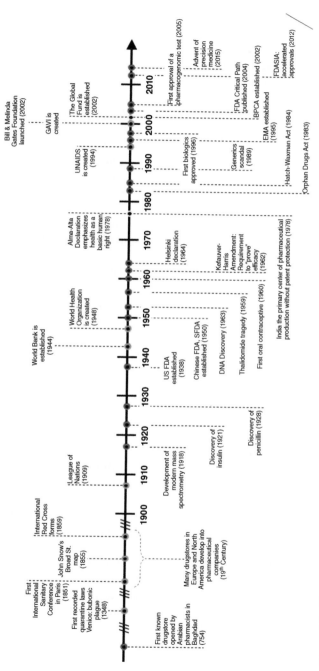

Figure 1.5 Timeline of events defining the pharmaceutical industry in the modern era (bottom panel) with the chronology of major global health milestones (top panel).

pharma companies and a host of corporate buyouts of the smaller firms. The industry is transformed by new DNA chemistries and technologies for analysis and computation.

Drugs for heart disease and AIDS become a feature of the 1980s, involving challenges to regulatory bodies and opportunities for faster drug approval. Managed care and health maintenance organization (HMOs) spread during the 1980s as part of an effort to contain rising medical costs; the development of preventative and maintenance medications became more important.

A new business atmosphere existed in the 1990s with mergers and takeovers, and there was a dramatic increase in the use of contract research organization (CROs) for clinical development and even basic R&D. Marketing also changed dramatically in the 1990s. The internet made possible direct purchase of medicines by consumers and of raw materials by drug producers, transforming the nature of business. In the United States, direct-to-consumer advertising proliferated on radio and TV because of FDA regulations in 1997 – liberalized requirements for the presentation of risks. There was a progression from hit-and-miss approach to rational drug discovery in laboratory design and natural product surveys.

The pharmaceutical industry continues to thrive, yet there are several challenges that may affect the industry's future growth. Drug prices are at an all-time high, R&D productivity has only just begun to climb again following a shortening in 2016/2017, and the pharmaceutical landscape is constantly changing with the rapid growth of biosimilars and disruptions of health technology.

There is still a major issue over high drug prices, particularly in the United States. Mounting pressures by patients, politicians, and regulatory bodies over drug pricing and reimbursement led to price freezes in 2018 and a proposal to introduce an "international pricing index" through Medicare – which would aim to reduce Medicare spending by 30%. The proposal was met widely with criticism due to concerns and in early 2019 several pharmaceutical companies hiked their prices up even further – an average of 6.3%. Biosimilars have made big waves in recent years and there is strong growth predicted across all markets, forecasting over 20% increases over the next five years. However, even though biosimilars are growing at an accelerated rate, the market is still dominated by small molecules with 76% of the market share.

Although biosimilars are a growing segment and threaten to take market share from small molecules, there are some challenges to their production. Although biosimilars will present competition for biologics, they represent significant savings to the consumer. In the United States, the projected cost savings from switching to biosimilars is expected to be between $40 and $250 billion within the next 10 years. This will go some way in combatting the drug price crisis and make life-saving medicines more affordable.

Technology trends are driving a shift toward patient-centric healthcare, as evidenced by wearable biometric devices and telemedicine. This trend is resulting in more informed patients who are likely to take a more active role in any treatment plan their doctor may prescribe. Patient-centric care can provide challenges and rewards for the pharmaceutical industry. The main challenge will be determining how to leverage the power of health technology and shifting focus from partnerships with the medical community to partnerships directly with the consumer.

References

Barrett, J.S. and Heaton, P.M. (2019). Real world data: an unrealized opportunity in global health? Submitted. *Clinical Pharmacology and Therapeutics* 106 (1): 57–59. PMID: 31188467, doi: https://doi.org/10.1002/cpt.1476.

Costa-Neto, E.M. (2012). Zoopharmacognosy, the self-medication behavior of animals. *Interfaces Científicas-Saúde e Ambiente* 1 (1): 61–72.

Engle, C. (2003). *Wild Health*, 24–38. Houghton Mifflin Press, ISBN-13: 978-0618340682, ISBN-10: 0618340688.

Jones, A.W. (2011). Early drug discovery and the rise of pharmaceutical chemistry. *Drug Test Analysis* 3: 337–344.

Krantz, J.C. Jr. (1966). New drugs and the Kefauver-Harris amendment. *The Journal of New Drugs* 6 (22): 77–79.

Mukherjee, P.K., Harwansha, R.K., Bahadur, S. et al. (2017). Development of Ayurveda – Tradition to trend. *Journal of Ethnopharmacology* 197: 10–24.

Pharmacopeia of the Peoples Republic of China (2015). 10 Anhui Science and Technology Press, ISBN 978-7-5067-7337-9.

Zhang, L. and Schuppan, D. (2014). Traditional Chinese Medicine (TCM) for fibrotic liver disease: hope and hype. *Journal of Hepatology* 61: 166–168.

Chapter Self-Assessments: Check Your Knowledge

Questions:

- Why is regulatory oversight necessary?
- How do generic drugs benefit patients?
- Why does the National Center for Complimentary and Integrative Health interested in re-evaluating the science behind many traditional medicine practices?
- How did the age of mergers and acquisitions and the evolution of healthcare affect the pharmaceutical industry?

Answers:

- The need for regulation and oversight historically came on the heels of a tragedy in which a lack of understanding, negligence, or simple greed instigated an event that was ultimately dangerous for patients. The main emphasis for regulatory oversight is to protect the public from unsafe medicines and practices.
- Generic drugs provide equivalent, safe alternatives to prescription medicines to patients at a lower price than their name-brand alternatives marked by the sponsor (innovator) company once patent life has expired.
- Many traditional medicine practices are still in use, some with successful outcomes. There is a great effort at present to consider the integration of these approaches with more modern medical treatments as healthcare considers a shift in emphasis from treatment to prevention strategies. This is especially true given the greater appreciation for holistic approaches to maintaining healthy homeostasis.
- With mergers and takeovers, a dramatic increase in the use of CROs for clinical development and even basic R&D occurs, influencing internal headcount and staffing models. Marketing also changed dramatically with the internet making a possible direct purchase of medicines by consumers and of raw materials by drug producers, transforming the nature of business.

Quiz:

1 True or false. Alchemists have had no real impact on pharmaceutical sciences because of their preoccupation with mysticism and conversion of metals into gold.

2 Common traditional medicines include all of the following except which:
   A New Orleans voo doo
   B Traditional Chinese Medicine (TCM)
   C Ayurvedic Medicine
   D Aboriginal Bush Medicine
   E "a" and "d" are excluded

3 The modern pharmaceutical industry traces its origin to two sources: _____ that moved into wholesale production of drugs such as morphine, quinine, and strychnine in the middle of the nineteenth century and _____ that established research labs and discovered medical applications for their products starting in the 1880s. (choose the best answer below):
   A Apothecaries and Dye, and Chemical companies
   B Early pharmacies and tea traders

    C Early pharmacies and traditional medicine companies
    D Apothecaries and traditional medicine companies

4 Key discoveries of the 1920s and 1930s, such as insulin and penicillin, became
    _____ and _____ creating the infrastructure for R&D and manu-
    facturing under one roof for a few companies. (Choose the best answer):
    A Discovered and produced
    B Developed and marketed
    C Developed and distributed
    D Mass-manufactured and distributed

2

The Modern Pharmaceutical Industry

Small, Medium, and Large PhRMA, Biotech, and Generics . . .
Jeffrey S. Barrett

Aridhia Digital Research Environment

Size and Other Things Matter

The size of individual companies is very much dependent upon the intended scope of services and/or products they intend to provide. The decision regarding scope, mission, and infrastructure is necessarily made by the company's leadership and potentially their board of directors. Historically, the growth of pharmaceutical companies was promoted by the necessity to manufacture and distribute goods and services to a larger market. Likewise, regulatory oversight and the need for quality controls and quality assurance mandated additional staffing considerations for pharmaceutical sponsors to comply with an ever-increasing demand for the demonstration of rigor and reproducibility of their processes. The modern world has also dictated growth in the ability to market, advertise, and sell products to a global economy. These factors contributed to the designation of the term "Large Pharma" to companies that exceeded the 100 000-employee mark by the mid-1980s.

Of course, there are many ways to assess the size of an organization. Physically, we can look at the number of employees, the geographic regions where the company has a presence, and the number of regions to which they market and distribute their goods. Financially, we can look at their earnings, investment in research and development (R&D), and other indices of profitability including market share as indicators of financial acumen and also of size. Productivity and efficiency related to size are also captured in the size and diversity of their pipeline in addition to the number of drug candidates at different phases of development.

Table 2.1 provides an illustration of the diversity among various companies that represent the pharmaceutical industry. The company attributes linked to size were chosen somewhat arbitrarily but selected to illustrate the diversity across various well-known and some lesser-known companies based on their intended

Fundamentals of Drug Development, First Edition. Edited by Jeffrey S. Barrett.
© 2022 John Wiley & Sons, Inc. Published 2022 by John Wiley & Sons, Inc.
Companion website: www.wiley.com/go/Barrett/FundamentalsDrugDevelopment

Table 2.1 Examples of representative pharmaceutical companies illustrating diversity across various product sectors and sizes.

Company	Focus	# Employees	R&D budget	# Drugs in pipeline	Earnings
Johnson & Johnson	Healthcare company operating in the pharmaceutical, medical devices, and consumer care sectors	>135 000	$10.8B (2018)	216	$81.6B (2018)
Merck	Pharmaceuticals, vaccines, and animal health products	~69 000	$9.75B (2018)	191	$42.3B (2018)
Pfizer	Biologics, vaccines, small-molecule medicines, and consumer products	~93 000	$8.0B (2018)	192	$53.6B (2018)
Gilead Sciences	Therapeutics against life-threatening diseases (HIV, hepatitis B, hepatitis C, and influenza)	~11 000	$3.1B (2018)	66	$22.1B (2018)
Takeda	Plasma-derived therapies and vaccines	~30 000	$3.5B (2018)	164	$15.9B (2018)
Teva	Generic drugs, active pharmaceutical ingredients, and proprietary pharmaceuticals	~9000	$1.2B (2018)	66	$18.9B (2018)
Novavax	Vaccines: novel products to prevent a broad range of infectious diseases	355	$172M (2018)	4	$21.2M (2018)
Sorrento Therapeutics	Antibody-centric emphasis – new therapies for cancer and chronic cancer pain	382	$170 (2017)	6	1 product on market-issued IPO for investors
Catalyst Biosciences	Treatment of hemophilia using high potency coagulation factors that promote blood clotting	11	$21.5M (2018)	2	0 (no products on market; looking for investors)
Karyopharm Therapeutics	Novel first-in-class drugs directed against nuclear transport and related targets for the treatment of cancer and other major diseases	353	$161M (2018)	4	$30.3M (2018)

Source: References: https://www.fiercebiotech.com/special-report/top-10-pharma-r-d-budgets-2018.https://www.globenewswire.com/news-release/2019/03/07/1749763/0/en/Catalyst-Biosciences-Reports-Fourth-Quarter-and-Full-Year-2018-Operating-Financial-Results-and-Provides-a-Corporate-Update.html.https://www.valuewalk.com/2018/03/sorrento-therapeutics-srne-short.https://investors.karyopharm.com/news-releases/news-release-details/karyopharm-reports-fourth-quarter-and-full-year-2018-financial.https://www.owler.com/company/novavax#overview; https://ycharts.com/companies/NVAX/r_and_d_expense.

scope of interest. It should be appreciated that while a strong correlation between the diversity of firms' development efforts and the success probability of individual projects exists, there is no effect of scale per se. Large firms' superior performance in drug development appears to be driven by returns to scope rather than returns to scale. Scope is confounded with firm fixed effects, however, suggesting an important role for inter-firm differences in the organization and management of the development function (Cockburn and Henderson 2001).

In this chapter, we will discuss broadly the differences in the various types of companies that drive the pharmaceutical industry from the perspective of their size, organization, infrastructure, and mission. Examples will be used to illustrate key differences and to quantify relationships that achieve some economy of scale (or not).

You Are What You Do

Each business opportunity will have some alignment of its intended scope of work defining its financial investment with a business plan that proposes a specific organizational structure including actual and/or projected headcount and budget requirements. Most smaller companies tend to anchor their organization to a target therapeutic area or technology or modality (new chemical entity small molecules, biologics or generic drugs, biosimilars, or vaccines). While larger pharmaceutical companies frequently expand the diversity of the modalities they support/provide, they also recognize that this comes at an infrastructure cost and does not always achieve an economy of scale.

Pharmaceutical companies handle the research, production and delivery of pharmaceutical drugs, devices, and/or vaccines to healthcare service providers (physicians, pharmacies, hospitals, etc.) and consumers. Pharmaceutical products must go through extensive research and testing (i.e. Research & Development), as well as follow regulations and obtain approval from government entities such as the Food and Drug Administration (FDA) before the products can reach their targeted market. Once products have been approved for large-scale distribution, pharmaceutical companies scale up manufacturing and align themselves with intermediaries to sell the drug through various channels (direct-to-consumer or through distribution channels such as pharmacies and hospitals). Figure 2.1 provides an example organization chart for a large pharmaceutical company with the R&D, operations, and commercial contributions to the business represented.

Figure 2.1 also serves as a baseline for discussing differences among different businesses within the umbrella of the pharmaceutical industry (e.g. generics, vaccines, biologics, and nutraceuticals) relative to pharmaceutical companies that

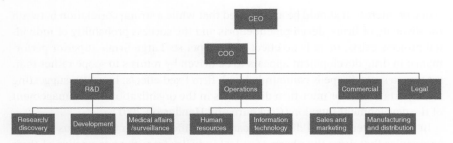

Figure 2.1   A representative organization chart for a large, full-service pharmaceutical company.

focus on small-molecule drug development. It is also an anchor for describing how outsourcing and/or partnering can influence the organizational structure and dedicated headcount requirements.

Generics

A generic drug is a product that compares to the pioneer, or reference, drug product (branded drug) in dosage form, route of administration, strength, quality, safety, and performance characteristics. The generic drug must have the same intended use as the pioneer product that serves as its prototype. The availability and utilization of generic alternatives to brand-name drugs have had a significant effect on cost savings for healthcare consumers.

The history of generic drugs can be traced to the Durham-Humphrey Amendment of 1951, which established two distinct categories of drugs: those that are unsafe to use without medical supervision and must be prescribed, and those that can be sold without a prescription. Despite the differentiation, multiple products continued to appear on the market in the early 1950s, which potentiated difficulties with inventory and drug counterfeiting. This led to efforts by the American Pharmaceutical Association (APhA, now referred to as the American Pharmacy Association) to pass anti-substitution resolutions and state legislation requiring pharmacists to dispense either the branded drug prescribed or a generic drug from a specific manufacturer unless only a generic name was provided. While these laws helped prevent the substitution of low-quality products, they limited opportunities for the manufacture of generic products of sufficient quality.

Eventually, the Kefauver-Harris Drug Amendments also required all manufacturers of related products to submit an abbreviated new drug application (ANDA) for products manufactured between 1938 and 1962. ANDAs contained information similar to that found in a pioneer drug application, with the exception of safety and efficacy. After 1962, the FDA established a new mechanism of proving safety and efficacy by allowing the "literature-based" new drug application (NDA). In response to rising costs and complaints about the influence of "Big Pharma," Congress in 1984 passed the Hatch-Waxman Act, which created a new regulatory track for generic drugs. As long as generic manufacturers could prove their drugs were bioequivalent to brand-name drugs, meaning they acted similarly in the body, they could get approved. It was a boon for generic drug makers, and in principle for the American public, allowing market competition to yield less expensive but equivalent drugs.

It is well appreciated today that generic medicines tend to cost less than their brand-name counterparts because they do not have to repeat animal and clinical (human) studies that were required of the brand-name medicines to demonstrate safety and effectiveness. In addition, multiple applications for generic drugs are often approved to market a single product; this creates competition in the marketplace, typically resulting in lower prices. The reduction in upfront research costs means that, although generic medicines have the same therapeutic effect as their branded counterparts, they are typically sold at substantially lower costs. When multiple generic companies market a single approved product, market competition typically results in prices about 85% less than the brand name. According to the IMS Health Institute, generic drugs saved the US healthcare system $1.67 trillion from 2007 to 2016. The modern generic industry relies heavily on the development of new drugs and drug approval to create future assets for generics post patent expiration.

Global regulatory authorities responsible for both new drug and generic drug approvals are likewise essential stakeholders in the process and well-being of both innovator and generic drug manufacturers. It should be noted that so close is this relationship that there are separate review divisions within the FDA to review the respective NDAs and ANDAs for generics though both adhere to the same standards and expectations. Likewise, the respective professional societies that support each industry – Pharmaceutical Manufacturers Association (PhMA) and Generic Pharmaceuticals Association (GPhA, now AAM, the Association for Accessible Medicines) – have similar missions and emphasis. Based on recent data from the Center for Justice and Democracy at New York Law School (https://www.thebalance.com/top-generic-drug-companies-2663110), 80% of all drugs prescribed are generic, and generic drugs are chosen 94% of the time when they are available.

Without the requirement to discover new chemical entities (NCEs) and prove safety and efficacy, generic drug makers obviously require less staffing in the R&D portion of the organization, specifically those engaged in drug discovery, clinical research, and clinical operations. For example, the largest generic pharmaceutical company Mylan Pharmaceuticals ($11.26 billion/year revenue from a global portfolio of more than 7500 products, including generic, branded generic, branded pharmaceutical, and OTC drugs) employs around 35 000 employees roughly one-third the size of the largest pharmaceutical innovator companies. This is a reasonable rule of thumb for the difference in staffing based on the absence of discovery and clinical efforts (roughly two-thirds reduction in staffing based on the difference in scope). This is admittedly imprecise but serves as a reasonable benchmark given that both industries market to a global market and must manufacture, sell, and distribute their products similarly. A more detailed look at generic drug development is provided in Volume 2 of this series, Chapter 22.

Biotech, Biologics, and Biosimilars

Biologics (or biologic drugs) refer to products produced from living organisms or containing components of living organisms. Biologic drugs include a wide variety of products derived from human, animal, or microorganisms by using biotechnology. Types of biologic drugs include vaccines, blood, blood components, cells, allergens, genes, tissues, and recombinant proteins [Declerck 2012]. Biologic products may contain proteins that control the action of other proteins and cellular processes, genes that control the production of vital proteins, modified human hormones, or cells that produce substances that suppress or activate components of the immune system. A biosimilar (also known as follow-on biologic) is a biologic medical product that is almost an identical copy of an original product that is manufactured by a different company. Biosimilars are approved versions of original "innovator" products and can be manufactured when the original product's patent expires. Reference to the innovator product is an integral component of the regulatory approval of a biosimilar before it can be marketed and prescribed to patients.

Many small-molecule drugs can be taken orally and tend to work in the body within cells. Since biologics are significantly larger in size, they are typically injected and interact within the body in the bloodstream or on the surfaces of cells, rather than within the cells. In contrast, small-molecule drugs are typically composed of only 20–100 atoms. Small biologics, such as hormones, are typically composed of 200–3000 atoms, while large biologics, such as antibodies, are typically composed of 5000–50 000 atoms.

Manufacturing processes for biologics differ greatly from the manufacturing processes for small-molecule drugs. Small-molecule drugs are generally synthesized using chemical reactions. Biologics, by comparison, are typically produced within specially engineered cells. Small molecules are well-characterized and can be easily purified and analyzed with routine laboratory tests. Biologics – especially larger biologics – tend to be produced as diverse mixtures of molecules that differ very slightly from one another, which make them difficult to characterize. It follows that the properties of the biologic often depend directly on the nature of the manufacturing process. Furthermore, proteins have unique structural organization patterns that affect the way that they work in the body; even biologics that are chemically the same may have different biological effects due to differences in structural folding. Due to both the size and sensitivity of biologics, these medicines are most frequently administered by injection, inhalation, or infusion into a patient's body. While small-molecule drugs can be swallowed and enter the human body without being noticed by the immune system, the same is not true of biologics. The large molecules of biologic medicines are always detected, and the human body's immune system must then decide whether to mount an immune response. Specifically, without precise design and administration, the patient's immune system may consider the biologic a foreign substance and take steps to neutralize and eliminate it.

Manufacturing complexities and their established linkage to clinical performance likewise obligate regulatory authorities to be rigorous in their reviews and establish a credible audit process to ensure routine performance expectations and ultimately maintain consumer confidence. For the pharmaceutical company producing biologics, they must likewise employ a staff that can not only deliver the necessary manufacturing controls but also pursue competitive advantages that protect the product from biosimilar competition and increase the strength and duration of relevant patents.

While technically vaccines and other large molecules and proteins such as heparin, low molecular weight heparin, and insulin fall into the category of biologics, the more recent emphasis on biologics has been focused on antibody (whole and fragment)-based therapies. Examples of approved traditional biologics include growth factors, insulin, erythropoietin (EPO), enzymes, interferon, and granulocyte colony-stimulating factor. Today, biologics represent a significant portion of most "big pharma" companies' portfolios, and companies originally focused on small molecules as a modality are diversifying to accommodate biologic drugs as part of their portfolio. It is projected that biologics will account for up to 30% of pharmaceuticals under development in the next few years.

Currently, the top 10 biologic therapies account for 36% of all biologic spending. This is far above the top 10 small molecules, which collectively hold only 20% of the original brand small-molecule market (Andrews et. al., 2015). The same concentration also applies in terms of therapeutic landscape. The three largest biologic therapy areas (autoimmune, diabetes, oncology) are worth $110B, over half of all biologic revenue. They are represented in nine of the top 10 biologics and are increasingly relevant due to their contribution of 70% of biologic growth since 2010.

The large therapy areas have dominated the biologic market as a result of a high number of strong launches into high-unmet-need indications and the lack of biologic entrance into other large disease areas. However, a change in the market is imminent. These areas are increasingly competitive, and the introduction of biosimilars will add further downward pricing pressure. Even so, the additional cost of manufacturing and development of biologics is outweighed by the high efficacy, improved probability of technical success, and patent protection suggesting that these development trends will continue in the short term.

Vaccines

Vaccines work by developing and enhancing the immunity of the body to naturally fight against serious infections and illnesses. Vaccines act by training and strengthening the immune system to develop resistance against antibodies and illnesses by imitating an infection to create a natural immune response. Most of the active types of vaccines remain in the system as antibodies even after they have completed their process of creating a stronger immune system. With respect to the classification of vaccines, there are five types of vaccines presently administered to infants and adults: live vaccines, inactive vaccines, toxoid vaccines, conjugate vaccines, and subunit vaccines.

The growth of the leading vaccine manufacturers is driven by the global economic recovery in recent years, which has led to a higher level of disposable income with more people realizing the importance of their infants and children to get vaccinated. While the order changes somewhat, the top 10 vaccine developers over the last five years include GlaxoSmithKline, Merck & Co, Sanofi, Pfizer, Novavax, Emergent BioSolutions, CSL, Innovio Pharmaceuticals, Bavarian Nordic, and Mitsubishi Tanabe.

With respect to the organization and infrastructure of vaccine companies, they are appreciably different from their drugs and biologics counterparts though they do tend to be smaller overall typically based on the size difference in the vaccine versus drugs portfolio but also based on the nature of their development programs (preclinical and clinical). A telling feature is the simple comparison of the number of employees within companies that maintain both drugs and vaccine business units. In a few cases, these are separate organizations with separate leadership up through the CEO in fact. Consider the comparison between Merck, Sanofi, and GSK drug versus vaccine businesses in Table 2.2.

Because vaccines are often targeted against infectious disease that is often easily spread among infected populations, the process of immunization requires coordination within each of the targeted geographic areas and must rely on local infrastructure to supply and distribute vaccines and manage immunization. The details of coordination of these events are the responsibility of the vaccine sponsor in conjunction with the individual governments and their local infrastructure. Likewise, there is tremendous diversity in how this is accomplished and the

Table 2.2 Comparison between drug and vaccine organizations for companies supporting both drug and vaccine research and development.

	Separate business?	Highest common line management	Number of employees [drugs vs. vaccines]	Co-located?
Merck	No	Executive VP and chief patient officer	69 000 vs. ~7000 [2018]	Yes – integrated with drug research; manufacturing separate
Pfizer[a]	No	R&D head	92 400 vs. 4000 [2018]	Yes – integrated with drug research; manufacturing separate
Sanofi	Yes [Sanofi Pasteur]	CEO	110 000 vs. 15 000 [2018]	No – no overlapping sites or facilities
GSK	No	CEO	98 462 vs. 17 000 [2018]	No – vaccine R&D headquarters in Maryland; drug R&D Hub in Pennsylvania

[a] https://www.pfizer.com/science/vaccines/research. *Source:* https://www.gsk.com/en-gb/about-us/corporate-executive-team/roger-connor.

difference in how vaccines are distributed, and immunization administered between the United States and sub-Saharan Africa is very striking. Both commercial and regulatory groups within these organizations must likewise understand the constraints of the geographic areas they hope to promote the product to and have personnel or contractors available to manage the process locally. In many cases, this is accomplished via outsourcing or partnering locally in order to manage expenses and limit the necessity of local full time employees (FTEs) at all corners of the world. Not surprisingly, most of the actual headcount is focused on regions where profits are highest.

Nutraceuticals, Natural Products, and Supplements

A nutraceutical product is a food or fortified food product that not only supplements the diet but also assists in treating or preventing disease (apart from anemia), so provides medical benefits. Nutraceuticals are not tested and regulated to the extent of pharmaceutical drugs. A natural product refers to anything that is produced by life and includes biotic materials (e.g. wood, silk), bio-based materials (e.g. bioplastics, cornstarch), bodily fluids (e.g. milk, plant exudates), and other natural materials that were once found in living organisms (e.g. soil, coal).

A more restrictive definition of a natural product is any organic compound that is synthesized by a living organism. When referring to supplements we usually mean dietary supplements and in this sense, a dietary supplement refers to a product (typically taken orally) that contains one or more ingredients (such as vitamins, minerals, herbs, or amino acids) that are intended to supplement one's diet and are not considered food. While the manufacturers of nutraceutical, natural products, and supplements may consider themselves separate in their product lines and marketing, with respect to the pharmaceutical industry they can be contrasted together. We will identify them as nutraceutical companies from the point of comparison, but this is made from the standpoint of their organizational structures only.

Nutraceutical companies are generally not considered to be part of the pharmaceutical industry in part because of the lack of regulatory oversight on their products and processes. Recent concerns about the content and quality of their products given the rise in popularity and uptake by consumers, however, may change that status over time. Also, some pharmaceutical companies have dabbled in nutraceuticals as an extension of their consumer product offerings further blurring the line between the two. As research into nutraceuticals that allude to the medical benefits of these products (e.g. St John's Wort [Sarris et al. 2012], Milk Thistle [Fried et al. 2012]) particularly when combined with actual drugs is suggested, so too will findings that suggest that the therapeutic window for these products may not be as broad as once previously believed and that quality standards [Lee et al. 2007] will likely need to be enforced in order to maintain consumer confidence.

Regarding their organizations, these companies conduct little actual research themselves although they may fund or outsource certain research in order to support their product quality and address any concerns about their clinical use. Likewise, their workforce is mostly concerned with commercial activities and focused on manufacturing and distribution of the product in addition to the expected sales, marketing, and advertising required. Some of the most prominent nutraceutical companies include Archer Daniels Midland (ADM) Company, Genomatica, Ajinomoto, and Naturex. Pharmaceutical companies with nutraceutical business units include GlaxoSmithKline and Bayer Healthcare AG. A list of nutraceutical-related companies can be easily found on the internet (https://pipecandy.com/list-of-nutraceutical-companies-usa) though given the dynamic and volatile nature of this industry it would be difficult to stay current with such developments unless that is your core business. What is relevant for the pharmaceutical industry is that some level of regulation is likely for these products in the near future. This necessity will both require these companies to hire additional research and regulatory staff and pull resources from FDA and global regulatory authorities to review and assess their progress.

Outsourcing and Partnering

One way that the pharmaceutical industry has been able to adjust its workforce allocation to the various tasks in R&D, operations, and commercial endeavors including manufacturing, sales, and distribution of products has been through outsourcing and partnering. Both are relatively recent activities for the industry, but both have had a tremendous impact on organizational efficiency and resourcing flexibility. Biopharmaceutical companies increasingly are partnering with a diverse set of healthcare stakeholders to address scientific and technological challenges, create greater efficiencies in R&D, and accelerate the discovery, production, and delivery of critical new treatments for patients in need.

Outsourcing broadly refers to a business practice in which a company hires another company or an individual to perform tasks, handle operations, or provide services that either would usually be executed or had previously been done by the company's own employees. Outsourcing in the pharmaceutical industry is a relatively common practice today though in the past firms preferred to rely on internal resources (certainly prior to 1970s). With respect to the scope of current outsourcing efforts, the range of functions targeted is quite broad with operations activities in the IT and HR space being common across all industries. Pharma-specific activities have created opportunities for large and expanding contract research organization (CRO) and contract manufacturing organization (CMO) industries. Established CROs (e.g. Parexel, Covance, Icon, PRA Health Sciences, PPD, etc.) and CMOs (e.g. Patheon, Catalent, Lonza, PCI, etc.) become preferred vendors for many pharmaceutical partners; legal, finance, and procurement groups within individual companies are keen to negotiate contracts, which guarantee favored status with respect to the availability of services, preferential payment options, and cheaper prices.

CROs were originally established to provide specific services, such as clinical trials, to allow pharmaceutical companies to focus their resources on proprietary R&D, while CMOs provided services for late-stage drug development. Driving forces for the emergence of CROs and CMOs included the increase in federal regulations for improving new drug safety and efficacy and the need to access manufacturing capacity. At the same time, biotechnology companies were developing early-stage drug leads and concepts using innovative technology. Pharmaceutical companies accessed these innovations by forming partnerships with the biotech companies early in the discovery of these new drug opportunities.

A Tufts study conducted in 2012 found that biopharmaceutical companies are increasingly forming partnerships with academic medical centers with the goal of identifying promising pathways for potential breakthrough therapies through basic research in medicine, as well as guiding their translation into clinical development of new medical products.

On the discovery and early research front, the completion of the Human Genome Project led to the identification of more than 120000 genes and up to 10000 new drug targets (Kelly 2015). Prior to genomics, only about 500 gene-based drug targets had been identified. The 6- to 20-fold increase in drug targets from today's current market has resulted in a need for more efficient methods to elucidate these new drug targets. With the information generated by the genomics revolution, detailed structural information about proteins, using expression, purification, and 3D-structural determination (proteomics), has also become available. This information can be used to more rapidly identify physiological drug targets by utilizing structural biology to design lead compounds rationally. In addition, recent advances in high-throughput screening/combinatorial chemistry, as well as the automation of existing technologies and the development of novel technologies, allow more rapid discovery and optimization of lead compounds. The explosion in the number of new targets requires pharmaceutical companies to partner with companies that can turn the information into new drugs.

Many companies have embraced the need for partnering as part of overall strategies to improve the quality of their portfolio and to gain an advantage over others in their competitive therapeutic areas – many examples exist (Sutton 2019). Throughout the past 10 years, biopharmaceutical companies have increasingly sought to address previously unmet medical needs by building on scientific advances in genomic and molecular medicine. For example, the era of personalized medicine is rapidly changing the way diseases are identified, patients are diagnosed, and treatment decisions are made. Collaboration across the biopharmaceutical R&D ecosystem has been essential in driving important scientific breakthroughs in novel diagnostics technology and in identifying molecular targets for the development of personalized medicines. Biopharmaceutical companies are committed to advancing targeted therapies and medicines to treat serious conditions and unmet medical needs. In 2013–2014, AstraZeneca established several strategic R&D outsourcing partnerships with academic organizations, including Academic Drug Discovery Consortium (ADDC), Medical Research Council Laboratory of Molecular Biology (MRC LMB), and Cancer Research UK (CRUK) Cambridge Institute. Meanwhile, AstraZeneca's global biologics research and development arm MedImmune has already been collaborating with The University of Texas MD Anderson Cancer Center toward developing immunotherapies against cancer. Pfizer has undertaken a similar strategy in the United States, having positioned many of its research and development facilities close to major bioscience hubs, such as San Francisco and La Jolla in California and Cambridge, Massachusetts. Bristol-Myers Squibb partnered with Allied Minds, a Boston-based group focused on the commercialization of academic research, to scour American universities for innovative drug discovery ideas. GlaxoSmithKline teamed up with the University of Leicester to develop novel drugs against blood cancer.

A common partnering objective is to progress a single asset through the R&D process, obtain approval, and launch. Today's nonasset-based partnerships diverge notably from that model – collaborative alliances may include three or more parties and are often composed of a mix of ecosystem stakeholders, including biopharmaceutical companies, academia, nonprofits, and government entities. Importantly, these partnerships feature shared control and decision-making, thus spreading both the potential risks and rewards. Nonasset-based partnership examples include the following:

- Joint Ventures (JVs): Two or more entities enter a collaboration wherein all involved parties agree to jointly contribute to R&D-related activities to achieve a specific objective. They typically involve joint governance and decision-making and sharing of accompanying risks and rewards.
- Consortium: Three or more parties pool resources and work together to achieve a common goal, such as accelerating scientific discovery in a particular disease area or technology. Some consortia include "pre-competitive" arrangements in which all players work together to solve problems and develop capabilities in areas where they would typically compete with each other.
- Other: Parties provide financial resources and/or marketing, educational, and promotional programs (e.g. company support of broader disease awareness efforts).

While it is tempting to imagine that all partnerships go well, it is simply not the case, and many end prematurely without good reason. Each potential partner brings a different set of values, priorities, resources, and competencies to a partnership. The challenge of any partnership is to bring these diverse contributions together, linked by a common vision in order to achieve sustainable development goals. The hallmark of good partnering includes setting clear expectations, considering your partner a part of your team, allowing the partnership room to grow, and communicating with honesty and transparency.

References

Andrews, A., Ralston, S., Blomme, E., and Barnhart, K. (2015). A snapshot of biologic drug development: challenges and opportunities. *Human and Experimental Toxicology* 34 (12): 1279–1285.

Cockburn, I.M. and Henderson, R.M. (2001). Scale and scope in drug development: unpacking the advantages of size in pharmaceutical research. *Journal of Health Economics* 20: 1033–1057.

Declerck PJ. Biologicals and biosimilars: a review of the science and its implications. *Generics and Biosimilars Initiative Journal (GaBI J)*. 2012;1(1):13–16. DOI: https://doi.org/10.5639/gabij.2012.0101.005.

Fried MW, Navarro VJ, Afdhal N, et al. Effect of silymarin (milk thistle) on liver disease in patients with chronic hepatitis C unsuccessfully treated with interferon therapy: a randomized controlled trial. *JAMA* 2012;308(3):274–282. doi:https://doi.org/10.1001/jama.2012.8265.

Kelly, E. (2015). *Business Ecosystems Come of Age.* Deloitte University Press.

Lee, J.I., Narayan, M., and Barrett, J.S. (2007). Analysis and comparison of active constituents in commercial standardized silymarin extracts by liquid chromatography–electrospray ionization mass spectrometry. *Journal of Chromatography B* 845: 95–103.

Sarris J, Fava M, Schweitzer I, Mischoulon D. St John's wort (Hypericum perforatum) versus sertraline and placebo in major depressive disorder: continuation data from a 26-week RCT. *Pharmacopsychiatry* 2012;45(7):275–278. doi: https://doi.org/10.1055/s-0032-1306348.

Sutton, S. (2019). How to find your secret source. *The Medicine Maker* (August 2019), Issue 56, pp. 22–29.

Chapter Self-Assessments: Check Your Knowledge

Questions:
- Why is there more inherent protection from generic competition for biologic drugs as opposed to small molecules?
- Do you think partnering in the pharmaceutical industry is a fad that will die out with concerns about intellectual property?
- Why is it inappropriate to only consider the number of employees when we consider the size of an organization? What is reasonable for consideration?
- Why aren't smaller companies just miniature versions of larger companies? Why is there no economy of scale?

Answers:
- Differences including the extent of IP, manufacturing complexities, and the uncertain linkage to clinical safety obligate greater regulatory scrutiny and likewise create more uncertainty, investment, and effort to produce an acceptable biosimilar.
- Biopharmaceutical companies increasingly are partnering with a diverse set of healthcare stakeholders to address scientific and technological challenges, create greater efficiencies in research and development (R&D), and accelerate the discovery, production, and delivery of critical new treatments for patients in need.
- There are many ways to assess the size of an organization. Physically, we can look at the number of employees, the geographic regions where the company has a presence, and the number of regions to which they market and distribute their goods. Financially we can look at their earnings, investment in R&D

and other indices of profitability including market share, as indicators of financial acumen but also of size. Productivity and efficiency related to size are also captured in the size and diversity of their pipeline in addition to the number of drug candidates at different phases of development.

- Most smaller companies tend to anchor their organization to a target therapeutic area or technology or modality (new chemical entity small molecules, biologics or generic drugs, biosimilars, or vaccines). While larger pharmaceutical companies frequently expand the diversity of the modalities they support/provide, they also recognize that this comes at an infrastructure cost and not does not always achieve an economy of scale.

Quiz:

1   Historically, the growth of pharmaceutical companies was promoted by the necessity to _____and _____ goods and services to a larger market. Choose the best answer:
   A market and advertise
   B manufacture and distribute
   C research and develop
   D promote and sell
   E none of the above is correct

2   True or false. Nutraceuticals are not tested and regulated to the extent of pharmaceutical drugs.

3   Some examples of non-asset-based partnership examples include which of the following (choose the best answer):
   A Joint ventures
   B acquisition
   C consortium
   D merger
   E none of the above is correct
   F a and c

4   When multiple generic companies market a single approved product, market competition typically results in prices about _____ % less than the brand name. Choose the correct answer:
   A 85
   B 10
   C 50
   D 95
   E none of the above is correct

3

Legal Considerations, Intellectual Property, Patents/Patent Protection, and Data Privacy
Jeffrey S. Barrett

Aridhia Digital Research Environment

Introduction

The US pharmaceutical industry consists of companies primarily engaged in researching, developing, manufacturing, and marketing drugs and/or vaccines for human or veterinary use. The industry consists of both brand name and generic drugs, with different companies specializing in each type and some with business units covering multiple modalities and marketing varied products under the same company or as a family of business units (e.g. Johnson & Johnson). Brand-name drugs are patent-protected formulations that are marketed under a specific brand name by the company that owns the patent. Generic drugs are often administered in place of the brand-name drug once the formulation's patent protection has expired, and other companies are free to produce the same drug.

Within the United States, brand-name pharmaceutical manufacturing generates approximately 70–80% of total drug revenue (based on recent cost estimates); generic drug manufacturing generates approximately 20–30% of total drug revenues. In terms of revenue growth, the generic industry has been outpacing brand-name drugs in recent years. Certain brand names and generic drugs are also approved for over-the-counter sales without a prescription. The US over-the-counter (OTC) pharmaceutical industry is estimated to grow at more than 10% annually for the next several years, with large pharmaceutical companies focusing more on OTC. The marketplace is always evolving, as is the global healthcare ecosystem in which the pharmaceutical industry resides.

For a company to sustain and grow its portfolio, it must continue to create a value proposition supporting its worth to the external world and the marketplace, engage in a variety of agreements with contractors, other companies, and partners, and protect its intellectual property in the form of patents, copyrights, and

Fundamentals of Drug Development, First Edition. Edited by Jeffrey S. Barrett.
© 2022 John Wiley & Sons, Inc. Published 2022 by John Wiley & Sons, Inc.
Companion website: www.wiley.com/go/Barrett/FundamentalsDrugDevelopment

trademarks among other assets. Likewise, there are numerous legal relationships that must be developed and navigated. The worth and value of a company are far beyond these agreements, however, and the reputation of the company contributes to the valuation as well. In this chapter, we discuss the value proposition of a company, expose the quantitative metrics typically assessed for this purpose, and consider the various legal agreements that many companies engage in while doing business. Finally, we will explore the various forms of intellectual property to understand the legal aspects of protecting it and ensuring that it is properly valued.

The Value Proposition for Drug Development and the Pharmaceutical Industry

The value proposition is extremely complex for those engaged in pharmaceutical research and development. Companies must target healthcare providers, payers, and patients with compelling value propositions that explain in clear, concise terms what a pharmaceutical product has to offer and why it is superior to competitors. At the same time, there is often intense competition among companies, and the value proposition can be used as a means to differentiate companies. One aspect of valuation is based on operational and financial performance metrics. There are four commonly accepted methods that are typically considered when assessing a pharmaceutical company's value (see Table 3.1).

Additional performance metrics that managers in the pharmaceutical industry may use to benchmark their performance against others in the industry include the following: sales, operating profit margin, R&D spending as a percentage of sales, core earnings per share, operating free cash flow as a percentage of sales, number of major regulatory approvals received, number of Phase 1–3 clinical trials, percent of sales due to recently released products, average development cost per product.

Of course, value needs to be defined in a broader way beyond financial metrics as well. Less quantifiable indicators beneficial to valuation would include evidence that a company can improve physicians' ability to practice while making economic sense for payers. The value proposition thus needs to articulate how an individual product will address the needs of each stakeholder. Physicians won't believe a value proposition that does not have strong evidence behind it, of course, and clinical data and experience is essential. Companies may struggle to develop an asset (e.g. drug or vaccine) with a strong, unique value proposition if they don't have the evidence to support it. Evidence in this context might include published clinical trial results, presentations at scientific meetings, and/or health economic data. The core element of the value proposition must be something competitors don't have, which often suggests riskier clinical trials. That may mean moving down a nonvalidated development pathway that incurs additional risk – companies tend to take a

Table 3.1 Common valuation methods used to assess pharmaceutical companies.

Valuation method	Description
Asset-based valuation	This method calculates a business's equity value as the fair market value of a company's assets less the fair market value of its liabilities. This approach is also sometimes referred to as a "cost-based approach"; that is, the business's value is equal to the cost of acquiring its physical assets. This approach is seldom used for a pharmaceutical company because its value is more closely related to intangible assets, R&D expenditure, and cash flows.
Income approach (capitalization of earnings)	This method is most applicable to companies that face predictable and constant growth in earnings and have an established record of operations. The business value under this method is equal to the cash flow projection for the next year divided by a capitalization rate (i.e. the appropriate discount rate less the predicted growth rate).
Income approach (discounted cash flow)	The value of equity utilizing this method is equal to the present value of free cash flows available to common equity holders over the life of the business. This method works well for both established companies with low growth rates as well as new companies with higher rates of growth and requires forecasting future cash flows.
Market approach	This method utilizes market indications of value based on metrics from guideline publicly traded pharmaceutical companies or privately held businesses. The financial metrics of public companies or those of private transactions, such as P/E[a], P/S[b], and EV/EBITDA[c], can be used to generate valuation multiples that are then used to calculate business value.

[a] The price-to-earnings ratio (P/E ratio) is the ratio for valuing a company that measures its current share price relative to its per-share earnings (EPS).
[b] The price-to-sales (P/S) ratio is a valuation ratio that compares a company's stock price to its revenues. It is an indicator of the value placed on each dollar of a company's sales or revenues.
[c] The EV/EBITDA ratio is a comparison of enterprise value and earnings before interest, taxes, depreciation, and amortization. This is a very commonly used metric for estimating business valuations. It compares the value of a company, inclusive of debt and other liabilities, to the actual cash earnings exclusive of the noncash expenses.

low-risk development approach, but that won't give them the evidence for a differentiated value proposition. A good marketing organization works with R&D during the earlier phases of product development and designs a product profile that supports a strong, differentiated value proposition while the product is in development. Likewise, it is essential to evolve the value proposition over the life cycle of a development candidate. While initially focusing on segments where the value proposition is strongest, the company needs to be able to migrate to different segments (markets, indications, etc. – see Chapters 6–8 regarding the TPP).

Legal Agreements Supporting the Business of a Pharmaceutical Company

There are several kinds of agreements that are a part of normal business operations for a pharmaceutical company and others that are meant to protect businesses from unfavorable legal situations. Some common agreements include partnership agreements, indemnity agreements, and nondisclosure agreements. All are relevant to the pharmaceutical industry and the companies that support it.

General business contracts are agreements that cover some of the most vital topics related to any business, including the structure of the company and protections available to shareholders. A wide variety of general business contracts are available, including partnership agreements, equipment leases, franchise agreements, and employment agreements (Ponzio et. al., 2011). Sales-related contracts make it easier to transfer titles when needed. A bill of sale is one of the most common sales-related contracts. With a bill of sale, two parties can transfer ownership of a piece of property. For instance, bills of sale are frequently used to transfer the title of an automobile. During normal operations, a business may use several sales-related contracts, such as purchase orders (agreements used when a business commits to purchasing a certain item), security agreements (used if a company has to put up an asset as collateral for a loan), and warranty (describes actions that would result in a contract being terminated).

Employment contracts are the third category of business contracts. These contracts are an important part of hiring employees and can protect both the company and employees. A general employment contract defines the relationship between the company and its employee and covers several topics: length of employment, compensation, benefits, and grounds for termination. If an employee leaves the company, they may be asked to sign a noncompete agreement, which would prevent them from seeking employment with a competitor for a set period of time. Some other types of employment contracts include independent contract agreements, consulting agreements, distributor agreements, and confidentiality agreements. All of these agreements come into play for a pharmaceutical company, and all require legal counsel or input in order to finalize.

Due Diligence – How Much to Tell, Share, and Show

Due diligence is an extensive process undertaken by a company in order to thoroughly and completely assess another organization's business, assets, capabilities, and financial performance. There are numerous aspects of due diligence analysis, and the process is not only relevant for the acquisition of a company but other assets, including specific technologies, drug or vaccine candidates at various stages of development, business units (e.g. consumer products, veterinary medicine, etc.) or even whole therapeutic areas or franchises within an organization. Many examples of each of these exist within the pharmaceutical industry over the past three decades.

The main types of due diligence inquiry are administrative, financial, human resources, asset, environmental, taxes, intellectual property, legal, customer, and strategic fit. In each of these areas, a company will assemble a sub-team to address the various aspects as part of the due diligence exercise. Table 3.2 provides a listing of the type of materials that are typically reviewed by the various sub-teams participating in the due diligence exercise.

Table 3.2 Common due diligence categories with the source materials reviewed as part of the process.

Due diligence category	Materials reviewed by sub-team
Administrative	Admin-related items such as facilities, occupancy rate, number of workstations, etc. The purpose is to verify the various facilities owned or occupied and determine whether all operational costs are captured in the financials.
Financial	Checks whether the financials provided in the Confidentiality Information Memorandum (CIM) are accurate or not. The purpose is to provide a thorough understanding of all the company's financials, including, but not restricted to, audited financial statements for the last three years, recent unaudited financial statements with comparable statements of the last year, the company's projections and basis of such projections, capital expenditure plan, schedule of inventory, debtors and creditors, etc.
Human resources	• Analysis of total employees, including current positions, vacancies, due for retirement, and serving notice period. • Analysis of current salaries, bonuses paid during the last three years, and years of service. • All employment contracts with nondisclosure, nonsolicitation, and noncompetition agreements between the company and its employees. • HR policies regarding annual leave, sick leave, and other forms of leave. • Analysis of employee problems, such as alleged wrongful termination, harassment, discrimination, and any legal cases pending with current or former employees. • Potential financial impact of any current labor disputes, requests for arbitration, or grievances pending. • List and description of all employee health benefits and welfare insurance policies or self-funded arrangements.

(Continued)

Table 3.2 (Continued)

Due diligence category	Materials reviewed by sub-team
Asset	Detailed schedule of fixed assets and their locations (if possible physical verification), all lease agreements for equipment, a schedule of sales and purchases of major capital equipment during the last three to five years, real estate deeds, mortgages, title policies, and use permits assessed.
Environmental	• List of environmental permits and licenses and validation of the same. • Copies of correspondence and notices with the EPA and state or local regulatory agencies. • Verification that the company's disposal methods are in sync with current regulations and guidelines. • Check for any contingent environmental liabilities or continuing indemnification obligations.
Taxes	Documentation of tax compliance and potential issues typically includes verification and review of the following: • Copies of all tax returns – including income tax, withholding, and sales tax – for the past three to five years. • Information relating to any past or pending tax audits of the company. • Documentation related to NOL (net operating loss) or any unused credit carryforwards of deductions or tax credits. • Any important, out-of-the-ordinary correspondence with tax agencies.
Intellectual property	• Schedule of patents and patent applications. • Schedule of copyrights, trademarks, and brand names. • Pending patents clearance documents. • Any pending claims case by or against the company with regards to violation of intellectual property.
Legal	• Copy of Memorandum and Articles of Association. • Minutes of Board Meetings for the last three years. • Minutes of all meetings or actions of shareholders for the last three years. • Copy of share certificates issued to Key Management Personnel. • Copy of all guarantees to which the company is a party. • All material contracts, including any joint venture or partnership agreements; limited liability company, or operating agreements. • Licensing or franchise agreements. • Copies of all loan agreements, bank financing agreements, and lines of credit to which the company is a party.

Table 3.2 (Continued)

Due diligence category	Materials reviewed by sub-team
Customer	• The company's top customers: customers who make the largest total purchases from the company and customers who are the "largest" in terms of their total assets. • Service agreements and corresponding insurance coverage. • Current credit policies; assess the efficiency of accounts receivable. • Customer Satisfaction Score and related reports for past three years. • List, with explanations, of any major customers lost within the past three to five years.
Strategic fit	Some of the key strategic fit issues that acquirers look at and evaluate: • Does the target company have important technology, products, or market access that the acquirer lacks and has need of or can make profitable use of? • Does the target company have key personnel that represents a substantial gain in human resources? • Assess operational and financial synergies benefits that can be expected from the target's integration with the acquirer. • If the target company is to be merged with the acquirer or another firm the acquirer already owns, examine the plan for merging and project how long the merger process will take and estimate the cost of implementing the actual process of merging the two firms. • Determine the best personnel from both the acquirer and the target to manage the merger process.

Other areas of due diligence research include IT networks, issues of stocks and/or bonds, research and development (R&D), and sales and marketing. Conducting thorough due diligence is critical to any successful acquisition, regardless of the category type. Without complete and intimate knowledge of the target company or its asset of interest, it is impossible to make the best-informed decisions on mergers and acquisitions. The history of the pharmaceutical industry is filled with examples of both good and bad mergers and acquisitions (see Chapter 1) so this is a process that senior managers take very seriously. In a proposed merger or a situation where shares of stock in the acquiring company constitute a major part of the purchase transaction, the target company may well look to perform its own due diligence on the potential acquiring company.

As mentioned previously, the usual procedure for due diligence planning and execution begins with the assignment of team members from relevant functional groups within an organization to form due diligence teams. A great deal of planning is involved in order to provide a thorough assessment as described above. An alternative, of course, is to outsource this effort (see also Chapter 11 on this topic). There are many external contract research organizations (CRO) that provide this service, so that team members and other vital staff are able to perform their normal operations. Of course, there are always concerns with having a proxy for such an important task, but there are experts with years of experience in healthcare business management, software engineering, interoperability, and regulatory management that can perform a comprehensive assessment of a potential opportunity, but it will still come at the expense of educating them. Figure 3.1 shows the approach of one such commercial due diligence CRO (https://pcpimaging.com/due-diligence) with a strong correlation to the approach laid out in Table 3.2. As with any contract partner, trust and experience usually contribute to the decision to outsource or not.

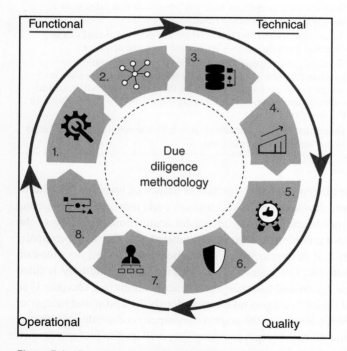

Figure 3.1 Due diligence methodology for complete end-to-end assessment of a potential acquisition or partnership from a commercial CRO.

Figure 3.1 (Continued)

Intellectual Property, Patents, and Patent Protection

Intellectual property refers to the exclusive rights granted by the State over creations of the human mind, in particular, inventions, literary and artistic works,
distinctive signs, and designs used in commerce. Intellectual property is divided
into two main categories: industrial property (IP) rights, which include patents,
utility models, trademarks, industrial designs, trade secrets, new varieties of
plants, and geographical indications; and copyright and related rights, which
relate to literary and artistic works.

IP rights are extremely important for the pharmaceutical industry. The use of the
IP system by subject matter experts (SMEs) in the pharmaceutical industry depends
largely on the business strategy of a company, its size, resources, innovative capacity, competitive context, and field of expertise. Research-based, innovation-led companies that seek to develop new drugs, improve or adapt existing drugs or develop
new pharmaceutical/medical equipment or processes tend to rely heavily on the
patent system to ensure they recover the investments incurred in research and
development. Companies that rely on licensing in or licensing out of pharmaceutical products will need to be knowledgeable about the patent system so that they are
able to negotiate fair and balanced licensing contracts. SMEs in the pharmaceutical
industry may use the wealth of information contained in patent documents as a
crucial input to their R&D work, to get ideas for further innovation, to ensure their
"freedom to operate" or to find out when a patent is due to expire opening the door
for the introduction of generics. Confidential information, protected as trade secrets,
is also important for many companies, as is the valuable know-how or undisclosed
test data relating to new or improved drugs (Nealey et al. 2015). Other examples of

trade secrets in the pharmaceutical industry could include R&D information, software algorithms, inventions, designs, formulas, ingredients, devices, and various methods. Understanding the trademark system is important for companies selling branded products. Industrial designs, plant variety protection, and copyright and related rights are generally less relevant to most SMEs in the pharmaceutical sector, but this could vary depending on the product line and strategy of each company.

A patent is an exclusive right granted by the State for an invention that is new, involves an inventive step (or is nonobvious) and is capable of industrial application (or useful). It provides its owner the exclusive right to prevent others from making, using, offering for sale, selling, or importing the patented invention without the owner's permission. A patent is a powerful business tool for companies to gain exclusivity in the market over a new product or process and develop a strong market position and/or earn additional profits through licensing. A patent is granted by the national or regional patent office. It is valid for a limited period of time, generally for 20 years from the filing date (or priority date) of the patent application, provided the renewal (or maintenance) fees are paid to keep the patent in force. In some countries, a longer period of protection may be obtained for pharmaceutical products to compensate for the loss of an effective period of protection due to delays in obtaining marketing approval from the relevant public health regulatory bodies. In return for the exclusive rights granted by a patent, the inventor is required to disclose his invention to the public in the patent application with sufficient detail to enable a person skilled in the relevant technology to practice the claimed invention. Patents, and in many countries patent applications, are disclosed to the public through publication in an official journal or gazette.

A trademark is a sign capable of distinguishing the goods or services produced or provided by one enterprise from those of other enterprises. Any distinctive words, letters, numerals, drawings, pictures, shapes, colors, logotypes, labels, or combinations that distinguish the origin of goods or services may be considered a trademark. Figure 3.2 shows some common pharmaceutical trademarks (company logos) that many will recognize.

Figure 3.2 Recognizable trademarks across the pharmaceutical industry.

In some countries, advertising slogans are also considered trademarks and may be registered as such at national trademark offices. An increasing number of countries also allow for the registration of less traditional forms of trademarks such as single colors, three-dimensional signs (shapes of products or packaging), audible signs (sounds), or olfactory signs (smells). Aside from the protection of logos and brand names, in some countries, companies in the pharmaceutical industry rely on trademark protection for the distinctive shape or color of pharmaceutical products (such as capsules or tablets) and product packaging.

Trademark protection can be obtained through registration or, in some countries, also through use. Even where trademark rights can be acquired through use, companies are well-advised to register a trademark by filing the appropriate application form with the national or regional trademark office. While the term of protection may vary, in a large number of countries, registered trademarks are protected for 10 years. Registration may be renewed indefinitely (usually for consecutive periods of 10 years) provided renewal fees are paid in time.

Broadly speaking, confidential business information that provides an enterprise a competitive edge may be considered a trade secret. The misappropriation, disclosure, or unauthorized use of such information is regarded as an unfair practice and a violation of the trade secret. Depending on the legal system, the protection of trade secrets forms part of the general concept of protection against unfair competition or is based on specific provisions or case law on the protection of confidential information. Confidential business information may benefit from protection as a trade secret as long as: it is not generally known or readily accessible, to circles dealing with that type of information; it has commercial value because it is secret, and it has been subject to reasonable steps by the rightful holder of the information to keep it secret (e.g. through physical and electronic control mechanisms or by entering into nondisclosure or confidentiality agreements).

In the field of pharmaceuticals, great importance is attached to the protection of undisclosed experimental data, which is required to be submitted for obtaining marketing approval of new drugs. Authorities in charge of marketing approval for new drugs are thus required to protect such data against unfair commercial use by competitors. Further, authorities should protect such data against disclosure, except where necessary to protect the public or unless steps are taken to ensure the protection of such data against unfair commercial use. The duration of data exclusivity varies from country to country but is often 10 years.

An industrial design is the ornamental or aesthetic aspect of a product. The design may consist of three-dimensional features, such as the shape or surface of a product, or of two-dimensional features, such as patterns, lines, or color. As an example, the appearance of an oral dosage form, including the size, shape, embossing, etc. (often referred to as the final market image) is chosen based on the target population and following some degree of market survey. In most countries, an industrial design must be registered in order to be protected under industrial design law. As a general rule, to be registrable, the design must be "new" or "original," and sometimes both. Once a design is registered, a registration certificate is issued. The duration of protection varies significantly from country to country but is generally of at least 10 years, as requested by the TRIPS Agreement (though renewals may be required to benefit from the full length of protection). In the pharmaceutical industry, industrial design protection may be used, for example, to obtain exclusivity over the design of medical equipment.

Copyright grants authors, artists, and other creators (e.g. software companies, multimedia producers, website designers) legal protection for their literary and artistic creations. "Related rights" are the rights granted to people who often play a creative role in communicating some types of works to the public, such as performers, producers of sound recordings, and broadcasting organizations. Copyright protection in the pharmaceutical industry may arise, for example, in relation to advertising campaigns, scientific publications, or other creative output.

The Modern World of Data Privacy, Protection, and Sharing

Data protection is an important aspect of the pharmaceutical industry throughout all stages of a product lifecycle – from innovation to exit. In the early stages of product development, manufacturers often seek to obtain trial data related to their therapeutic areas and franchises to provide a competitive edge; security issues arising during licensing and collaborative activities in the later stages of product development; after development and approval, patients seek to protect their privacy surrounding the use of a drug.

The Health Insurance Portability and Accountability Act (HIPAA), passed by Congress in 1996, has been a source of controversy in the data protection sphere. HIPAA requires confidential treatment of protected health information yet provides for disclosure of it in certain instances, such as when related to treatment, payment, or healthcare operations. Another issue is social media. Pharmaceutical manufacturers want to utilize social media tools to promote their products and provide information on health and diseases, but the industry is hesitant to proceed due to a lack of consistent guidance from the Food and Drug Administration (FDA) regarding enforcement in the social media sphere, as well as the onus of monitoring and reporting adverse events. Other data protection issues surround transactional due diligence, i.e. the practice of gathering information regarding a transaction with a company. Buyers and licensees need to have the requisite information to make an informed decision regarding the deal (e.g. licensing a pharmaceutical product) but, at the same time, the selling or licensing company has an interest in protecting its valuable information in case the deal falls through – and avoid having provided a third party with unfettered access to protected information (Ponzio et al. 2011). Similarly, companies have varied policies regarding corporate records retention, that is, the duration that corporate records must be preserved (Restaino et al. 2011). Record preservation is required in order to comply with regulatory obligations, but preservation beyond the required period, while valuable, must be weighed against the large costs of preserving vast quantities of information and the potential liability of maintaining confidential information. Finally, litigants can demand to see protected information if pertinent to a legal dispute. Companies do not want their protected information leaked out and try to protect against disclosure by using confidentiality agreements to ensure the protection of the data that must be disclosed in litigation.

A milestone in data privacy is the relatively recent (25 May 2018) adoption of the General Data Protection Regulation (GDPR) – the European Union's (EU) new data protection/data privacy regulation. GDPR was created to reform EU data privacy law for the digital age with the following objectives: promote transparency, so individuals understand what personal data is collected and how it is used, expand privacy rights and give individuals greater control over the use of their personal data, and require companies to uphold these privacy rights and impose penalties for noncompliance. GDPR applies to any organization established in the EU as well as any organization (regardless of where established) offering goods or services to or monitoring EU residents. Personal data is defined as any information that can identify a living person and distinguish a person from others and can include name, email/postal addresses, telephone number, government ID number, picture, date of birth, social security number, criminal acts/ records, ethnic origin, genetic information, physical or mental health information, sexual orientation, and biometric data for the purpose of uniquely identifying a natural person. A key principle for GDPR is that personal data be processed lawfully, fairly, and in a transparent manner in relation to the data subject.

GDPR is very explicit in the definition of conditions that define this principle. It is stated in Article 6 of GDPR that processing shall be lawful only if and to the extent that at least one of the following applies: (1) the data subject has given consent to the processing of his or her personal data for one or more specific purposes; (2) processing is necessary for the performance of a contract to which the data subject is party or in order to take steps at the request of the data subject prior to entering into a contract; (3) processing is necessary for compliance with a legal obligation to which the controller is subject; (4) processing is necessary in order to protect the vital interests of the data subject or of another natural person; (5) processing is necessary for the performance of a task carried out in the public interest or in the exercise of official authority vested in the controller; and (6) processing is necessary for the purposes of the legitimate interests pursued by the controller or by a third party, except where such interests are overridden by the interests or fundamental rights and freedoms of the data subject, which require protection of personal data, in particular where the data subject is a child. Article 5 further explains that personal data can only be processed for specific, explicitly declared, and legitimate purposes, for example, to detect, assess, understand and prevent adverse reactions and to identify, and take actions to reduce the risks of and increase the benefits from medicinal products for the purpose of safeguarding public health (GVP module VI.C.6.2.2.10).

A key component of GDPR is the requirement that personal data cannot be re-used for other purposes. It is stated that personal data shall be adequate, relevant, and limited to what is necessary in relation to the purposes for which they are processed (emphasis on data minimization) – sponsors are guided to not collect more personal data than needed. This is in stark contrast to the practices of the past when companies would accumulate personal data and conduct additional analyses (often referred to as "secondary use") sometimes outside the specified user in the Informed Consent. Additional laws are certain to follow, and the penalties of noncompliance to GDPR are severe both in financial and reputational metrics. It is clear that the industry must modernize its practices with respect to communications and data management in order to comply with GDPR and future privacy laws.

References

GDPR (2018). https://eugdpr.org.

Nealey, T., Daignault, R.M., and Cai, Y. (2015). Trade secrets in life science and pharmaceutical companies. *Cold Spring Harbor Perspectives in Medicine* 5: a020982.

Ponzio, T.A., Feindt, H., and Ferguson, S. (2011 1). License compliance issues for biopharmaceuticals: special challenges for negotiations between companies and non-profit research institutions. *LES Nouvelles* 46 (3): 216–225.

Restaino, D., Gladstone-Restaino, L., and Shaw, M. (2011). Data protection in the pharmaceutical industry: concerns and considerations. *International In-house Counsel Journal* 5 (17): 1. ISSN 1754-0607 print/ISSN 1754-0607.

Chapter Self-Assessments: Check Your Knowledge

Questions:
- Why is it important for a pharmaceutical company to create a value proposition?
- What performance metrics are commonly used to judge a company relative to the competition?

- What is due diligence and why is it important?
- What does intellectual property refer to and why is it important to pharmaceutical companies?

Answers:

- For a company to sustain and grow its portfolio, it must create a value proposition supporting its worth to the external world and the marketplace, engage in a variety of agreements with contractors, other companies and partners, and protect its intellectual property in the form of patents, copyrights, and trademarks among other assets. The worth and value of a company is far beyond agreements, however, and the reputation of the company contributes to the overall valuation.

- Performance metrics that managers in the pharmaceutical industry may use to benchmark their performance against others in the industry include the following: sales, operating profit margin, R&D spending as a percentage of sales, core earnings per share, operating free cash flow as a percentage of sales, number of major regulatory approvals received, number of Phase 1–3 clinical trials, percent of sales due to recently released products, average development cost per product.

- Due diligence is an extensive process undertaken by a company to thoroughly and completely assess another organizations business, assets, capabilities, and financial performance. There are numerous aspects of due diligence analysis, and the process is not only relevant for the acquisition of a company but other assets including specific technologies, drug or vaccine candidates at various stages of development, business units (e.g. consumer products, veterinary medicine, etc.) or even whole therapeutic areas or franchises within an organization.

- Intellectual property refers to the exclusive rights granted by the State over creations of the human mind inventions, literary and artistic works, distinctive signs and designs used in commerce. Intellectual property is divided into two main categories: industrial property rights, which include patents, utility models, trademarks, industrial designs, trade secrets, new varieties of plants, and geographical indications; and copyright and related rights, which relate to literary and artistic works. Industrial property (IP) rights are extremely important for the pharmaceutical industry. The use of the IP system by subject matter experts (SMEs) in the pharmaceutical industry depends largely on the business strategy of a company, its size, resources, innovative capacity, competitive context, and field of expertise. Research-based, innovation-led companies that seek to develop new drugs, improve, or adapt existing drugs or develop new pharmaceutical/medical equipment or processes, tend to rely heavily on the patent system to ensure they recover the investments incurred in research and development.

Quiz:

1 Which of the following is not a common valuation method used to assess pharmaceutical company's value? Choose the best answer.
   A Maximum likelihood method
   B Asset-based valuation
   C Income approach
   D Market approach
   E Both a and b

2 True or false. If an employee leaves a company, they may be asked to sign a non-compete agreement, which would prevent them from seeking employment with a competitor for a set period of time.

3 A patent is granted by the national or regional patent office. It is valid for a limited period of time, generally for _____ years from the filing date (or priority date) of the patent application, provided the renewal (or maintenance) fees are paid to keep the patent in force. Choose the best answer to fill in the blank.
   A 5
   B 10
   C 20
   D 25
   E None are correct

4 The acronym GDPR stands for . . .. (Choose the best answer)
   A Good data is pretty reasonable
   B Global Data Privacy Regulation
   C General Data Privacy Regulation
   D General Data Protection Regulation
   E None are correct

4

Global Regulatory Landscape
Jeffrey S. Barrett

Aridhia Digital Research Environment

Introduction – The Necessity of Regulation

As discussed in the introductory chapter, the growth and development of the pharmaceutical industry go hand in hand with the recognition of the need for regulatory oversight and the foundation of governing bodies who would take on that responsibility framing the requirements for the evolving industry and proposing a process by which they could ensure that patients receiving and prescribers recommending new medicines would be both protected and informed. The history of regulation for the industry is still recent. It was not until 1938 that a newly enacted US Food, Drug, and Cosmetic Act subjected new drugs to premarket safety evaluation for the first time (Junod 2014). This required regulators in the United States to review both preclinical and clinical test results for new drugs. Although the law did not specify the kinds of tests that were required for approval, the new authority allowed drug officials to block the marketing of a new drug formally or delay it by requiring additional data. The act also gave regulators limited powers of negotiation over scientific study and approval requirements with the pharmaceutical industry and the medical profession. The creation of the US Food and Drug Administration (USFDA) was, of course, a milestone itself and represented the first organization of its kind to have such a role. This event did precipitate the creation of a global regulatory community; however, it was not until several years later that China (https://www.sfdachina.com), Europe (https://www.ema.europa.eu/en), and Japan (https://www.pmda.go.jp/english) created similar organizations and still later that they coordinated amongst each other in any meaningful way. National agencies now exist in most parts of the world including many developing countries. Some smaller countries benefit from regional agencies, and of course, organizations such as ICH and the World Health Organization (WHO) provide an umbrella for

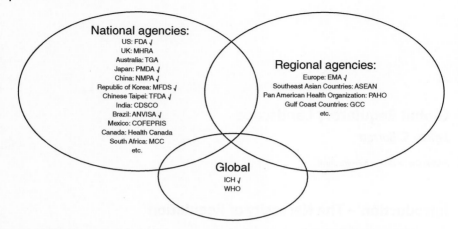

Figure 4.1 Global regulatory environment landscape with national, regional, and global organizations identified (√ symbol denotes ICH member).

global regulatory guidance. Figure 4.1 illustrates the overlap of national, regional, and global regulatory organizations at present. It should be appreciated that the landscape is always changing; efforts to harmonize such as ICH serve a key role in ensuring that regulatory guidance is standardized as much as possible.

Unfortunately, the evolution of regulatory science and the laws that support, modify and fund the underlying organizations is often indexed upon tragedies that occurred during drug development, but it is also important and relevant to note that it was because these organizations existed that there could be an appropriate response to these tragedies often in the form of new requirements and regulations. Many of these regulations were legally mandated by the governments managing and funding the various regulatory agencies. Figure 4.2 shows the timeline of critical legislation that governed many of the key milestones in drug development that evolved over time in the United States (upper chart) and in the rest of the world (lower chart) (Dunne et al. 2013).

In this chapter, we will review the primary global regulatory authorities representing the economies that drive the industry, describing their current organizational structure and procedural expectations (Pezzola and Sweet 2016). The reference point for these descriptions and differences in structure or procedure will be the USFDA, the oldest and largest regulatory organization, which is still viewed as the standard-bearer for regulatory science. Other organizations, particularly those that govern the largest populations, will be discussed as well, and we will review how the collective process works for both drugs and vaccines. Finally, we will describe efforts to harmonize regulatory science and make it easier for drug and vaccine developers to coordinate regulatory filings and obtain market access globally.

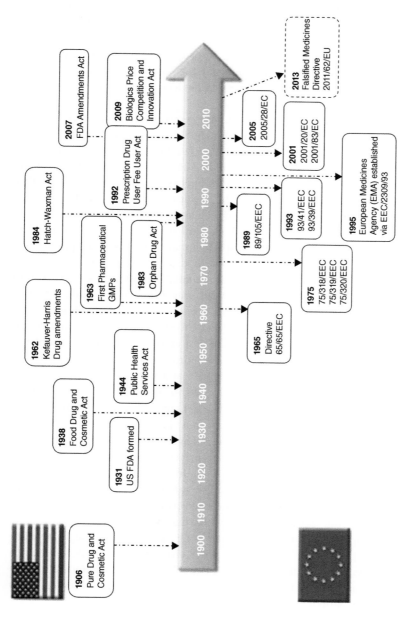

Figure 4.2 Timeline history of pharmaceutical regulations-timeline of significant legislations in the twentieth and twenty-first centuries (original and unedited figure from [Dunne et al. 2013], license to reproduce figure through creative commons, https://creativecommons.org/licenses/by/2.0).

The USFDA

FDA or USFDA is a federal agency of the US Department of Health and Human Services, one of the US federal executive departments. The FDA is responsible for protecting and promoting public health through the control and supervision of food safety, tobacco products, dietary supplements, prescription, over-the-counter drugs, vaccines, biopharmaceuticals, blood transfusions, medical devices, electromagnetic radiation emitting devices, cosmetics, animal (foods and feed), and veterinary products.

In 1927, the Bureau of Chemistry's regulatory powers were reorganized under a new Department of Agriculture (USDA) body, the Food, Drug, and Insecticide organization. This name was shortened to the FDA three years later. The FDA was empowered by the US Congress to enforce the Federal Food, Drug, and Cosmetic Act, which serves as the primary focus for the agency; the FDA also enforces other laws, notably Section 361 of the Public Health Service Act and associated regulations, many of which are not directly related to food or drugs. These include regulating lasers, cellular phones, condoms, and control of disease on products ranging from certain household pets to sperm donation for assisted reproduction. The FDA is led by the Commissioner of Food and Drugs, appointed by the president with the advice and consent of the Senate. The Commissioner reports to the Secretary of Health and Human Services. The FDA has its headquarters in unincorporated White Oak, Maryland. The agency also has 223 field offices and 13 laboratories located throughout the 50 states, the US Virgin Islands, and Puerto Rico as of 2020. In 2008, the FDA began to post employees to foreign countries, including China, India, Costa Rica, Chile, Belgium, and the United Kingdom.

The US Congress increased FDA's authority and mandated formal rules for drug evaluation in response to precipitating events, notable cases of widespread adverse drug reactions. Historically, legislative interventions in the United States were predicated on the notion that patients must be protected by the state from the worst ravages of free-market capitalism. Congress and the FDA expected government control over premarket testing to protect patients otherwise open to abuses by industry and the medical profession. In the 1980s and 1990s, however, patients represented by disease-based organizations sought greater access to drugs and speedier approvals. At the same time, critics warned that the country's competitive standing depended on the pharmaceutical and biotech sectors. A strict boundary between testing and marketing – established by legislative initiatives and implemented rigorously by FDA officials – then was softened to allow for greater access to new medicines. Regulation of pharmaceuticals in the United States has followed an overall progression from the medical profession to a new consumer/patient oversight model.

EMA

The European Medicines Agency (EMA) is an agency of the European Union (EU) in charge of the evaluation and supervision of medicinal products (EMA 2020). Prior to 2004, it was known as the European Agency for the Evaluation of Medicinal Products or European Medicines Evaluation Agency (EMEA). The EMA was set up in 1995, with funding from the EU and the pharmaceutical industry, as well as indirect subsidy from member states, its stated intention to harmonize (not replace) the work of existing national medicine regulatory bodies. The hope was that this plan would not only reduce the €350 million annual cost drug companies incurred by having to win separate approvals from each member state but also that it would eliminate the protectionist tendencies of sovereign states unwilling to approve new drugs that might compete with those already produced by domestic drug companies.

The EMA replaced the Committee for Proprietary Medicinal Products and the Committee for Veterinary Medicinal Products, though both were reborn as the core scientific advisory committees. The agency was located in London prior to the United Kingdom's vote for withdrawal from the EU, relocating to Amsterdam in March 2019. The EMA has seven scientific committees (Committee for Medicinal Products for Human Use (CHMP), Pharmacovigilance Risk Assessment Committee (PRAC), Committee for Medicinal Products for Veterinary Use (CVMP), Committee for Orphan Medicinal Products (COMP), Committee on Herbal Medicinal Products (HMPC), Committee for Advanced Therapies (CAT), and Pediatric Committee (PDCO)) and several working parties and related groups which conduct the scientific work of the agency. The committee's evaluations of marketing-authorization applications submitted through the centralized procedure provide the basis for the authorization of medicines in Europe. These groups can be consulted by the agency's scientific committees on scientific issues relating to their field of expertise. The groups are made up of members who have expertise in a scientific field, selected from the list of European experts maintained by the agency.

EMA committees each have their own rules of procedure. To carry out a scientific assessment, usually a committee appoints a rapporteur to prepare an assessment report, which the committee will consider and eventually adopt as part of a scientific opinion or recommendation. For certain procedures, a "co-rapporteur" also prepares an assessment independently from the rapporteur. An assessment team supports the rapporteur and co-rapporteur with the necessary expertise and resources. The EMA secretariat provides technical, scientific, and administrative support for each assessment. Rapporteurs and co-rapporteurs can establish multinational assessment teams by including experts from other member states as well as their own. This is intended to mobilize the best expertise for medicines evaluation regardless of where experts are geographically based. For more information, see the European medicines regulatory network. A peer-review process provides additional quality assurance of certain scientific assessments. EMA committees try to reach their conclusions by consensus whenever possible, but if not, the committee holds a vote. A more current review of the EMA mission and recent initiatives can be found in the paper by Nicotera (Nicotera et al. 2019).

PMDA (Japan)

Japan's Ministry of Health, Labor, and Welfare (MHLW) is the regulatory body that oversees food and drugs in Japan, which includes creating and implementing safety standards for medical devices and drugs. In conjunction with the MHLW, the Pharmaceutical and Medical Device Agency (PMDA) is an independent agency that is responsible for reviewing drug and medical device applications (PMDA 2020). The PMDA works with the MHLW to assess new product safety, develop comprehensive regulations, and monitor post-market safety. The Pharmaceuticals and Medical Devices Agency (Japanese: 独立行政法人医薬品医療機器総合機構, Dokuritsugyōsei hōjin iyakuhin'iryōkikisōgōkikō) is an Independent Administrative Institution responsible for ensuring the safety, efficacy, and quality of pharmaceuticals and medical devices in Japan. It is similar in function to the Food and Drug Administration in the United States or the Medicines and Healthcare products Regulatory Agency in the United Kingdom. It was founded on 1 April 2004 with headquarters in Tokyo, Japan. Details of the PMDA organizational structure, function, and mission can be found in English on their website, https://www.pmda.go.jp/english and in a previously published overview (Nagasaka 2020).

Current Japan PMDA regulations are laid out in the Pharmaceuticals and Medical Devices Act (PMD Act), also known as the Act on Securing Quality, Efficacy, and Safety of Pharmaceuticals, Medical Devices, Regenerative and Cellular Therapy Products, Gene Therapy Products, and Cosmetics. The PMD Act affects all aspects of Japanese medical product registration, including in-country representation, certification processes, licensing, and quality assurance systems. The PMD Act came into force on 25 November 2014 and replaced the Pharmaceutical Affairs Law (PAL). Key features of the regulations include the following: (1) some Class III medical devices are able to undergo third-party certification, (2) medical software programs are independently regulated, (3) manufacturers are required to be registered rather than be licensed, and (4) quality management systems (QMS) are streamlined. QMS inspection is conducted on the Marketing Authorization Holder and is conducted per product family, not on individual products. The PMDA reviews new drugs, generic drugs, over-the-counter (OTC) drugs ("behind-the-counter" [BTC] drugs as referred to in Japan [equivalent to OTC drugs in the United States]), and quasi-drugs, and conducts re-evaluations of previously approved drugs. Orphan drugs and other priority drugs are given priority reviews in accordance with their clinical significance.

NMPA (China)

The National Medical Products Administration (NMPA) (Chinese: 国家药品监督管理局) (formerly the China Food and Drug Administration, or CFDA) was founded based on the former State Food and Drug Administration (SFDA) (NMPA/CFDA 2020). The original organization was founded in 1950; in March 2013, the former regulatory body was rebranded and restructured as the China Food and Drug Administration, elevating it to a ministerial-level agency. In 2018, as part of China's 2018 government administration overhaul, the name was changed to "National Medical Products Administration" and merged into the newly created State Administration for Market Regulation. The headquarters are in Xicheng District, Beijing.

As in other geographic regions, the Chinese regulatory authority is beholden to the central government that provides modifications and improvements to their institutions via the legislature. Likewise, recent governmental oversight has provided a mechanism for NMPA to evolve further. On 26 August 2019, China's Standing Committee of National People's Congress (NPC) adopted a significant revision of the Drug Administration Law (DAL). The newly adopted DAL (Revised DAL) went into effect on 1 December 2019. The Revised DAL was the first overhaul of the DAL since 2001. Perhaps, the most significant feature of the Revised DAL is the adoption of a nationwide marketing authorization holder (MAH) system. This system links marketing licenses directly to the products, permitting flexibility in designing contract manufacturing and distribution arrangements. The Revised DAL addresses several other significant issues, including encouraging drug innovation, facilitating the drug approval process, improving drug traceability and pharmacovigilance, and amending the definition of counterfeit drugs.

One key difference between the Chinese and other regulatory systems is the review and approval of Traditional Chinese Medicine (TCM) in addition to drugs, vaccines, and devices. TCM is a branch of traditional medicine that is said to be based on more than 3500 years of Chinese medical practice that includes various forms of herbal medicine, acupuncture, cupping therapy, gua sha, massage (tui na), bonesetter (die-da), exercise (qigong), and dietary therapy. While the practice of TCM has been maintained for generations, it is only recently that it has come under the same kind of scrutiny as conventional drug development.

Likewise, the practice has also grown to be influenced by modern Western medicine. Many governments including the United States have similarly enacted laws to regulate TCM practice though they do not review and approve TCM.

Everybody Else

As previously mentioned, smaller and developing countries often lack the infrastructure and/or technical staff to manage regulatory oversight in a meaningful way and choose to either participate in regional agencies that review submissions within a common or identifiable region or adopt the review policies of a larger entity (e.g. EMA or FDA) assuming those decisions are applicable to patients in their counties. Neither approach is necessarily optimal, but both offer some level of scrutiny and oversight beyond the capabilities of many low- or middle-income countries (LMIC) that would struggle to provide adequate staffing of an internal agency.

While there are mechanisms in place to get drug, device, and vaccine submissions reviewed and approved around the world, the standards are different, and despite the appreciation of the need for harmonization, there are areas left behind. Recently Pezzola et al. (Pezzola and Sweet 2016) surveyed developing states around the world to assess the extent of standards in place and adhered to with surprising results. The authors found remarkable resistance to the implementation of global pharmaceutical norms for quality standards in developing states and in regulatory infrastructure. Human capacity across many developing countries remains limited. Most notably, variation among states is stark. Countries that have been leaders in establishing global norms do not appear to have influenced their neighbors in establishing regional patterns. Finally, in contrast to traditional theories of international norms diffusion, global standard-setters such as the United States or EU appear to have surprisingly little influence on the standard-setting.

How It Works (or doesn't) for Drugs

Not surprisingly, perhaps, the regulatory requirements of various countries of the world vary from each other. Therefore, it is challenging for the companies to develop a single drug that can be simultaneously submitted in all the countries for approval. The regulatory strategy for product development is essentially to be established before the commencement of developmental work in order to avoid major surprises after the submission of the application. The role of the regulatory authorities is to ensure the quality, safety, and efficacy of all medicines in circulation in their country. It not only includes the process of regulating and monitoring the drugs but also the process of manufacturing, distribution, and promotion of it. One of the primary challenges for regulatory authority is to ensure that the pharmaceutical products are developed as per the regulatory requirement of that country. This process involves the assessment of critical parameters during product development (Handoo et al. 2012).

Governments have the responsibility to protect their citizens. Likewise, it is the responsibility of individual national governments to establish regulatory authorities with strong guidelines for quality assurance and drug regulations in their respective territories. Somewhat parallel with the ongoing harmonization and movement toward creating a common market for medicines inside the EU, the need for wider harmonization was felt by officials from Japan, the EU, and the United States during the International Conference of Drug Regulatory Authorities (ICDRA) organized by WHO. Informal discussions led to a need for the harmonization of requirements relating to the new innovative drugs and also subsequently paved the way for the establishment of the International Conference on Harmonization of Technical Requirements for the Registration of Pharmaceuticals for Human Use (discussed in more detail later). The driving force behind these efforts has been the increase in global trade in pharmaceutical products and growth in the complexity of technical regulations related to drug efficacy, safety, and quality.

Controversy persists about the differences in US and EU regulatory processes, costs, and the time it can take for a drug candidate to proceed from concept to approval under the regulations of each. A frequently held assertion is that slower FDA approval processes deprive American citizens of effective drugs that are available to Europeans, and critics have characterized FDA processes as "slow, risk averse, and expensive" (van Norman 2016). However, the Institute of Medicine determined that current FDA premarketing procedures for medical devices are insufficient to assure device safety, particularly those approved largely on their similarity to previously cleared "predicate" devices, rather than on prospective, randomized clinical trials. In the EU, concerns abound that drugs may be approved too quickly, to the detriment of patient safety. In recent years, there have been calls to tighten approval processes and to establish regulatory consistency between the FDA and the EU. Efforts include recent legislation in the US Congress to facilitate release in the United States of drugs that have already achieved European approval. Proposed changes to regulations of the European Commission (EC) regarding device approval are under discussion but are vigorously opposed by both industry and patient groups insisting that it will impede the availability of innovative therapies to the public.

In general, however and using the US system as a baseline, for registration of research of new chemical entities (NCEs) for human phase testing, pharmaceutical sponsors begin the process by applying for permission to engage in such testing. This application (IND [Investigational New Drug] in the United States) assimilates specific requirements on preclinical evaluation with emphasis on safety (toxicology experiments especially) and some indication that there is a rationale for the drug candidate to work in human patients (pharmacology experiments and target product profile) along with a plan for human phase testing (Investigator's Brochure and Phase 1 protocol). The regulatory authority receiving the application will have the responsibility to review the application and respond regarding permission to proceed within an expected finite time period. See Chapters 6 and 7 for a detailed discussion of the US process. Essentially, this basic procedure is followed by each country's regulatory agency with some minimal variation. Pending approval of the IND equivalent, human phase testing proceeds (Phases 1–3), and approval is granted or not (pending the review and assessment of the new drug application (NDA in the United States)).

How it Works for Vaccines

The assessment, licensure, control, and surveillance of biological medicinal products including vaccines are major challenges for national regulatory authorities confronted by a steadily increasing number of novel products, complex quality concerns, and new technical issues arising from rapid scientific advances. While national and regional regulatory authorities manage the specific guidance, tracking, and review of submission materials in accordance with the local governing authority that has jurisdiction over the review and approval of vaccine candidates, the WHO also has a role. Through its consultative approach, WHO identifies and consolidates current consensus opinions on key regulatory issues and communicates them to national authorities and manufacturers through guidance documents addressing both general issues and specific products. Through this mechanism, national regulatory authorities are informed on the scientific background needed to assess critical issues and are advised on which regulatory approaches and methodologies have been found to be optimal for ensuring the global supply of uniformly high, quality, and efficacious biological medicinal products. The WHO also provides guidance documents with respect to vaccine development and quality in addition to guidance on immunization standards. Much of the guidance on immunization is focused on access, particularly for LMIC member states. Increased access to biotherapeutic products was recently identified as a global public health priority, articulated in resolution WHA67.21 of the World Health Assembly, for example. The resolution calls on WHO to provide more support to the member states to regulate biotherapeutics and make them accessible to their populations.

Regarding the history of vaccine regulation, at the end of the nineteenth century, several vaccines for humans had been developed. They were smallpox, rabies, plague, cholera, and typhoid vaccines. However, no regulation of vaccine production existed. On 1 July 1902, the US Congress passed "An act to regulate the sale of viruses, serums, toxins, and analogous products," later referred to as the Biologics Control Act (even though "biologics" appears nowhere in the law). This was the first modern federal legislation to control the quality of drugs. This act emerged in part as a response to 1901 contamination events in St. Louis and Camden involving the smallpox vaccine and diphtheria antitoxin. The act created the Hygienic Laboratory of the US Public Health Service to oversee the manufacture of biological drugs. The Hygienic Laboratory eventually became the National Institutes of Health. The act established the government's right to control the establishments where vaccines were made. The United States Public Service Act of 1944 mandated that the federal government issue licenses for biological products, including vaccines. After a poliovirus vaccine accident in 1954 (known as the Cutter incident), the Division of Biologics Standards was formed to oversee vaccine safety and regulation. Later, the DBS was renamed the Bureau of Biologics, and it became part of the Food and Drug Administration. It is now known as the Center for Biologics Evaluation and Research (CBER). In the EU, the EMA supervises the regulation of vaccines and other drugs. A committee of the WHO makes recommendations for biological products used internationally, and many countries have adopted the WHO standards.

The process of vaccine development is analogous to drug development, with the starting point being exploratory and preclinical development focused on vaccine candidate selection in line with the target indication. Following preclinical development, assuming a viable candidate has been identified, the sponsor, usually a private company, submits an application for an IND to the USFDA or complementary authority depending on the geographic region of interest. The sponsor describes the manufacturing and testing processes, summarizes the laboratory reports, and describes the proposed human study. An institutional review board representing an institution where the clinical trial will be conducted must approve the clinical protocol. The FDA then has 30 days to approve the application. Once the IND application has been approved, the vaccine is subject to the traditional three phases of clinical testing (Phases 1–3). Assuming the candidate passes the various development stage gates and after the completion of a successful Phase 3 trial, the vaccine developer will submit a Biologics License Application (BLA) to the FDA. Then the FDA will inspect the factory where the vaccine will be made and approve the labeling of the vaccine. After licensure, the FDA will continue to monitor the production of the vaccine, including inspecting facilities and reviewing the manufacturer's tests of lots of vaccines for potency, safety, and purity. The FDA has the right to conduct its own testing of manufacturers' vaccines. A variety of systems monitor vaccines after they have been approved. They include Phase 4 trials, the Vaccine Adverse Event Reporting System, and the Vaccine Safety Datalink. Phase 4 trials may be optional studies that drug companies conduct after a vaccine is released. Likewise, the manufacturer may continue to test the vaccine for safety, efficacy, and other potential uses. The CDC and FDA established The Vaccine Adverse Event Reporting System (VAERS) in 1990. The goal of VAERS, according to the CDC, is "to detect possible signals of adverse events associated with vaccines." (A signal, in this case, is evidence of a possible adverse event that emerges in the data collected.) About 30 000 events are reported each year to VAERS. Similar systems are available in other organizations and parts of the world (e.g. MSAEFI [Monitoring System for Adverse Events] by CDC and Adverse Events following Immunization [AEFI] in Canada).

Harmonization

The realization that it was important to have an independent evaluation of medicinal products before they are allowed on the market was reached at different times in different regions. However, in many cases, the realization was driven by tragedies, such as that with thalidomide in Europe in the 1960s. For most countries, whether they had initiated product registration controls earlier, the 1960s and 1970s saw a rapid increase in laws, regulations, and guidelines for reporting and evaluating the data on safety, quality, and efficacy of new medicinal products. The industry, at the time, was becoming more international and seeking new global markets; however, the divergence in technical requirements from country to country was such that industry found it necessary to duplicate many time-consuming and expensive test procedures in order to market new products, internationally.

Harmonization of regulatory requirements was pioneered by the EC in the 1980s, as the EC, Europe moved towards the development of a single market for pharmaceuticals. The success achieved in Europe demonstrated that harmonization was feasible. At the same time, there were discussions between Europe, Japan, and the United States on possibilities for more globalized harmonization. It was, however, at the WHO Conference of Drug Regulatory Authorities (ICDRA), in Paris, in 1989, that specific plans for action began to materialize. Soon afterward, the authorities approached the International Federation of Pharmaceutical Manufacturers and Associations (IFPMA) to discuss a joint regulatory-industry initiative on international harmonization, and ICH was conceived.

The birth of ICH took place at a meeting in April 1990, hosted by EFPIA in Brussels. Representatives of the regulatory agencies and industry associations of Europe, Japan, and the United States met, primarily, to plan an International Conference, but the meeting also discussed the wider implications and terms of reference of ICH. At the first ICH Steering Committee meeting of ICH, the Terms of Reference were agreed upon. It was decided that the Topics selected for harmonization would be divided into Safety, Quality, and Efficacy to reflect the three criteria: the basis for approving and authorizing new medicinal products. One of the many benefits of ICH involvement as a harmonizer of regulatory standards is their creation of the common technical document (CTD) (FDA ICH M4 Guidance 2001). The CTD is a set of specifications for an application dossier for the registration of medicines and designed to be used across Europe, Japan, and the United States. It is an internationally agreed format for the preparation of applications regarding new drugs intended to be submitted to regional regulatory authorities in participating countries. It was developed by the EMA (Europe), the FDA (United States), and the MHLW (Japan). The CTD is maintained by the International Conference on Harmonization of Technical Requirements for Registration of Pharmaceuticals for Human Use (ICH). The CTD is divided into five modules: (1) Administrative and prescribing information, (2) Overview and summary of modules 3–5, (3) Quality (pharmaceutical documentation), (4) Preclinical (Pharmacology/Toxicology), and (5) Clinical – efficacy and safety (Clinical Trials). Detailed subheadings for each Module are specified for all jurisdictions. The contents of Module 1 and certain subheadings of other Modules will differ based on national requirements. After the United States, EU, and Japan, the CTD has been adopted by several other countries including Canada and Switzerland.

In the postapproval arena, the Mutual Recognition Agreement (MRA) between FDA and EU allows drug inspectors to rely upon information from drug inspections conducted within each other's borders. Under the Food and Drug Administration Safety and Innovation Act, enacted in 2012, FDA has the authority to enter into agreements to recognize drug inspections conducted by foreign regulatory authorities if the FDA determined those authorities are capable of conducting inspections that met US requirements. FDA and the EU have collaborated since May 2014 to evaluate the way they each inspect drug manufacturers and assess the risk and benefits of mutual recognition of drug inspections. The benefits of MRA are twofold: (1) it yields greater efficiencies for United States and EU regulatory systems by avoiding duplication of inspections, and (2) it enables reallocation of resources towards inspection of drug manufacturing facilities with potentially higher public health risks across the globe.

FDA continues to perform some inspections in EU countries with capable inspectorates, such as product manufacturing assessment inspections to support marketing approval decisions. However, FDA expects to perform fewer routine surveillance inspections in EU countries with a capable inspectorate. FDA is collaborating with the following inspectorates it has assessed as capable and is reviewing their recent inspection reports and related information in determining each manufacturer's suitability for the US market in lieu of an FDA site inspection. FDA completed its capability assessment of all EU inspectorates as of July 2019.

The emergence of a consumer model of regulation poses several critical questions about the longer-term role of government, industry, the medical profession, and citizens. The era of paternalistic medicine has passed, but the notion that patients can act as consumers and make appropriate decisions concerning medical treatment poses countervailing risks of its own. This should be very clear now in the wake of the COVID-19 pandemic when the president of the United States openly recommended off-label use of a drug approved for malaria for an unassessed (by common regulatory standards) infectious disease indication (KFF 2020). Simply stated, a better accommodation among key players needs to be struck to foster the safe use of pharmaceuticals. The precise form of this accommodation will necessarily vary from one country to the next, which holds out the possibility for additional policy learning from future cross-national comparisons. Likewise, it is fair to say that the global regulatory landscape is dynamic and ever-changing.

References

Dunne S, Shannon B, Dunne C and Cullen W. A review of the differences and similarities between generic drugs and their originator counterparts, including economic benefits associated with usage of generic medicines, using Ireland as a case study. *BMC Pharmacology and Toxicology* 2013, 14:1 (accessed 22 March 2020).

European Medicines Agency (EMA) (2020). https://www.ema.europa.eu/en (accessed 22 March 2020).

Guidance for Industry, ICH M4: Organization of the CTD (2001) U.S. Department of Health and Human Services Food and Drug Administration Center for Drug Evaluation and Research (CDER) Center for Biologics Evaluation and Research (CBER) https://www.fda.gov/cber/gdlns/m4ctd.pdf.

Handoo, S., Arora, V., Khera, D. et al. (2012). A comprehensive study on regulatory requirements for development and filing of generic drugs globally. *International Journal of Pharmaceutical Investigation* 2 (3): 99–105. https://doi. org/10.4103/2230-973X.104392. PMID: 23373001; PMCID: PMC3555014.

Junod, S.W. (2014). US Food and Drug Administration. https://www.fda.gov/ media/110437/download.

KFF Trump Continues To Promote Off-Label Use Of Drug For COVID-19 Despite Expert Opinion Of Unproven Effectiveness (2020). https://www.kff.org/cb771f3 (accessed 23 April 2020).

Nagasaka, Satoru (2020). An Overview of Pharmaceutical and Medical Device Regulation in Japan (PDF). Morgan, Lewis & Bockius LLP. https://www. morganlewis.com/-/media/files/publication/outside-publication/article/overview_ pharma_device_reg.ashx (accessed 22 March 2020).

National Medical Products Administration (NMPA/CFDA) (2020). https://www. sfdachina.com (accessed 22 March 2020).

Nicotera, G., Sferrazza, G., Serafino, A., and Pierimarchi, P. (2019). The Iterative Development of Medicines Through the European Medicine Agency's Adaptive Pathway Approach. *Frontiers in Medicine (Lausanne)* 6: 148. https://doi. org/10.3389/fmed.2019.00148. PMID: 31316991; PMCID: PMC6610487.

Pezzola, A. and Sweet, C.M. (2016). Global pharmaceutical regulation: the challenge of integration for developing states. *Global Health* 12: 85. https://doi.org/10.1186/ s12992-016-0208-2.

Pharmaceuticals and Medical Devices Agency – PMDA (2020). https://www.pmda. go.jp/english (accessed 22 March 2020).

Van Norman, G. (2016). Drugs and devices: comparison of European and U.S. approval processes. *JACC: Basic to Translational Science* 1 (5): 399–412.

Chapter Self-Assessments: Check Your Knowledge

Questions:
- Why is the government involved with the regulation of drugs in the US?
- How does the EMA manage its geographic boundaries with respect to regulatory review?
- How is the World Health Organization (WHO) involved with the regulation of vaccines?
- The ICH-created common technical document is touted as one of the organizations most successful contributions toward harmonization. What is the CTD and how does it help with global regulatory submissions?

Answers:
- The U.S. Congress provides FDA's authority via legislation and mandates formal rules for drug evaluation in response to precipitating events, notably cases of widespread adverse drug reactions. Historically, legislative interventions in the United States were predicated on the notion that patients must be protected by the state from the worst ravages of free-market capitalism. Congress and the FDA expected government control over pre-market testing to protect patients otherwise open to abuses by industry and the medical profession.
- EMA committees each have their own rules of procedure. To carry out a scientific assessment, usually, a committee appoints a rapporteur to prepare an assessment report, which the committee will consider and eventually adopt as part of a scientific opinion or recommendation. For certain procedures, a 'co-rapporteur' also prepares an assessment independently from the rapporteur. An assessment team supports the rapporteur and co-rapporteur with the necessary expertise and resources. The EMA secretariat provides technical,

scientific, and administrative support for each assessment. Rapporteurs and co-rapporteurs can establish multinational assessment teams by including experts from the other Member States as well as their own. This is intended to mobilize the best expertise for medicines evaluation regardless of where experts are geographically based.

- Through its consultative approach, WHO identifies and consolidates current consensus opinions on key regulatory issues and communicates them to national authorities and manufacturers through guidance documents addressing both general issues and specific products. Through this mechanism, national regulatory authorities are informed on the scientific background needed to assess critical issues and are advised on which regulatory approaches and methodologies have been found to be optimal for ensuring the global supply of uniformly high, quality, and efficacious biological medicinal products. The WHO also provides guidance documents with respect to vaccine development and quality in addition to guidance on immunization standards.
- The CTD is a set of specifications for an application dossier for the registration of medicines and designed to be used across Europe, Japan, and the United States. It is an internationally agreed format for the preparation of applications regarding new drugs intended to be submitted to regional regulatory authorities in participating countries. It was developed by the European Medicines Agency (EMA, Europe), the Food and Drug Administration (FDA, United States), and the Ministry of Health, Labor and Welfare (Japan). The CTD is maintained by the International Conference on Harmonization of Technical Requirements for Registration of Pharmaceuticals for Human Use (ICH). The Common Technical Document is divided into five modules: (1) Administrative and prescribing information, (2) Overview and summary of modules 3–5, (3) Quality (pharmaceutical documentation), (4) Preclinical (Pharmacology/Toxicology), and (5) Clinical – efficacy and safety (Clinical Trials). Detailed subheadings for each module are specified for all jurisdictions. The CTD conserves the time and cost of submission preparation by avoiding region-specific submission requirements.

Quiz:

1 The acronym for the Japanese regulatory authority is PMDA. What does the acronym stand for? Choose the best answer. a
   A Pharmaceutical and Medical Device Agency
   B Pharmaceutical Manufacturers Drug Agency
   C Product and Medical Development Agency
   D Pharmaceutical and Medical Development Agency
   E None are correct

2 True or false. The EMA currently has its headquarters in London, UK. False

3 True or false. The National Medical Products Administration (NMPA) in China also reviews the submission of traditional Chinese medicine (TCM) applications. True

4 The CDC and FDA established The Vaccine Adverse Event Reporting System (VAERS) in 1990. The goal of VAERS, according to the CDC, is to "_____ _____." Choose the best answer.

   A monitor vaccine performance on the marketplace
   B monitor safety signals related to vaccine usage
   C detect possible signals of adverse events associated with vaccines
   D detect counterfeit vaccine adverse effects
   E None are correct

5

Phases of Drug Development and Drug Development Paradigms

Jeffrey S. Barrett

Aridhia Digital Research Environment

Introduction – The Necessity of Establishing Clinical Efficacy and Evolution of Human Phase Testing

In a recent account, Dr. Suzanne White Junod provided a short history of the Food and Drug Administration (FDA) (Junod 2014) and reviewed some of the prehistory establishing the need for clinical testing. In this review, she provided the following historical anecdote.

> *"The Babylonians reportedly exhibited their sick in a public place so that onlookers could freely offer their therapeutic advice based on previous and personal experience. The first mention of a paid experimental subject came from Diarist Samuel Pepys who documented an experiment involving a paid subject in a diary entry for November 21, 1667. He noted that the local college had hired a 'poor and debauched man' to have some sheep blood 'let into his body'. Although there had been plenty of consternation beforehand, the man apparently suffered no ill effects."*

The anecdote provides a reminder of a time not that long ago when there was no process for drug development, and the regulations that guided the conduct of experimental testing and the generation of data to support product registration were not in place. What has ensued, thankfully, is a well-established and evolving process to guide sponsors to generate data that provides the necessary evidence to support the marketing of new medicines allowing regulators to approve these drug candidates with some degree of confidence. Along the way, this process has evolved into well-defined and somewhat distinct stages that involve multidisciplinary teams and collaborations with development partners, academic collaborators, and regulatory scientists that participate in what we refer to as modern drug development.

Fundamentals of Drug Development, First Edition. Edited by Jeffrey S. Barrett.
© 2022 John Wiley & Sons, Inc. Published 2022 by John Wiley & Sons, Inc.
Companion website: www.wiley.com/go/Barrett/FundamentalsDrugDevelopment

Drug development then is the process of bringing new medicines to the market once a lead compound has been identified through the process of drug discovery. It includes preclinical research on microorganisms and animals, filing for regulatory statutes, such as via the US Food and Drug Administration (USFDA) for an investigational new drug to initiate clinical trials on humans, and may include the step of obtaining regulatory approval with a new drug application to market the drug. This process proceeds in a sequential manner, with drug sponsors attempting to quickly generate the data to convince regulators that their drug candidate is safe and efficacious in the intended target populations to secure market access while preserving as much of their patent life as possible. In this chapter, we will review the stages of modern drug development, evaluate past, current and potentially future paradigms aimed at streamlining the process and review tools such as the target product profile (TPP) and integrated product development plan (IPDP) used by sponsors to outline and refine their development programs.

Stages of Development

The commonly appreciated stages of drug development include discovery, product characterization, formulation development, delivery, packaging development, pharmacokinetics and drug disposition, preclinical toxicology evaluation, additional preclinical evaluation leading to IND application, bioanalytical testing, clinical development including clinical trials culminating in a regulatory submission (NDA for the USFDA) if successful. These stages are distinct with some overlap and generally but not entirely sequential. They are broadly defined in Table 5.1, but additional detail is provided in distinct chapters (e.g. Chapters 6 through 9 focused on Discovery through Phase 3 clinical testing) later in the text.

The Historical Paradigm

There are many schematics that organize the product stages in a sequential manner illustrating the duration of the individual phase and occasionally the cost, usually on a chronologic scale that illustrates the cumulative time in development. In this way, one can compare development time in the context of the 20-year patent life granted to a drug candidate, process, or use (Roses 2008), assessing return-on-investment (ROI). Figure 5.1 provides one such representation. This figure provides an ordering of the development stages into sequential steps necessary for a drug to progress through the research and development (R&D) pipeline. The phases are shown in chronologic order, along with the approximate times taken within each stage. The most crucial step is the proof of concept for

Table 5.1 Drug development stages that form the basis of current paradigms.

Stage	Focus
Discovery	Target identification – choosing a biochemical mechanism involved in a disease condition. Drug candidates discovered in academic and pharmaceutical/biotech research labs are tested for their interaction with the drug target. Typically, 5000–10000 molecules for each potential drug candidate are subjected to a rigorous screening process which can include functional genomics and/or proteomics as well as other screening methods. Once interaction with the drug target is confirmed, validation of the target by checking for activity versus the disease condition for which the drug is being developed ensues. One or more lead compounds are usually selected.
Product characterization	Promising candidate molecules must be characterized – size, shape, strengths, and weaknesses, preferred conditions for maintaining the function, toxicity, bioactivity, and bioavailability must be determined. Characterization studies will undergo analytical method development and validation. Early-stage pharmacology studies help to characterize the underlying mechanism of action of the compound.
Formulation, delivery, packaging development	Drug developers must devise a formulation that ensures the proper drug delivery. Formulators look ahead to clinical trials at this phase of the drug development process. Drug formulation and delivery are refined continuously until, and even after, the drug's final approval. Scientists determine the drug's stability – in the formulation itself and for all the parameters involved with storage and shipment, such as heat, light, and time. The formulation must remain potent and sterile, and it must also remain safe (nontoxic). It may also be necessary to perform leachables and extractables studies on containers or packaging.
Pharmacokinetics and drug disposition	Pharmacokinetic (PK) and ADME (Absorption/Distribution/Metabolism/Excretion) studies provide useful feedback for formulation scientists. PK studies yield parameters such as AUC (area under the curve), Cmax (maximum concentration of the drug in blood), and Tmax (time at which Cmax is reached). Data from animal PK studies are compared to data from early-stage clinical trials to check the predictive power of animal models.

(Continued)

Table 5.1 (Continued)

Stage	Focus
Preclinical toxicology testing and IND application	Preclinical testing analyzes the bioactivity, safety, and efficacy of the formulated drug product. This testing is critical to a drug's eventual success and, as such, is scrutinized by many regulatory entities. During the preclinical stage of the development process, plans for clinical trials and an investigative new drug (IND) application are prepared. Studies taking place during the preclinical stage should be designed to support the clinical studies that will follow. The main stages of preclinical toxicology testing are acute, repeat dose, genetic toxicity, reproductive toxicity, carcinogenicity, and toxicokinetic studies.
Bioanalytical testing	Bioanalytical laboratory work and method development support most of the other activities in drug development. The bioanalytical effort is key to the proper characterization of the molecule, assay development, developing optimal methods for cell culture or fermentation, determining process yields, and providing quality assurance and quality control for the entire development process. It is also critical for supporting preclinical toxicology/pharmacology testing and clinical trials.
Phase 1 clinical development (focus on human pharmacology)	Thirty days after a sponsor has filed its IND, it may begin a small, Phase 1 clinical trials unless the FDA places a hold on the study. Phase 1 studies are used to evaluate safety, pharmacokinetics, and tolerance, generally in healthy volunteers. These studies include initial single-dose studies, dose-escalation, and short-term repeated-dose studies (see Chapter 7).
Phase 2 clinical development (therapeutic exploratory, POC)	Phase 2 clinical studies are small-scale trials to evaluate a drug's preliminary efficacy and side-effect profile in typically 100–250 patients. Additional safety and clinical pharmacology studies are also included in this category, and its primary goal is to establish Proof-of-concept (POC) – see Chapter 8.

| Phase 3 clinical development (therapeutic confirmatory) | Phase 3 studies are large-scale clinical trials designed to confirm safety and efficacy in large patient populations. While Phase 3 studies are in progress, preparations are made for submitting the Biologics License Application (BLA) or the new drug application (NDA). BLAs are currently reviewed by the FDA's Center for Biologics Evaluation and Research (CBER). NDAs are reviewed by the Center for Drug Evaluation and Research (CDER) – see Chapter 9. |
| Phase 4 (postmarketing and surveillance commitments) | Clinical research is conducted after a drug has been approved. Due to the modest size of developmental programs, evaluation of a drug's toxicity profile and overall understanding of its safety can only partially be determined prior to approval. The understanding at the approval of an NCE's toxicity profile and overall benefit–risk is best considered provisional. FDA often imposes obligations on drug manufacturers, as a condition of FDA approval, to conduct one or more Phase 4 post-marketing studies to fill important data gaps. |

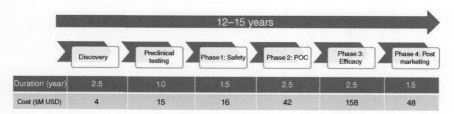

	Duration (year)	Cost ($M USD)

(schematic flow: Discovery → Preclinical testing → Phase 1: Safety → Phase 2: POC → Phase 3: Efficacy → Phase 4: Post marketing, spanning 12–15 years)

	Discovery	Preclinical testing	Phase 1: Safety	Phase 2: POC	Phase 3: Efficacy	Phase 4: Post marketing
Duration (year)	2.5	1.0	1.5	2.5	2.5	1.5
Cost ($M USD)	4	15	16	42	158	48

Figure 5.1   Representative drug discovery and development schematic.

efficacy, which is early on in drug development but can be considered an essential end to a drug discovery program of five to seven years' duration. There are four clinical phases, as mentioned previously as well. This effort is a large hurdle for academia and many biotechnology companies, which is why most of the development activity still resides in "Big Pharma."

After passing the preclinical safety hurdles, a molecule ready for development could be worth many times the cost of the safety /toxicology investment (or not). The return on investment through this step can be quite high for those who can take the risk (Paul et al. 2010). This is the financial incentive for companies to adopt a "quick kill" approach where they adhere to Go/No-go criteria and define quantitative stage gates by which they will judge the progress of development of a drug candidate. In this way, the attributes that the project team and senior leadership have agreed to will be required to be met or surpassed, or the compound will be removed from development consideration. The sooner this can be determined in the development phase, the better from a financial point of view, hence the "quick kill" designation.

While the duration of time in each phase has been reduced somewhat due to the influence of innovations such as high-throughput screening (Macarron et al. 2011) and in silico modeling (Sieburg 1990), these improvements haven't necessarily improved the number or quality of drug candidates in the manner initially envisioned. The cost per phase has escalated; the average cost per phase shown in Figure 5.1 is based on 2008 estimates and most assuredly has gone up since then. Likewise, there are those who feel that the current paradigm could benefit from some further adjustment and innovation.

The Search for Greater Efficiency – Innovation Driving Paradigm Shifts

Over the past three decades, the number of new molecular entities approved by the FDA has averaged 20–30 drugs per year, except for a peak in the mid-1990s that experienced a doubling of this rate. This modest productivity cannot be explained by lack of funding, as the research budgets of government- and industry-funded programs have increased threefold to fivefold over the same time period. Considering the historical perspective of drug discovery and the role of serendipity, it can be argued that the current emphasis on translational research diverts scientists from pursuing basic-science studies that give rise to fundamental discoveries. In many cases, retro-translational research (from clinic to basic science) is necessary before disease processes can be understood well enough for scientists to develop therapeutics. Ultimately, a balance of disease-oriented and basic-science research on fundamental processes is optimal though the question remains of how best to strike this balance within the timing constraints of drug development and the competitive marketplace.

Declining or stagnant R&D productivity has led many to suggest that the current paradigm for drug discovery and development requires disruptive innovation (i.e., a transformation of the pharmaceutical industry driven by new technology, new business models, or policy decisions that improve therapy and create value for patients and society in a way that could not be achieved through other means) to break out of a current crisis by identifying and rapidly bringing new discoveries to market (see Paul et al. 2010; FitzGerald 2011; Munos and Chin 2011; Scannell et al. 2012; Elkins et al. 2013; Bowen and Casadevall 2015). But how does one choose an alternative paradigm that is testable and generalizable to all therapeutic areas? Do companies really avoid the herd mentality and blaze a trail or not? In reality, much of the current approach is dictated to a large extent by regulatory requirements defined by risk management principles. Many have called for FDA reform (Kesselheim et al. 2016), but this is not a clear solution either, and leaving the protection of the American public to politicians would not seem to be prudent.

New Paradigm Proposals

Many opinions exist on how the current drug development paradigm can be modified to reduce the time and cost of development. Most of the proposals call for increased investment in discovery science to select better candidates linked to better defined targets. On the development side, there is the feeling that increased collaboration with shared risk and development costs, such as the conduct of platform trials, will further reduce cost and improve efficiency while approving better drugs with well-defined therapeutic windows and target populations. All of these are laudable goals, but how do we get there?

On the drug discovery front, the enormous progress in the development of new methods in the field of molecular biology and computer science is currently unprecedented, and the drug design and discovery should soon be able to construct a virtual drug with all the desired chemical, physical and biological properties to survive the rigors of clinical testing before doing a single chemical reaction. Drug design and discovery may be used to rationally construct a drug "blueprint" for each individual for tailored therapy based on our genetic makeup. Proteomics and emerging fields like chemogenomics and metabolomics, along with more accessible DNA chips, should eventually be able to construct protein chips that have the ability to perform high throughput structural genomics to unravel the conformations of all relevant proteins in specific disease processes. While various researchers are working to generate protein microarray, other alternative strategies involving HPLC, 2-D gel electrophoresis, and mass spectrometry are providing attractive alternative methods for protein analysis.

On the development front, novel collaboration forums highlighting the input of the patient and payer communities are evolving (Hernandez et al. 2015) at the same time. The Patient-Centered Outcomes Research Institute (PCORI) created the National Patient-Centered Clinical Research Network (PCORnet), a coordinated, interoperable "network of networks" comprising 13 Clinical Data Research Networks (CDRNs) and 18 Patient-Powered Research Networks (PPRNs). Each CDRN represents a collaboration among existing health systems, including academic health centers, community hospitals, health plans, inpatient and outpatient hospitals and providers, Federally Qualified Health Centers, pediatric hospitals and providers, integrated delivery systems, electronic health record companies, and regional Health Information Exchanges. In contrast, PPRNs typically comprise organizations of patients, families, and advocates affected by a particular medical condition that have coalesced, often in collaboration with academic researchers, into research networks dedicated to addressing issues relevant to their care and outcomes.

One of the more novel concepts that these collaboration models permit is the ability to conduct a pragmatic clinical trial by embedding the study within usual care, recruiting a diverse patient population with minimal eligibility criteria, promoting the continuation of usual care without standardized treatment protocols, and relying on electronic data collection with reduced need for costly primary data collection. In one such trial (ADAPTABLE trial, [Hernandez et al. 2015]), the primary composite outcome of interest – death, hospitalization for nonfatal myocardial infarction, or stroke – and a primary safety endpoint of major bleeding complications were chosen with input from patients. The trial sought to recruit 20 000 high-risk patients with heart disease with an expected cost of less than $1000 per participant, an amount far below that of a typical trial of this scope. By leveraging electronic health record data collected during usual patient care, ADAPTABLE aimed to reduce the burdens that traditional processes for research data collection impose on patients, clinicians, and practices.

Moving away from generalizations and looking at specific paradigm proposals offers a different perspective. One such new paradigm proposed by Danhof et al. (Danhof et al., 2018) describes a shift towards a pathology-based era of systems therapeutics with implications for the future of drug research. In this concept, specialists are needed not only from the pharmaceutical sciences, the biological sciences, or the medical sciences but also from fields such as engineering, computational sciences, and robotics. Danhof postulates that the progress in cell biology, genetics, molecular, and systems pharmacology, is evidence that the next paradigm shift in drug research is unfolding. The adoption of "biological network transduction models," evaluating drug effects as the result of multiple interactions in a biological network, has yielded unprecedented opportunities to understand the functioning of biological systems, to identify the molecular mechanisms of drug action, and to design therapeutic strategies aimed at modifying disease processes rather than controlling symptoms. It is expected that these transitions in science and drug development will shape new avenues for an avalanche of advanced treatments reaching patients in the years to come. But how does this opportunity become a reality with respect to current R&D practices?

Some proposals for new drug development paradigms have been linked to specific therapeutic areas (e.g. antibiotics, HIV, infectious disease, and Alzheimer's disease), recognizing bottlenecks with current development paths and/or leverage to be gained by both pre-competitive and open collaboration among drug sponsors. This is often a necessity in small target populations such as pediatric oncology or rare diseases. Some example proposals are summarized below.

Antibiotic drug development is challenging for a variety of reasons including resistance development, financially unfavorable treatment paradigms (most effective if they're used sparingly), and large Phase 3 trials required. New proposals for antibiotic development have been proposed suggesting a shift in the baseline paradigm (Farha and Brown 2019). In one such proposal, a paradigm change based on repurposing existing drugs for antimicrobial agents is described (see Figure 5.2). The authors discuss modifications to early drug development, enabling screening platforms for antimicrobial discovery, and present encouraging findings of novel antimicrobial therapeutic strategies. Also covered are the general advantages of repurposing over de novo drug development and challenges of the strategy, including scientific, intellectual property, and regulatory issues.

The main point of the antibiotic repurposing proposal is that development time and cost can be reduced by essentially skipping Phase 1 and Lead Optimization since, in theory, these would have been completed previously.

To a certain extent, the HIV therapeutic area has already experienced a shift in the current, traditional development paradigm from the baseline scenario illustrated in Figure 5.1. Given the urgent need to improve the therapeutic options for HIV patients during the height of the AIDS epidemic, both the FDA and the European Medicines Agency (EMA) supported so-called "Early Access Programs" (EAPs). Since 1992, the FDA introduced the "Priority Review" or "Fast Track," designed to make available new drugs for the treatment of serious or life-threatening diseases (conditions associated with morbidity that have a significant impact on specific factors, such as survival or day-to-day functioning) without therapeutic alternatives. For these drugs, the "breakthrough designation" can be expected

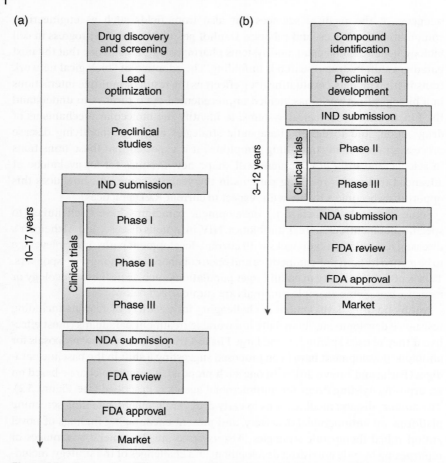

Figure 5.2 Conventional (a) versus modified (b) drug development paradigm proposed for repurposing existing drugs with antimicrobial activity shortening the development timelines and the cost of drug development (Farha and Brown 2019).

(see Chapter 4). The FDA's "Fast track" imposes on the pharmaceutical company lower standards than the regular procedure. Similarly, in the European context, specific regulatory procedures, including approval under exceptional circumstances as well as conditional and accelerated approval, have been introduced in order to accelerate the marketing authorization of a new drug. With such procedures, the marketing authorization application can be based on incomplete clinical data (even data from Phase 2 studies), and its evaluation can be reduced from 210 to 150 days if the applicant provides sufficient justification for an accelerated assessment. Depicted schematically in Figure 5.3 (Scavone et al. 2019), the

Regular drugs development process

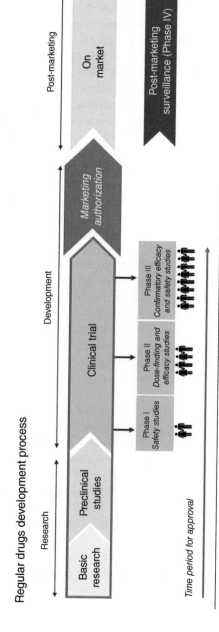

Accelerated approval process

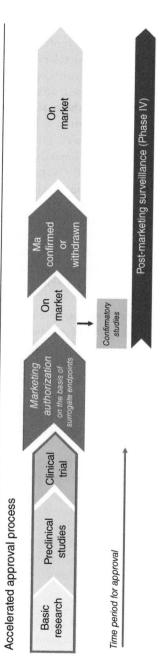

Figure 5.3 Traditional vs. accelerated development and approval process utilized for HIV drug development (Scavone et al. 2019).

accelerated approval process also suggests earlier market access with the expectation that confirmatory and Phase 4 commitments begin on the heels of the approval. While it most assuredly represents a paradigm shift, the accelerated approval process is more about risk tolerance in the face of an unmet medical need.

Finally, oncology is yet another therapeutic area where new development paradigms have been proposed. Most of the incentives for the proposals in oncology are not only related to the modest survival gains from recently approved medicines but also the length of time and cost of development. While some of the proposed paradigms project greater efficiency during discovery stage activities, most are focused on a breakthrough approval paradigm similar to that established for HIV. Wagstaff (Wagstaff 2017) summarized the opinions of leading figures in academic drug development that addressed the issue in an article in Cell published (Workman et al. 2017), which was widely covered in the mass media, including an editorial in the UK newspaper, The Times. Under the title, "How much longer will we put up with $100,000 cancer drugs?" the authors, from top centers in the US, UK, and the Netherlands, called for "the formation of new relationships between academic drug discovery centers and commercial partners, which can accelerate the development of truly transformative drugs at sustainable prices."

One of the key concepts discussed by the academic oncologists was the necessity of promoting public–private partnerships aimed at enabling the academic sector to generate more innovative high-risk ideas and then also do much of the work to "derisk" them. The intention is for academic partners to define and/or refine the patient population, select the appropriate biomarker and advise on the availability of a prototype drug (e.g. for a comparator in Phase 3 trial). One concern is that Pharma sponsors would still propose a conventional, large Phase 3 trial model and payback to the pharmaceutical companies based on the maximum the market will bear.

The novel part of the proposal is based on the suggestion that academic partners collaborate with generic drug makers that are used to working with lower profit margins with the idea that highly innovative but derisked drugs from academic drug discovery and development (Figure 5.4) can be further developed, manufactured and submitted for regulatory approval by the generic sponsor. The sentiment presumes that many of the drug candidates will have a strong mechanistic rationale and an associated biomarker of response such that the registration trials can be small and the success rate much higher than in traditional pharma trials.

Regulatory bodies are also open to novel ways for oncology drug approval. The EMA recently launched an adaptive licensing program enabling companies to obtain marketing authorization approval based on small well designed, biomarker-supported trials. Its early days for these proposals and uncertainty regarding their

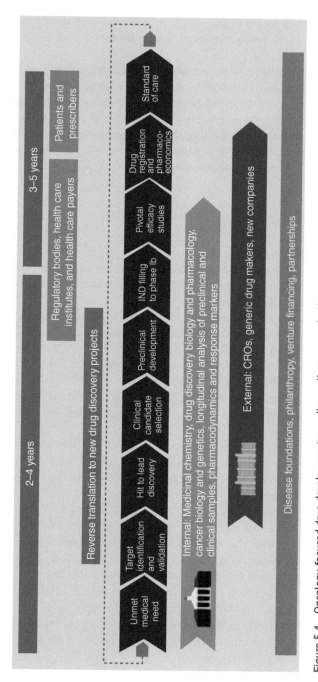

Figure 5.4 Oncology-focused drug development paradigm reliant on academic and non-profit partnerships to re-risk early-stage development and partnership with generic drug manufacturers (Workman et al. 2017).

uptake and little track record to judge them by at the moment, but hopefully, there will continue to be a steady stream of proposals focused on innovation and efficiency.

While the current drug development paradigm has been well entrenched for the past three decades, there is no doubt that the cost of drug development and the Phases 2 and 3 attrition rates could be improved. Collaboration is a key component of the future solution, and a "one-size-fits all" solution is unlikely. There should be great optimism that a future solution can be implemented with more broad participation from all key stakeholders and hopefully with the best interests of patients at the forefront of these proposals.

The TPP and IPDP

Two important documents that a sponsor uses to capture its current thinking around a drug candidate's potential and development plans are the TPP and IPDP. Both are produced by the early development project teams, and both evolve as a drug candidate is developed.

The TPP is a format for a summary of a drug development program described in terms of labeling concepts. Both research and commercial goals are captured in the TPP. A TPP is often shared with the appropriate FDA review staff to facilitate communication regarding a particular drug development program. Submission of a TPP is voluntary. The published FDA guidance (FDA TPP Guidance 2007) describes the purpose of a TPP, its advantages, and its optimal use. It also provides guidance on how to complete a TPP and relates case studies that demonstrate a TPP's usefulness.

The TPP constitutes an important evaluation tool in "gate reviews" if such reviews are enabled by the organization (Breder et al. 2017). Likewise, there are standard templates often used to facilitate TPP creation. One such template appears below in Table 5.2.

An additional useful guide often utilized during initial TPP creation is the TPP summary of efficacy template shown in Table 5.3. As the table depicts, it is essential that the sponsor thinks about and quantifies as best as possible criteria that define the target, minimally acceptable, and optimal targets for key endpoints generated from the drug candidate's evaluation. Such criteria assist the sponsor in adhering to Go/No-go criteria and help regulatory authorities evaluate the appropriate level of risk tolerance and benefit: risk especially since they are able to give guidance from the perspective of seeing multiple sponsors' proposals.

Table 5.2   General statement template used for TPP creation.

Project name	(Name)
Project description	Summary description of the product
Project category	Is the project an additional indication for an existing drug or a new project?
Strategic fit and value	How well does this drug/biologic fit with the core expertise and capabilities of the company?
Value to patients	What is the specific value of this drug/biologic to patients? Does it offer therapeutic, safety, or ease of use advantages over existing or upcoming drugs/biologics
Company's competitive position	Does the company have a competitive advantage?
Company's IP position	Brief summary of the IP position regarding this drug
Rationale for success	Brief summary as to why the developing team believes that this product would be successful
Factors for success	Brief statement as to the company's core competencies and market conditions that would drive a successful outcome
Key risk factors	Brief statement identifying possible risks
Consequences for not pursuing the project	What would happen if this project were not pursued?
Possible alternatives to this project	Are there any alternatives to this project?

The TPP may contain additional elements regarding product design and formulation (including purity, contaminants, storage conditions, and shelf life), any delivery system associated with the drug, projected dates of submissions, regulatory approval and launch, cost of goods, pricing, market size and target, optimistic and minimal conditions. Once completed, the TPP is revised at key stage gates including the end of Phases 1 and 2. Figure 5.5 provides an example of a TPP for Long-Acting/Extended Release Antiretrovirals which was discussed at a past workshop (Workshop on Development of Long-Acting/Extended Release Antiretrovirals, Boston, 2014, https:// longactinghiv.org/content/target-product-profile-future-laer-arvs). Many such examples exist and are easily accessed as starting points for TPP creation.

The FDA strongly advocates the use of a TPP, although it does not mandate it. The FDA has prepared a template included in recent draft guidance (http://www.fda. gov/downloads/Drugs/GuidanceComp lianceRegulatoryInformation/Guidances/ ucm080593.pdf). For each element of the label, the template proposes the following items: Target (language in the Package Insert that the sponsor hopes to achieve),

Table 5.3 TPP summary of efficacy template.

	Primary indication				
	Safety				
	Clinical	Non-clinical	Drug interactions	Precautions	Contraindications
Optimistic	>Target if fewer and less severe AE profile or = Target	Laboratory or other findings similar to those observed for the same class or similar classes of compounds that have been approved	>Target if fewer and less severe interactions or = Target	>Target if no or fewer precautions or = Target	>Target if no or fewer contraindications or = Target
Target	Target safety is usually equivalent to the known safety of the same class or similar classes of compounds that have been approved		Interactions similar to those observed for the same class or similar classes of compounds that have been approved	Precautions similar to those observed for the same class or similar classes of compounds that have been approved	Contraindications similar to those observed for the same class or similar classes of compounds that have been approved
Minimal	= Target (<Target would be acceptable if risk/benefit ratio is favorable)	= Target (<Target would be acceptable if risk/benefit ratio is favorable)	=Target (<Target acceptability criteria should be explained)	= Target (<Target acceptability criteria should be explained)	= Target (<Target acceptability criteria should be explained)

	Minimum	Base Case	Optimum
Dosing frequency	• Q Week	• Q Month	• ≥Q 2 Month
Route	• IV	• IV/IM/SC	• IM/SC
Safety profile	• Similar to EFV	• No systemic allergic reactions • Similar to raltegravir	• No mitochondrial toxicity • Similar to FTC/3TC • Removable by hemofiltration
Metabolism • includes drug-drug interactions	• No preference	• Not metabolized by CYP3A4 • CYP3A4 inhibition preferred to induction	• Not a substrate for CYP3A4 • No effect on CYP3A4 or glucuronidation
Tissue penetration (viral suppression)	• LN • Genital tract	• CNS • LN • Genital tract	• CNS • LN • Genital tract
Storage	• 2 yr, refrigeration acceptable	• 3 yr at 20–25 °C; or • 2 yr at ≥40 °C	• ≥3 yr at ≥40 °C
Resistance profile	• Similar to EFV • No cross resistance	• Infrequent • No cross resistance	• None • Protects other ARV's
Cost of goods	• POC	• Affordable for RLC's	• <EFV

Figure 5.5 Target product profile for long-acting/extended-release antiretrovirals.

Annotations (summary information regarding completed or planned studies), and Comments (section to provide clarity. The TPP template links each labeling concept to a specific study or other sources of data.

The IPDP is an internal document produced by the sponsor that captures the detailed steps and procedures envisioned by the project team required to advance the drug candidate through development while defining the stage gates and development milestones that the candidate should achieve to advance. Many of these milestones are defined relative to the development stage (i.e. Phases 1–3 milestones). The document is authored by the various project team representatives with input from their functional leadership and reviewed by senior leaders within R&D and commercial parts of the organization. The IPDP will include critical assumptions and assessments of time and budget constraints informed by competitive intelligence and the marketplace. A sponsor may choose to share the IPDP with development partners or others outside their organization, but this most certainly will be restricted by confidentiality agreements. A representative template for an IPDP table of contents is provided in Table 5.4.

Both the TPP and IPDP are critical to the planning efforts that pharmaceutical companies engage in while developing new chemical entities in pursuit of market access. They capture much of the decision-making that happens through the process as defined by the current drug development paradigm. If the current paradigm evolves, it will likely be captured by both the TPP and IPDP.

Table 5.4 IPDP table of contents template.

Section	Author(s)	Purpose
Cover page/summary	Project manager	Specifies the candidates name and/or company ID, the IPDP Version and Date, the current development phase, next stage gate, investment total until next stage gate, change control and version, and the listing of the functional team members
Objectives statement and integrated development plan executive summary	All functional team members	Statement of objectives, current stage-gate status, and scientific rational
Background on disease burden, interventions and product development	Pharmacology and clinical team members	Summary of unmet medical need, size, and details of target population, disease progression including current standard of care.
Next stage gates, go/no-go criteria, and success metrics	All functional team members	Summary of stage gates and progression criteria along with identification of responsible parties (internal and external)
Assumptions used to determine development timelines	All functional team members	Assessment of key dependencies with respect to timelines and verification of candidate attributes through experimentation
End-to-end integrated development timeline	Project manager	Gantt chart of the development candidate, back-ups and competition,
High impact integrated project risks	Each functional group represented	Listing of key risk factors, ranked, and quantified by potential impact to the program
Future landscape and competitors	Clinical and commercial	Marketplace surveillance and summary along with economic indicators, reimbursement concerns, feedback from payers.
Key stakeholders and partnering strategy for development and hand-off	Business development, finance	Identification of any hand-offs for development, technology transfers, or funding sources with timelines and milestones
Functional domain strategy	Each functional team member	Strategy summaries for non-clinical and clinical development, regulatory, quantitative sciences, CMC, and other key contributors. Section also includes a listing of proposed studies, target dates, and expected milestones.
Appendices	Functional team	Current draft of TPP and/or competitor's TPP

References

Bowen, A. and Casadevall, A. (2015). Increasing disparities between resource inputs and outcomes, as measured by certain health deliverables, in biomedical research. *Proceedings of the National Academy of Sciences* 112 (36): 11335–11340.

Breder, C.D., Du, W., and Tyndall, A. (2017). What's the regulatory value of a target product profile? *Trends in Biotechnology* 35 (7): 576–579.

Danhof, M., Klein, K., Stolk, P. et al. (2018). The future of drug development: the paradigm shift towards systems therapeutics. *Drug Discovery Today* 23 (12): 1990–1995. https://doi.org/10.1016/j.drudis.2018.09.002.

Elkins, R.C., Davies, M.R., Brough, S.J. et al. (2013). Variability in high-throughput ion-channel screening data and consequences for cardiac safety assessment. *Journal of Pharmacological and Toxicological Methods* 68 (1): 112–122.

Farha, M.A. and Brown, E.D. (2019). Drug repurposing for antimicrobial discovery. *Nature Microbiology* 4: 565–577.

FitzGerald, G.A. (2011). Re-engineering drug discovery and development. *LDI Issue Brief* 17 (2): 1–4.

Guidance for Industry and Review Staff Target Product Profile (2007). A Strategic Development Process Tool, U.S. Department of Health and Human Services Food and Drug Administration Center for Drug Evaluation and Research (CDER). https://www.fda.gov/media/72566/download.

Hernandez AF, Fleurence RL, Rothman RL. The ADAPTABLE Trial and PCORnet: shining light on a new research paradigm. *Annals Internal Medicine* 2015; 163:635–636. doi: https://doi.org/10.7326/M15-1460.

Junod, S.W. (2014). US Food and Drug Administration. https://www.fda.gov/media/110437/download.

Kesselheim, A.S., Avorn, J., and Sarpatwari, A. (2016). The high cost of prescription drugs in the United States: origins and prospects for reform. *JAMA* 316 (8): 858–871. https://doi.org/10.1001/jama.2016.11237.

Macarron, R., Banks, M.N., Bojanic, D. et al. (2011). Impact of high-throughput screening in biomedical research. *Nature Reviews Drug Discovery* 10 (3): 188–195. https://doi.org/10.1038/nrd3368. PMID 21358738.

Munos, B.H. and Chin, W.W. (2011). How to revive breakthrough innovation in the pharmaceutical industry. *Science Translational Medicine* 3 (89): 89cm16–89cm16.

Paul, S.M., Mytelka, D.S., Dunwiddie, C.T. et al. (2010). How to improve R&D productivity: the pharmaceutical industry's grand challenge. *Nature Reviews Drug Discovery* 9 (3): 203–214.

Roses, A.D. (2008). Pharmacogenetics in drug discovery and development: a translational perspective. *Nature Reviews Drug Discovery* 7: 807–817.

Scannell, J.W., Blanckley, A., Boldon, H., and Warrington, B. (2012). Diagnosing the decline in pharmaceutical R&D efficiency. *Nature Reviews Drug Discovery* 11 (3): 191–200.

Scavone, C., di Mauro, G., Mascolo, A. et al. (2019). The new paradigms in clinical research: from early access programs to the novel therapeutic approaches for unmet medical needs. *Frontiers in Pharmacology* 10: 111.

Sieburg, H.B. (1990). Physiological Studies in silico. *Studies in the Sciences of Complexity* 12: 321–342.

Wagstaff, A. (2017). *A Breakthrough Business Model for Drug Development*, 56–63. Cancer world. March/April.

Workman, P., Draetta, G.F., Schellens, J.H.M., and Bernards, R. (2017). How much longer will we put up with $100 000 cancer drugs?'. *Cell* 168: 579–583.

Chapter Self-Assessments: Check Your Knowledge

Questions:
- What does the term "quick kill" refer to and why is it relevant to drug development?
- What is the incentive for suggesting a new drug development paradigm is warranted?
- What is a target product profile (TPP) and why is it relevant to drug development?
- What is an integrated product development plan and how is it used?

Answers:
- The term "quick kill" refers to the approach where a company adheres to strict Go/No-go criteria with well-defined quantitative stage gates by which they will judge the progress of development of a drug candidate. Attributes developed by the project team and approved by senior leadership are required to be met or surpassed or the compound is removed from development consideration. The sooner this can be determined in the development phase, the better from a financial point of view for the company, hence the "quick kill" designation.
- Declining or stagnant research and development (R&D) productivity has led many to suggest that the current paradigm for drug discovery and development requires disruptive innovation (i.e. a transformation of the pharmaceutical industry driven by new technology, new business models, or policy decisions that improve therapy and create value for patients and society in a way that could not be achieved through other means) to break out of a current crisis by identifying and rapidly bringing new discoveries to market.

- The TPP is a format for a summary of a drug development program described in terms of labeling concepts. Both research and commercial goals are captured in the TPP. The TPP constitutes an important evaluation tool in "gate reviews", if such reviews are enabled by the organization.
- The Integrated product development plan (IPDP) is an internal document produced by the sponsor that captures the detailed steps and procedures envisioned by the project team required to advance the drug candidate through development while defining the stage gates and development milestones that the candidate should achieve to advance. Many of these milestones are defined relative to the development stage (i.e. Phase 1, 2, and 3 milestones). The document is authored by the various project team representatives with input from their functional leadership and reviewed by senior leaders within R&D and Commercial parts of the organization. The IPDP will include critical assumptions and assessments of time and budget constraints informed by competitive intelligence and the marketplace.

Quiz:

1 What is the current estimate of the time to develop a new drug from Discovery through Phase 3 and submission?
   A 10–12 years
   B 12–15 years
   C 15–17 years
   D 17–20 years
   E None of the above are correct

2 True or false. Over the past 3 decades, the number of new molecular entities approved by the FDA has averaged 20–30 drugs per year, except for a peak in the mid-1990s that experienced a doubling of this rate.

3 Select the best choice to fill in the blanks. "Many opinions exist on how the current drug development paradigm can be modified to reduce the time and cost of development. Most of the proposals call for increased investment in _____ to select _____ linked to better _____."
   A Manufacturing innovation, more efficient processes, product quality
   B Innovative trial designs, patients more likely to respond to proposed therapy, trial outcomes
   C Discovery Science, better candidates, defined targets
   D Real-world data, more informative patient populations, probability of technical success

4 With respect to the IPDP, which group authors the part of the document that describes the next phase stage gates, "Go/No Go" criteria, and success metrics? Choose the best answer.

A Clinical Pharmacology

B Project management

C The Project Team Leader in consultation with senior management

D All functional groups contribute to this section

6

Drug Discovery and Preclinical Development
Jeffrey S. Barrett

Aridhia Digital Research Environment

Drug Discovery – Introduction

The discovery phase includes the early aspect of drug research, which is designed to confirm pharmacologic targets, identify an investigational drug candidate, and perform initial experiments that allow scientists to rank and select candidates for preclinical evaluation. This first stage of the process takes approximately three to six years. By the end, researchers hope to identify one or more promising drug candidates to further study in the lab and animal models and then in people. Researchers work to identify biological targets for a potential medicine. A drug target is a molecular structure in the body that, when it interacts with a potential drug compound, produces a clinical effect (treatment or prevention of a disease, for example). The investigators conduct studies in cells, tissues, and animal models to determine whether the target can be influenced by a drug candidate. Target validation is crucial to help scientists identify the most promising approaches before going into the laboratory to develop potential drug candidates, increasing the efficiency and effectiveness of the research and development (R&D) process.

In actuality, this phase consists of two distinct segments: an initial discovery phase, followed by a development phase. These two phases differ significantly from each other with respect to scope, challenges, and approaches. Differences notwithstanding, discovery and development must be integrated into a coherent whole for the process to be successful. Accordingly, much thought has been devoted to ensuring scientific, logistical, and organizational aspects of such integration are taken into consideration and optimized.

After learning more about the underlying disease pathway and identifying potential targets, researchers then seek to narrow the field of compounds to one lead compound – a promising molecule that could influence the target and,

Fundamentals of Drug Development, First Edition. Edited by Jeffrey S. Barrett.
© 2022 John Wiley & Sons, Inc. Published 2022 by John Wiley & Sons, Inc.
Companion website: www.wiley.com/go/Barrett/FundamentalsDrugDevelopment

potentially, become a medicine. This is often referred to as lead or candidate selection. This is accomplished in a variety of ways, including creating a molecule from living or synthetic material, using high-throughput screening techniques to select a few promising possibilities from among thousands of potential candidates, identifying compounds found in nature, and using biotechnology to genetically engineer living systems to produce disease-fighting molecules. High-throughput screening (HTS) refers broadly to methods and approaches that permit efficient and rapid assessment of compound attributes so that they can be ranked and prioritized as part of the Leads Optimization phase within drug discovery. The details of what constitutes HTS will be discussed later in this chapter.

Even at this early stage, investigators are already thinking about the final product and how it will be administered to patients (e.g. whether it is taken in pill form, injected, or inhaled). In turn, they must also consider the formulation (the design of dosage forms) of medicine and how easily it can be produced and manufactured.

The history of drug discovery describes a process driven by chemistry but guided by pharmacology and the clinical sciences, with drug research contributing more to the progress of medicine during the past century than any other scientific factor. The advent of molecular biology and, in particular, genomic sciences has had a deep impact on drug discovery. Recombinant proteins and monoclonal antibodies have greatly enriched our therapeutic toolset. Genomic sciences, combined with bioinformatic tools, have allowed us to dissect the genetic basis of multifactorial diseases and to determine the most suitable points of attack for future medicines, thereby increasing the number of treatment options (Drews 2000). Many milestones in drug discovery have been observed through the impact of new technologies (the discovery of penicillin and other antibiotics encouraged many drug companies to establish departments of microbiology and fermentation units), the influence of molecular biology (the potential to understand disease processes at the molecular, genetic level and to determine the optimal molecular targets for drug intervention), the process of target identification and validation and combinatorial chemistry (large numbers of hypothetical targets incorporated into *in vitro* or cell-based assays and exposed to large numbers of compounds representing numerous variations on a few chemical themes or, more recently, fewer variations on a greater number of themes in high-throughput configurations) and high-throughput screening (the generation of a high degree of structural diversity within a library). Figure 6.1 provides a schematic representation of the modern drug discovery process focusing only on the elements critical to target validation and candidate selection. As we will see, this is only part of the activities that occur during this phase.

In this chapter, we will describe the various elements of the discovery phase, exposing the contributions of many functional groups that aid in delivering drug candidates suitable for human phase testing as well as complete the extensive

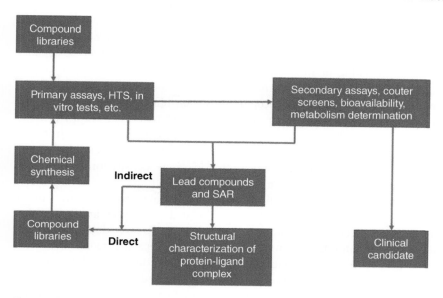

Figure 6.1 Schematic of typical Discovery phase activities used in the process of target identification and early candidate selection.

documentation required to submit an investigational new drug (IND) application to the US Food and Drug Administration (USFDA) or equivalent regulatory authority. In addition, we will define the key components of the discovery phase version of the target product profile (TPP) and discuss the milestones for a successful discovery candidate that is to move forward into Phase 1 (Kennedy 1997; Nikitenko 2006). Throughout each section, we will expose the timing and cost of the various activities as well.

Product Characterization

Product characterization is the foundation for all formulation and process development. Pharmaceutical scientists use a variety of specific analyses to understand the behavior of products in various stages of development to understand the impact of the individual components and process conditions. The primary goal of product characterization is to ensure a product's safety, purity, identity, and potency. The harmonized guideline Q6B, "Specifications: Test Procedures and Acceptance Criteria for Biotechnological/Biological Products, from the International Conference on Harmonization (ICH)" specifically calls for the determination of physicochemical properties, biological activity, immunochemical properties, purity, and impurity.

Requirements are greater for therapeutic proteins. For protein-based biotherapeutic drugs to achieve their desired effect in the patient, which occurs by interacting with receptor proteins, with other proteins, or with other targets, to enhance or inhibit signaling or interactions in the body, they must have the correct sequence, the correct size, the correct structure, and the correct post-translational modifications so that they are recognized by their binding partner. In addition, therapeutic proteins must be biologically active and have the correct physiochemical and immunochemical properties to elicit the correct response from the patient. Given the complexity of therapeutic protein production, it's vital to monitor all aspects of the protein to ensure that its sequence, structure, purity, and stability are correct and consistent.

Product characterization reveals the biochemical and biophysical nature of the product as well as the nature of product-related substances and impurities. Thorough product characterization is a necessary precursor to determining critical quality attributes (CQAs) and the associated analytical methods that, in turn, can be used as in-process controls and specifications and for stability testing. Process characterization focuses on understanding and defining the operating and design spaces for the process to achieve a product with consistent CQAs. It is advised by FDA that a product in development be analyzed at all stages during the processing pathway, including during discovery, upstream processing, downstream processing, and right through to the final product. This thorough analysis is meant to ensure that the product is consistent, active, stable, safe, and pure from start to finish.

Analysis of all processing steps also allows any changes to the protein or other problems, such as contamination, to be detected immediately, thereby allowing for determination of where the problem is occurring and enabling faster investigations to be performed. If an analysis is only carried out at one stage of a process, it is difficult to pinpoint the source of the problem. The advancement criteria supporting early product characterization (e.g. identification of actives, etc.) can be completed in approximately 12 months in most cases by performing activities in parallel. Beginning activities supporting the confirmation of hits prior to a "Go" decision is roughly the starting point in terms of timing. The accumulated project cost associated with a "No Go" decision at this point is estimated to be $1.46 million (Strovel et al. 2016).

Formulation, Delivery, Packaging Development

A large component of the early effort in product development is focused on supplying adequate amounts of the candidate molecules to support the required testing. The total amount of drug required depends upon its activity, effect, and physiochemical properties. Of course, if a compound survives early phase testing, the requirements for drug supply increase dramatically to full requirements for clinical phase evaluation. Typical requirements are 1 kg for Phase 1, 50–100 kg for Phase 2, and up to 1 ton for Phase 3. Commercial demands for the drug entity, assuming the candidate is approved, are projected by the project team based on the demand profile with high dose drug requirements (based on g/day estimates) potentially exceeding thousands of tons per year (low dose drugs can be in the 100 kg/yr range by contrast).

Initial in vitro biological activity tests can be easily carried out with milligram quantities, while animal toxicity studies and in particular pharmaceutical development, can easily elevate the active substance demand to a kilogram level. Chemical synthesis for the purpose of hit identification is usually performed on a fraction of millimolar scale, without any consideration for process development (Brodniewicz and Grynkiewicz 2010). For the drug lead and drug candidate level, the active drug substance has to be examined in detail, particularly in terms of impurity formation, and then optimized. Drug substance stability in time and under stress also has to be determined. This involves the development and validation of analytical methods and identification of critical parameters of the synthetic process, which can frequently be derived from academic knowledge or published information. Analytical specifications for an active pharmaceutical ingredient (API) can be changed during development, but it should start at a reasonable level of chemical purity. For a generic drug, there is a customary requirement of 99.8% of HPLC purity with no single unknown impurity crossing 0.1% level. For new drug candidates, especially at the preclinical study period, more flexible standards are possible, e.g. with no individual purity present above 0.5%. At the same time, based on advances in analytical techniques with coupled detection methods, it is reasonable to assume that a designation of "unknown impurity" is not acceptable and should not be included in the specification. Likewise, while there are obvious needs for packaging and delivering from a clinical supplies perspective, these also come into play for animal pharmacology and toxicology testing as well. In the context of non-human investigations, this role is typically maintained with the driving function, pharmacology, and safety assessment/toxicology, for example, while human phase evaluation will rely on clinical operations with coordination from Project Management and the originating therapeutic area clinical groups.

The entire process of drug discovery and preclinical development is summarized for the eventual purpose of filling a new drug application, in a standardized form of Common Technical Document (CTD), containing five modules: (1) Regional and administrative information (concerning applying organization), (2) overviews and summaries, (3) quality, (4) nonclinical study reports, and (5) clinical study reports. Module 3, dealing with the quality of a drug substance and a drug product, is in its first part of particular interest to any project devoted to new drug design, discovery, and development. Since final drug active substance (frequently described as API, short for active pharmaceutical ingredient) has to be exhaustively and meticulously examined and its properties fully characterized, in particular in respect to stability and content of impurities, a legitimate question arises, when pharmaceutical quality requirements become critical within a pathway of biological testing.

As the drug supply batch sizes grow, so do the chemical manufacturing requirements necessitating the sourcing of pilot plant production facilities and ultimately a commercial-grade manufacturing facility. Early considerations for these scale-up activities also happen within the Discovery/Preclinical phase, with the probability of technical success (POTS) weighing heavy on the decision to commit CMC resources on early-stage drug candidates. Regarding the cost of early formulation development, the estimated investment for the development of an acceptable clinical dosage form, delivery, reconstitution, stability testing (up to one year), and GMP quality studies is typically $500 000 to $1 M USD (Strovel et al. 2016).

Pharmacokinetics and Drug Disposition

Successful drugs must be absorbed into the bloodstream, distributed to the proper site of action in the body, metabolized efficiently and effectively, successfully excreted from the body, and demonstrated to be not toxic in the tests performed. Normally performed in living cells, in animals, and via computational models, these studies help researchers prioritize lead compounds early in the discovery process.

Preclinical drug metabolism and early-stage pharmacokinetic (PK) trials in animals continue to be a staple of the discovery and preclinical development phase as sponsors seek to characterize and rank their early drug candidates as well as engage in the planning of initial trials in humans. A number of primers on basic pharmacokinetic-pharmacodynamic (PK/PD) characterization can be found to fill in gaps in understanding (Gunaratna 2000). Typically, the first PK question is: What does the body do to a new drug candidate (essentially the pharmacokinetic characterization of the drug molecule and any active species)? This process begins with the route of administration and the release characteristics of the delivery system before the compound is absorbed. By following (and measuring) the concentrations of the drug over time in the relevant compartment (usually blood), it is then possible to determine the critical PK parameters: $t\frac{1}{2}$ of absorption, Tmax, Cmax, and the $t\frac{1}{2}$ of the elimination phase, and they are under the concentration-time profile (AUC). From these values, the intrinsic clearance can be determined. Once absorption begins, drug distribution starts. Distribution is affected by many factors such as ionization and protein binding, and it is sometimes passive; pumps and transporters may also affect distribution (e.g. across the blood–brain barrier). Some drugs are partitioned into fat, and others may be sequestered into specific sites or tissues. At some point during distribution, metabolism or biotransformation begins. Where and how drugs are biotransformed is important early knowledge, and the route of elimination (e.g. feces, urine, bile) must also be determined. Figure 6.2 provides an idealized concentration-time profile of an immediate-release formulation of a drug administered orally along with a representation of where along with the profile the critical aforementioned parameters are measured. While the shape of the profile may change when modified-release formulations or any other extravascular (situated or occurring outside the vascular system) route is administered, the basic processes are followed for non-systemically (non-iv administration) drugs, absorption, followed by distribution then elimination (via metabolism and/or excretion).

Knowledge of pharmacokinetics provides a mechanism to assess the feasibility of achieving a clinically effective but non-toxic dose as well as the means to modulate dose to adjust for situations in which certain sub-populations may require more or less exposure (e.g. pediatrics, elderly or organ-impaired patients).

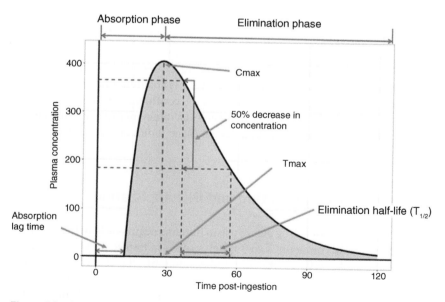

Figure 6.2 Idealized concentration-time profile for an orally administered drug.

Pharmacology Testing and Animal Models

Early pharmaceutical research drew on existing animal models that were used in experimental physiology, extending established scientific traditions of using animals in research. New potential medicines were not directed at a specific target, such as a cell receptor, as they are today. Rather, the effect of medicines was measured in relation to the general physiological response of an animal, such as changes in blood pressure. This method of screening for potentially beneficial effects of medicines used large numbers of animals and was inefficient and cumbersome. As pharmaceutical research expanded in the 1950s and 1960s, the use of animals expanded in parallel. In the 1980s, novel techniques, improved facilities, computer technology, and new materials became available and were integrated into the research and development process. The use of alternatives to solely animal-based research and development, such as cultured cells, also expanded.

Some (Shanks et al. 2009) maintain that credible evidence regarding the predictability of animal models, especially with respect to toxicology and pathophysiology to predict human outcomes, is not as compelling as we're often led to believe, and indeed the data in question seems to support this sentiment. Likewise, whether animals can be used to predict human response to drugs and other chemicals is still a contentious issue and one that has promoted considerations for *in silico* methodologies to replace animal testing at some point. It is easy to show that when one empirically analyzes animal models using scientific tools, they fall far short of being able to predict human responses. This is not surprising considering what we have learned from fields such as evolutionary and developmental biology, gene regulation and expression, epigenetics, complexity theory, and comparative genomics. Still, it is apparent that in vitro and in silico predictive tools are evolving with greater confidence in their predictive value (PBBPK FDA Guidance 2018), there is still a reluctance to trust entirely on their utilization, and regulatory guidance on these topics doesn't substitute for regulatory regulation.

Preclinical Toxicology Testing and IND Application

Establishing the safety of a drug before use in humans begins early in the development process, as lead compounds go through a series of tests to provide a preliminary assessment of safety. Therapeutic indices quantify the relative safety of a drug and can be estimated from the cumulative quantal dose-effect curves of a drug's therapeutic and toxic effects. Figure 6.3 illustrates the idealized relationship between therapeutic indices derived from animal toxicology studies and

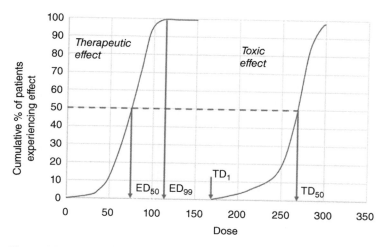

Figure 6.3 Schematic integrating toxicologic endpoints (TD_1, TD_{50}) onto idealized human therapeutic window illustrating how toxicology results influence early margin of safety calculations and FTIM dose prediction.

doses used in the calculation of safety indices utilized in first-time-in-man (FTIM) dose projections described in the typical IND protocol.

TD_1 = toxic dose to 1% of the population; TD_{50} = toxic dose to 50% of the population; ED_{50} = effective dose to 50% of the population; ED_{99} = effective dose to 99% of the population.

The therapeutic ratio is a ratio [TD_{50}/ED_{50}] of the dose at which 50% of patients experience the toxic effect to the dose at which 50% of patients experience the therapeutic effect. A therapeutic ratio of 2.5 means that approximately 2.5 times as much drug is required to cause toxicity in half of the patients as is needed to produce a therapeutic effect in the same proportion of patients. However, this ratio of toxic to therapeutic dose may not be consistent across the entire dose range if the dose-effect curves for the therapeutic and toxic effects are not parallel.

The goal of drug therapy is to achieve the desired therapeutic effect in all patients without producing toxic effects in any patients. Therefore, an index that uses the lowest toxic and highest therapeutic doses are more consistent with this goal than is the therapeutic ratio. The certainty safety factor (CSF) is the ratio of TD_1/ED_{99}. A CSF > 1 indicates that the dose effective in 99% of the population is less than the dose that would be toxic in 1% of the population. If the CSF < 1, there is an overlap between the maximally effective (ED_{99}) and minimally toxic (TD_1) doses. Unlike the therapeutic ratio, this measure is independent of the shapes of the cumulative quantal dose-effect curves for the therapeutic and toxic effects. The standard safety margin $\{[(TD1 - ED99)/ED99] \times 100\}$ also uses TD_1 and ED_{99}

but is expressed as the percentage by which the ED_{99} must be increased before the TD_1 is reached.

Regulatory authorities offer much guidance on the subjection of human dose projection based on animal toxicology trials, and specific guidance on a human equivalent dose (HED) offers multiple approaches for sponsors to arrive at a reasonable starting dose for FTIM studies (FDA Guidance 2005; Nair and Jacob 2016). Still, complementing toxicology assessment with pharmacology trials provides an early view of the therapeutic window as described earlier, and sponsors are always confronted with the dilemma of how much uncertainty they can tolerate at these critical junctures of development. Risk assessment plays a critical role in this endeavor, with risk tolerance varying widely among sponsors.

Before any clinical trial can begin, companies must file an investigational new drug (IND) application with the FDA. The application includes the results of the preclinical work, the candidate drug's molecular structure, details on how the investigational medicine is thought to work in the body, a listing of any potential side effects as indicated by the preclinical studies, and manufacturing information. The IND also provides a detailed clinical trial plan that outlines how, where and by whom the studies will be conducted. All INDs are submitted to the FDA and proceed after 30 days if there is no additional feedback or restriction given from the agency.

In addition to the IND application, all clinical trials must be reviewed, approved, and monitored by the institutional review board (IRB) or ethics committee (EC) at the institutions where the trials will take place. The IRB/EC has the responsibility to protect research participants and has the right to disapprove the study protocol or require changes before approving the planned clinical trials and allowing any participants to enroll. This process includes the development of appropriate informed consent documents, which will be required from all clinical trial participants.

With regard to specific toxicology requirements necessary to support an NDA, multiple types of studies must be completed prior to approval. Many of these need to be planned and completed prior to the IND application. FDA and all global regulatory authorities provide specific timelines regarding the nature and timing of toxicology trials required to support human phase testing (FDA Guidance 2010; Parasuraman 2011). Table 6.1 summarizes the various toxicology study types required to support early drug candidates along with their implications for future submission.

In general, acute, repeat-dose, and genetic toxicity trials are conducted prior to IND submission, and given the duration of carcinogenicity studies, these may be initiated at risk shortly after the FTIM study. Additional details of the individual toxicity study designs can be found in regulatory guidance documents and recent reviews (Parasuraman 2011).

Table 6.1 Toxicology requirements to support regulatory submission of an NDA or BLA.

Study type	Description	Purpose/Impact
Acute studies	Examine the effects of one or more doses administered over a period of up to 24h. Usually, at least two mammalian species are tested.	The goal is to determine toxic dose levels and observe clinical indications of toxicity; determine doses for repeated dose studies in animals and Phase 1 studies in humans.
Repeated dose studies	Note, may also be referred to as subacute, sub-chronic, or chronic. The specific duration should anticipate the length of the clinical trial that will be conducted on the new drug. Again, two species are typically required.	The studies are necessary to support multiple-dose testing in human (Phase 1 and beyond) trials.
Genetic toxicity studies	Procedures such as the Ames test (conducted in bacteria) detect genetic changes. DNA damage is assessed in tests using mammalian cells such as the mouse micronucleus test. The chromosomal aberration test and similar procedures detect damage at the chromosomal level.	These studies assess the likelihood that the drug compound is mutagenic or carcinogenic. The results of these studies are contained in a separate section of the drug label as well.
Reproductive toxicity studies	Segment I reproductive tox studies look at the effects of the drug on fertility. Segment II and III studies detect effects on embryonic and post-natal development.	Reproductive tox studies must be completed before a drug can be administered to women of child-bearing age.

(Continued)

Table 6.1 (Continued)

Study type	Description	Purpose/Impact
Carcinogenicity studies	The conventional test for carcinogenicity is the long-term rodent carcinogenicity bioassay. Studies observe test animals for a major portion of their life span for the development of neoplastic lesions during or after exposure to various doses of a test substance by an appropriate route of administration. The study is usually conducted using two species – rats and mice of both sexes. The animals are dosed by oral, dermal, or inhalation exposures based upon the expected type of human exposure. Dosing typically lasts around two years. Certain features are monitored throughout the study, but the key assessment resides in the full pathological analysis of the animal tissues and organs when the study is terminated.	Carcinogenicity studies are usually needed for drugs intended for chronic or recurring conditions. They are time-consuming and expensive and must be planned for early in the preclinical testing process.
Toxicokinetic studies	These are typically similar in design to PK/ADME studies, except they use much higher dose levels. They examine the effects of toxic doses of the drug and help estimate the clinical margin of safety.	There are numerous FDA and ICH guidelines that give a wealth of detail on the different types of preclinical toxicology studies and the appropriate timing for them relative to IND and NDA or BLA filings.

Bioanalytical Testing

Bioanalysis is a term generally used to describe the quantitative measurement of a compound (drug) or its metabolite in biological fluids, primarily blood, plasma, serum, urine, or tissue extracts (Pandey et al. 2010). The need for sound bioanalytical methods is well understood and appreciated in the discovery phase and during the preclinical and clinical stages of drug development. It is generally accepted that sample preparation and method validation are required to demonstrate the performance of the method and the reliability of the analytical results. The acceptance criteria should be clearly established in a validation plan prior to the initiation of the validation study. The reliability of analytical findings is a matter of great importance in pharmacologic and toxicologic experiments and, of course, in the context of clinical evaluation. Likewise, the importance of validation, at least of routine analytical methods, can therefore hardly be overestimated (Tiwari and Tiwari 2010). The developed assay should be sufficiently rugged that it provides opportunities for minor modifications and/or ease of adaptability to suit other bioanalytical needs such as applicability to a drug–drug interaction study, toxicokinetic study as well as characterization of the plasma levels of the metabolites.

Initially, in the discovery stage, the aim of bioanalysis is often only to provide reasonable values of either concentration and/or exposure which would be used to form a scientific basis for lead series identification and/or discrimination amongst several lead candidates. Therefore, the aim of the bioanalysis at this stage is typically to develop a simple, rapid assay with significant throughput to act as a great screening tool for reporting some predefined parameters of several lead contenders across all the various chemical scaffolds. Likewise, at this stage there is little emphasis on a rigorous assay validation as would be expected for clinical stage testing. See Figure 6.4 for complimentary formulation and bioanalytical activities required to support PK/PD characterization.

The initial method of analysis developed during the discovery phase of the molecule, with some modifications, may sometimes serve as a method of choice, to begin with as the candidate enters the preclinical development stage. Since the complexity of development generally tends to increase as the lead candidate enters the toxicological and clinical phase of testing, it naturally calls for improved methods of analytical quantization, improvement in selectivity and specificity, and employment of sound and rugged validation tools to enable estimation of PK parameters that would also aid in the decision-making of the drug molecule's advancement in the clinic in addition to safety and tolerability data gathered at all phases of development. Additionally, it becomes necessary to quantify active metabolite(s) in both animals and humans.

The analytical requirements for method development and validation become much stricter as a compound moves forward into clinical phase testing. There is

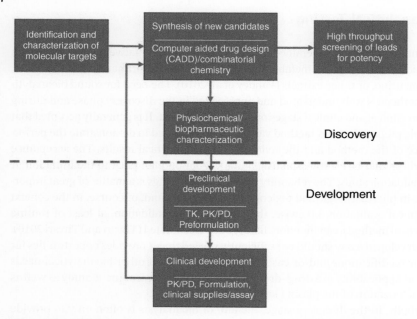

Figure 6.4   Different stages of discovery and development with respect to bioanalytical support expectations.

clear guidance on the expectations from regulatory authorities on the expected quality and rigor associated with the requisite method development and validation that needs to be in place prior to the initiation of sample analysis during clinical phase testing (FDA Guidance 2001). Another activity, even at an early stage, that the bioanalytical group would engage in would be the consideration of outsourcing method development, validation, and sample analysis to a contract research organization (CRO) depending on the availability of internal resources to fulfill this expectation.

Candidate Selection and Ranking Criteria

A drug candidate suitable for clinical testing is expected to bind selectively to the receptor site on the target, to elicit the desired functional response of the target molecule, and to have adequate bioavailability and biodistribution to elicit the desired responses in animals and humans; it must also pass formal toxicity evaluation in animals. The path from lead to clinical drug candidate represents the most idiosyncratic segment of drug discovery and development. While each program is unique, setbacks are common, and the TPP will define overall criteria that brackets an acceptable drug candidate for the proposed indication and population; those engaged in the discovery and preclinical evaluation will recommend specific criteria from which compounds may be compared and ranked such that a lead candidate and acceptable "backup" candidates can be chosen (Lipinski et al. 1997; Hefti 2008). Typical criteria used to evaluate and rank early development candidates often include but are not limited to the following: stability under a variety of conditions, half-life, bioavailability, metabolic pathway and DDI (drug-dependent interaction) potential, systemic and organ toxicity, estimated safety window, the potential for genotoxicity or cardiotoxicity, in vitro receptor affinity and effects and efficacy in animal models.

Lead investigational compounds that survive the initial screening may be further "optimized" or altered to make them more effective and safer. By changing the structure of a compound, scientists can give it different properties. For example, they can make a compound less likely to interact with other chemical pathways in the body, thus reducing the potential for side effects. Hundreds of different variations or "analogs" of the initial leads are produced and tested. The resulting compound is the candidate drug which will undergo years of further testing and analysis before potentially being reviewed and assessed for approval by the USFDA.

The Early TPP

The TPP affects all research activities during candidate selection and lead optimization, including focused compound design in order to reach the set TPP standards and planning of a screening cascade in order to maximize the number of testing cycles on key TPP parameters. Some salient TPP properties such as toxicological risks, predicted human dosing, and pharmaceutical properties can only be effective, and practically, assessed for the first time in a project timeline during early drug development. TPP definition and compliance have, therefore, far-reaching effects across the drug discovery–drug development value chain: they dictate which compounds are made in the first place, which compounds will be selected for clinical development, and ultimately which compounds will be successful at the end of the development cycle.

When considering the importance of the TPP to early drug development, it shouldn't be surprising that all of its parameters are essentially surrogates of clinical readouts, each characterized by its own uncertainty and variability based on the underlying data and methods used. Although major advances have been made in predicting human pharmacokinetics from animal data, there are still occasions for surprises in Phase 1 pharmacokinetic studies due to the intrinsic variability of human absorption, metabolic, and excretion properties, especially with compounds characterized by low-to-moderate bioavailability. When it comes to predicting pharmacological efficacy and toxicity, the current dismal clinical attrition statistics and the corresponding breakdown as to the primary reason for failure are sobering reminders of to what little extent we can predict clinical performance, although having clinically-validated biomarkers and genetics evidence for a given target can help to mitigate these risks. Furthermore, the various TPP parameters cannot be dealt with in isolation but are intimately connected. Integration of TPP parameters so as to provide clinically useful estimates such as starting dose, dose frequency, and therapeutic windows adds an additional layer of complexity and uncertainty during early drug development. Given these premises, early drug development is where the multidisciplinary nature of drug discovery and development makes the biggest impact. Successful integration of scientific data from disciplines such as medicinal chemistry, process chemistry, pharmacology, toxicology, and pharmaceutics requires discipline experts to work seamlessly as a team, fluent in each other's vocabulary, able and willing to challenge and support each other. Their ability to proactively anticipate and address TPP-related issues, to master the interdependencies between TPP parameters, and to distill diverse inputs into actionable plans and schedules is as important to success as the quality of the scientific data generated and the validity of the therapeutic hypotheses being tested.

Milestones for Discovery

A successful discovery and preclinical development campaign will provide the following as experimental evidence of a validated target, candidates which are "druggable" to the extent that they can be incorporated into commercially-viable formulations consistent with the early target product profile, lead and backup candidates that are suitable to move forward toward an IND and ultimately a credible and defendable IND across all functions that contribute to the application. These represent the primary milestones of the phase of course, and several may be in various stages of completion as discovery and preclinical evaluation ensues (e.g. target validation and "druggability" determination). As the IND is the hard endpoint for successful human phase entry (start of Phase 1), many of the other factors roll up into the IND.

The usual time spent in drug discovery for a particular compound is two to three years with an estimated cost of 4 M USD (DiMasi et al. 2016). An additional one to two years are spent in preclinical testing at the cost of approximately 15 M USD (DiMasi et al. 2016). Of course, these are estimates only and are based on aggregate analysis. As with later stages of development, it is in the sponsor's best interest to filter out compounds that don't meet the specified TPP criteria and also not advance compounds ranked too far below the lead molecule.

References

Brodniewicz, T. and Grynkiewicz, G. (2010). Preclinical drug development. *Acta Poloniae Pharmaceutica – Drug Research* 67 (6): 579–586.

DiMasi, J.A., Grabowski, H.G., and Hansen, R.W. (2016). Innovation in the pharmaceutical industry: new estimates of R&D costs. *Journal of Health Economics* 47: 20–33.

Drews, J. (2000). Drug discovery: a historical perspective. *Science* 287 (5460): 1960–1964.

Guidance for Industry (2001). *Bioanalytical Method Validation*. Rockville, MD: US Department of Health and Human Services, FDA, Center for Drug Evaluation and Research. Food and Drug Administration.

Guidance for Industry (2005). *Estimating the Maximum Safe Starting Dose in Initial Clinical Trials for Therapeutics in Adult Healthy Volunteers*. U.S. Department of Health and Human Services Food and Drug Administration Center for Drug Evaluation and Research (CDER) https://www.fda.gov/media/72309/download.

Guidance for Industry (2010). *M3(R2) Nonclinical Safety Studies for the Conduct of Human Clinical Trials and Marketing Authorization for Pharmaceuticals*. U.S. Department of Health and Human Services, Food and Drug Administration, Center for Drug Evaluation and Research (CDER) and Center for Biologics Evaluation and Research (CBER) https://www.regulations.gov/document?D=FDA-2008-D-0470-0013.

Guidance for Industry (2018). *Physiologically Based Pharmacokinetic Analyses — Format and Content*. U.S. Department of Health and Human Services Food and Drug Administration Center for Drug Evaluation and Research (CDER) https://www.fda.gov/media/101469/download.

Gunaratna, G. (2000). Drug metabolism and pharmacokinetics in drug discovery – a primer for bioanalytical chemists, part I. *Current Separations* 19 (1): 17–23.

Hefti FF. Requirements for a lead compound to become a clinical candidate. *BMC Neuroscience* 2008, 9(Suppl 3): S7 doi:https://doi.org/10.1186/1471-2202-9-S3-S7.

Kennedy, T. (1997). Managing the drug discovery/development interface. *Drug Discovery Today* 2: 436–444.

Lipinski CA, Lombardo F, Domini BW, Feeney PJ. Experimental and computational approaches to estimate solubility and permeability in drug discovery and development settings. *Advanced Drug Delivery Reviews* 1997; 23:3–25. doi: https://doi.org/10.1016/S0169-409X(96)00423-1.

Nair, A. B., & Jacob, S. (2016). A simple practice guide for dose conversion between animals and human. *Journal of basic and clinical pharmacy*, 7(2), 27–31. doi:https://doi.org/10.4103/0976-0105.177703.

Nikitenko, A.A. (2006). Compound scale-up at the discovery-development interface. *Current Opinion Drug Discovery and Development* 9: 729–740.

Pandey S, Pandey P, Tiwari G and Tiwari R. Bioanalysis in drug discovery and development. Pharm Methods. 2010 Oct-Dec; 1(1): 14–24.

Parasuraman, S. (2011). Toxicological screening. *Journal of Pharmacology and Pharmacotherapeutics* 2 (2): 74–79. https://doi.org/10.4103/0976-500X.81895. PMID: 21772764; PMCID: PMC3127354.

Shanks, N., Greek, R., and Greek, J. (2009). Are animal models predictive for humans? *Philosophy, Ethics, and Humanities in Medicine* 4: 2. https://doi. org/10.1186/1747-5341-4-2. PMID: 19146696; PMCID: PMC2642860.

Strovel, J., Sittampalam, S., Coussens, N.P. et al. (2016). Early Drug Discovery and Development Guidelines: For Academic Researchers, Collaborators, and Start-up Companies. 1 May 2012. In: (eds. Sittampalam, G.S., Grossman, A., Brimacombe, K. et al.). Assay Guidance Manual [Internet]. Bethesda (MD): Eli Lilly & Company and the National Center for Advancing Translational Sciences; 2004-. https://www. ncbi.nlm.nih.gov/books/NBK92015.

Tiwari, G. and Tiwari, R. (2010). Bioanalytical method validation: An updated review. *Pharmaceutical Methods* 1 (1): 25–38. https://doi.org/10.4103/2229-4708.72226. PMID: 23781413; PMCID: PMC3658022.

Chapter Self-Assessments: Check Your Knowledge

Questions:
- What does product characterization mean and why is it relevant to initiate this effort in the drug discovery phase?
- What is the therapeutic ratio and how is it defined?
- What are carcinogenicity studies and what is their purpose?
- What are the important milestones for the Discovery phase of development?

Answers:
- Product characterization reveals the biochemical and biophysical nature of the product as well as the nature of product-related substances and impurities. Thorough product characterization is a necessary precursor to determine critical quality attributes (CQAs) and the associated analytical methods that, in turn, can be used as in-process controls and specifications and for stability testing. Process characterization focuses on understanding and defining the operating and design spaces for the process to achieve a product with consistent CQAs. It is advised by FDA that a product in development be analyzed at all stages during the processing pathway, including during discovery, upstream processing, downstream processing, and right through to the final product. This thorough analysis is meant to ensure that the product is consistent, active, stable, safe, and pure from start to finish.

- The therapeutic ratio is a ratio [TD50/ED50] of the dose at which 50% of patients experience the toxic effect to the dose at which 50% of patients experience the therapeutic effect. A therapeutic ratio of 2.5 means that approximately 2.5 times as much drug is required to cause toxicity in half of the patients than is needed to produce a therapeutic effect in the same proportion of patients. However, this ratio of toxic to therapeutic dose may not be consistent across the entire dose range if the dose-effect curves for the therapeutic and toxic effects are not parallel.
- Carcinogenicity studies are usually needed for drugs intended for chronic or recurring conditions. The conventional test for carcinogenicity is the long-term rodent carcinogenicity bioassay. Studies observe test animals for a major portion of their life span for the development of neoplastic lesions during or after exposure to various doses of a test substance by an appropriate route of administration. The study is usually conducted using two species – rats and mice of both sexes. The animals are dosed by oral, dermal, or inhalation exposures based upon the expected type of human exposure. Dosing typically lasts around two years. Certain features are monitored throughout the study, but the key assessment resides in the full pathological analysis of the animal tissues and organs when the study is terminated.
- Successful discovery and preclinical development campaign will provide the following as experimental evidence of a validated target, candidates which are "druggable" to the extend that they can be incorporated into commercially-viable formulations consistent with the early target product profile, lead and backup candidates that are suitable to move forward toward an IND and ultimately a credible and defendable IND across all functions that contribute to the application. These represent the primary milestones of the phase of the course, and several may be in various stages of completion as discovery, and preclinical evaluation ensues (e.g. target validation and "druggability" determination).

Quiz:

1 True or false. The accumulated project cost associated with a "No Go" decision at the end of the candidate selection/Discovery Phase is estimated to be $1.46 billion USD.

2 Which of the following are likely to demand the largest quantities of drug supply?
   A Hit identification studies
   B Commercial supplies
   C Phase 2 trial supplies
   D in vitro biologic activity experiments
   E None of the above

3 Which of the following is NOT included in an IND?
   A the results of the preclinical work
   B the candidate drug's molecular structure
   C details on how investigational medicine is thought to work in the body
   D a listing of any potential side effects as indicated by the preclinical studies
   E manufacturing information
   F All are included

4 Choose the best answer to fill in the blanks. A drug candidate suitable for clinical testing is expected to bind selectively to the receptor site on the target, to elicit the desired functional response of the target molecule, and to have adequate _____ and _____ to elicit the desired responses in animals and humans.
   A absorption and bioavailability
   B bioavailability and biodistribution
   C absorption and biodistribution
   D bioavailability and rapid elimination
   E None are correct

7

Phase 1

Jeffrey S. Barrett

Aridhia Digital Research Environment

Introduction, Phase 1 Defined

Phase 1 in the drug development process represents the initial foray into human clinical evaluation. The goal of Phase 1 drug development is to provide an initial assessment of the safety and tolerability of a drug candidate while also assessing the pharmacokinetics and pharmacodynamics of the relevant biologically active species over the range of doses studied. Typically, but not always, Phase 1 is conducted entirely in healthy volunteers reserving patient evaluation for later stages, assuming safety and activity milestones have been met. The exceptions to the healthy volunteer study populations are usually based on agents in whom the drug actions would constitute an unacceptable risk (e.g. chemotherapeutic agents to treat cancer) or when the target population represents an acceptable risk-benefit setting for Phase 1 (e.g. healthy elderly) and healthy volunteers would not permit an appropriate population to judge Phase 1 (e.g. pregnant women or HIV-infected patients). Phase 1 clinical investigation is beyond a single trial and addresses both acute (single dose) and chronic (multiple doses) conditions in addition to lifestyle factors (e.g. food effect, drug interaction potential studies), formulation development (bioavailability and bioequivalence trials), and special population trials. The various study types and designs are covered in detail in later sections of the chapter.

Beyond the conduct of the initial human trials, there are, of course, many other complementary activities from other parts of a pharmaceutical organization essential for a successful Phase 1 program. The objectives of this chapter include the identification of these key activities that occur during Phase 1, exposing the key milestones and decision criteria evaluated during Phase 1 compound progression, and examining the compound attributes which define a successful drug candidate across various modalities. We will also explore several therapeutic areas that deviate somewhat outside the conventional Phase 1 approach based on the target populations in question and illustrate how Phase 1 deliverables appear in the product label of approved medicines. Finally, we will expose the cost of the Phase 1 activities that roll up into the overall drug development budget.

Who Is Doing What – The Early TPP and Phase 1 Milestones

The early phase project teams are comprised of scientists the evaluate multiple compounds of interest within a target therapeutic area and mechanism of action. The diversity of these scientists is broad and includes pharmacologists, chemists, biologists, statisticians, and clinicians. In addition, early evaluations by marketing and commercial colleagues may be commonplace as well (see Chapter 11 for details of Project Team composition). The activities of these project team members are focused on (1) developing a suitable early formulation to be used in the first-time-in man (FTIM) Phase 1 trial, (2) evaluating the toxicology of potential drug candidates in both animals and humans, (3) proposing and defending the choice of relevant analytes and biomarkers to measure, and (4) proposing, designing and evaluating early human phase trials which define the therapeutic window of target drug candidates. These activities will likely be spread over multiple drug candidates with the intention of choosing one or more to move forward into Phase 2 testing, assuming they pass the requisite target product profile (TPP) milestones laid out for Phase 1.

As mentioned in the previous chapter, the TPP provides the sponsor's development ideas for a target candidate include criteria and milestones that define a minimally acceptable and target candidate. In order to achieve these metrics, a number of discrete activities are undertaken by various pharmaceutical scientists and the functional groups that support them. Table 7.1 below provides a list of critical activities undertaken during Phase 1, identifies the group responsible, and the impact of the activity of Phase 1 milestones.

While the TPP elements that describe the intended route of administration, formulation, and dosing frequency are specified by commercial interests and influenced by the existing marketplace with respect to approved drugs and drugs in development, the actual data that defines these attributes for a development candidate is generated in Phase 1 (Breder et. al. 2017). In addition, reference to these attributes is made in the drug monograph and package insert that a sponsor produces in collaboration with regulatory authorities for prescribing physicians and patients. The early TPP that relies on Phase 1 input will be most dependent on the following: (1) an estimate of the relevant moiety's (the active drug substance) half-life to confirm whether the compound achieves its target dosing frequency, (2) a safety profile that is at least as good as or better than the current standard of care, (3) identification of a biomarker that is suitable to track the

Table 7.1 Critical Phase 1 activities that define TPP attributes and critical milestones necessary for advancement to Phase 2.

Phase 1 deliverable	Group responsible	Impact on Phase 1 milestone
Validated, GLP analytical methods for all relevant moieties and biomarker	Bioanalytical and/or biomarker group (could be CRO as well)	• FTIM and other Phase 1 studies • Needed to define basic PK/PD and therapeutic window as well as interpretation of any AEs or ADRs
Assessment of PK/PD and guidance on sampling scheme	DMPK and/or clinical pharmacology	• FTIM and other Phase 1 studies • Needed to define basic PK/PD and therapeutic window as well as interpretation of any AEs or ADRs; suitability of dosing frequency requirement of TPP
Phase 1 site selection	Clinical operations	• Necessary before any trial can be initiated; includes identification of PI and staffing requirements
Phase 1 study design and protocol development	Clinical pharmacology and biostatistics	• Necessary before any trial can be initiated; includes sample size projection and statistical analysis plan • Included in the IND submission
Provision of dosing material and/or placebo for trials	CMC and clinical operations	• Necessary before any trial can be initiated; provided to site in time • Needs to be GLP/GXP compliant
Provision of suitable Phase 1 formulation	Formulations and/or CMC groups	• Necessary before any trial can be initiated • Needs to be GLP/GXP compliant • Composition disclosed in IND
Provision of adequate preclinical safety package (final reports from completed toxicology trials)	Toxicology/safety assessment	• Requirement for IND submission; necessary for acceptance of IND and earliest start date for FTIM study

actions of the active entities to support and defend dosing recommendations, (4) minimal or acceptable interactions that would allow competitive marketing against the current standard of care or avoid a black-box warning, and (5) an indication that a final market image formulation was achievable at the time of Phase 3 initiation. Table 7.2 provides additional details on why and how these Phase 1 deliverables are essential milestones for passage to Phase 2.

Table 7.2 Phase 1 milestones essential for inclusion in TPP and progression to Phase 2.

Phase 1 milestone	TPP attribute	Impact
Relevant PK, especially half-life estimates	Target dosing frequency	Typically, oral immediate-release formulations are targeted for once daily dosing; short half-lives require more frequent dosing and may suggest a No-Go based on the market
Safety profile vs. dose and exposure compared to placebo	Acceptable therapeutic window	Market access, being listed on formulary, and being reimbursed are all based on similar or superior safety relative to the current standard of care
Biomarker identified	Justification for dose and regimen	The NDA requires a detailed and referenced justification for dose that is also referenced in the product label
Initial DDI evaluation and screening completed	Acceptable DDI profile	Most drugs are metabolized to a form that can be eliminated from the body. Other drugs and supplements can compete/interact with metabolic pathways yielding unwanted consequences (high or lower than anticipated drug levels). Depending on this result and the vulnerability of the target population, this can be a competitive advantage or disadvantage
Formulation suitable to support FTIM study	Pre-formulation studies suggest a market image is attainable	A final market image will specify the size, shape, coating, and other physical features of an immediate release dosage form even though the FTIM study will be based on a "powder-in-bottle" formulation. Some drug substance properties and dosing requirements may suggest that this image in unattainable

As Table 7.2 confirms, multiple groups supporting early phase project teams are generating data to pursue the evaluation of critical milestones during Phase 1. Successful completion of this phase also requires the assessment of how Phase 1 deliverables compare against the initial desired TPP attributes laid out by the project team before Phase 1 was initiated (André and Foulkes 1998). Such comparison would be made for every development compound within a class and thus be used as a means of comparing and ranking compounds deemed as "lead" and "backup." While it is desirable to have several candidates at the early stages, there is clearly the intention to select a single candidate for later-stage testing, especially Phase 3 given the considerable cost involved. Pharmaceutical sponsors often refer to the desire to make a "quick kill" when warranted in early phase testing with the expectation of improved development efficiency and containing development costs. To be clear, "quick kill" in this context refers to eliminating the compound from consideration as a development candidate for the intended therapeutic area and stopping further development activities. It could also mean disbanding the project team if there are no backup compounds in the pipeline.

Regulatory Hurdles to Human Phase Testing

In order for this stage (Phase 1) to begin, several milestones must be achieved to warrant the expectation of an acceptable risk prior to human exposure to a new chemical entity (NCE). While general agreement on these milestones is understood based on historical summits on the ethics of clinical investigation (WHO, Helsinki agreement, etc.), the specific elements and the evidence that adequate preclinical data exists is governed by the relevant regulatory review authorities (e.g. FDA in the United States, Health Protection Branch in Canada, EMA in Europe). While there is not perfect agreement across all regulatory authorities regarding the numerous requirements for drug sponsors, the requirements for human phase testing are reasonably broadly accepted around the world. The requirements that must be demonstrated by sponsors are embedded in the review processes established by various regulatory authorities. In the United States, the FDA governs this dialogue with the investigational new drug (IND) application process.

Luckily, there are guidance documents that advise sponsors, for example, "Content and Format of Investigational New Drug Applications (INDs) for Phase 1 Studies of Drugs, Including Well-Characterized, Therapeutic, Biotechnology-derived Products" provided by FDA (see references at the end of the chapter, FDA Guidances 1995–2019b). Other regulatory authorities publish similar documents, of course. The current IND application content requirements are defined in Table 7.3.

The General Investigation Plan is essentially the roadmap of how the sponsor plans to investigate and develop the NCE. A good plan provides a detailed assessment of the drug target as well as key assumptions regarding the presumed mechanism of action along with initial ideas regarding how the sponsor plans to establish proof-of-mechanism and proof-of-concept in patients (deliverables for

Table 7.3 Primary sections comprising the IND application as defined by the US FDA (Code of Federal Regulation reference).

- Cover sheet (FDA Form-1571) [21 CFR 312.23(a)(1)]
- Table of contents [21 CFR 312.23(a)(2)]
- Introductory statement and general investigational plan [21 CFR 312.23(a)(3)]
- Investigator's brochure [21 CFR 312.23(a)(5)]
- Protocols [21 CFR 312.23(a)(6)]
- Chemistry, manufacturing, and control information [21 CFR 312.23(a)(7)]
- Pharmacology and toxicology information [21 CFR 312.23(a)(8)]
- Previous human experience with the investigational drug [21 CFR 312.23(a)(9)]
- 21 CFR 312.23(a)(10), (11) and (b), (c), (d), and (e)

the Phase 2 program). While the plan is not expected to represent the absolute, final development plan, its value is centered on establishing a credible rationale for why and how a particular NCE is suitable to be evaluated for the proposed indication(s).

The investigator's brochure (IB) is a compilation of the clinical and nonclinical data on the investigational product(s) that are relevant to the study of the product(s) in human subjects (see Table 7.4 for template IB table of contents). Its

Table 7.4 Template table of contents of an investigational brochure for a hypothetical compound.

Section	Purpose
Title page	Provides the sponsor's name, the identity of each investigational product and the release date.
Confidentiality statement	A statement instructing the investigator/recipients to treat the IB as a confidential document for the sole information and use of the investigator's team and the IRB/IEC.
Table of contents	
Summary	As described
Introduction	
Physical, chemical, and pharmaceutical properties and formulation	A description of the investigational product substance(s), and a summary of the relevant physical, chemical, and pharmaceutical properties. A description of the formulation(s) to be used, including excipients, should also be provided and justified if clinically relevant. Instructions for the storage and handling of the dosage form(s) should also be given. Any structural similarities to other known compounds should be mentioned.

Table 7.4 (Continued)

Section	Purpose
Nonclinical studies introduction	The information provided may include the following, as appropriate if known/available: • Species tested, number and sex of animals in each group, unit dose (e.g. milligram/kilogram [mg/kg]), dose interval, route of administration, duration of dosing, information on systemic distribution, duration of post-exposure follow-up. • Results, including the following aspects: – Nature and frequency of pharmacological or toxic effects, severity or intensity of pharmacological or toxic effects, time to onset of effects, reversibility of effects, duration of effects, and dose-response
Effects in humans introduction	
Pharmacokinetics and product metabolism in humans	A summary of the pharmacokinetics of the investigational product(s) including the following, if available: Pharmacokinetics (including metabolism, as appropriate, and absorption, plasma protein binding, distribution, and elimination); Bioavailability of the investigational product (absolute, where possible, and/or relative) using a reference dosage form; Population subgroups (e.g. gender, age, and impaired organ function); Interactions (e.g. product-product interactions and effects of food); Other PK data (e.g. results of population studies performed within trial(s).
Toxicology	A summary of toxicological effects found in relevant studies conducted in different animal species under the following headings where appropriate: Single dose; Repeated dose; Carcinogenicity; Special studies (e.g. irritancy and sensitization); Reproductive toxicity; Genotoxicity (mutagenicity)
Effects in humans introduction	A discussion of the known effects of the investigational product(s) in humans including information on pharmacokinetics, metabolism, pharmacodynamics, dose-response, safety, efficacy, and other pharmacological activities. Where possible, a summary of each completed clinical trial should be provided. Information should also be provided regarding results of any use of the investigational product(s) other than from in clinical trials.

(Continued)

Table 7.4 (Continued)

Section	Purpose
Pharmacokinetics and product metabolism in humans	A summary of information on the pharmacokinetics of the investigational product(s) should be presented, including the following, if available: Pharmacokinetics (including metabolism, as appropriate, and absorption, plasma protein binding, distribution, and elimination). Bioavailability of the investigational product (absolute, where possible, and/or relative) using a reference dosage form. Population subgroups (e.g. gender, age, and impaired organ function). Interactions (e.g. product-product interactions and effects of food). Other pharmacokinetic data (e.g. results of population studies performed within clinical trial(s).
Safety and efficacy	Information about the investigational product's/products' (including metabolites, where appropriate) safety, pharmacodynamics, efficacy, and dose-response obtained from preceding trials in humans (healthy volunteers and/or patients). In cases where several clinical trials have been completed, summaries of safety and efficacy across multiple trials by indications in subgroups may provide a clear presentation of the data. Important differences in adverse drug reaction patterns/incidences across indications or subgroups should be discussed.
Marketing experience	The IB should identify countries where the investigational product has been marketed or approved. Any significant information arising from the marketed use should be summarized (e.g. formulations, dosages, routes of administration, and adverse product reactions). The IB should also identify all the countries where the investigational product did not receive approval/registration for marketing or was withdrawn from marketing/registration.
Summary of data and guidance for the investigator	This section should provide an overall discussion of the nonclinical and clinical data and should summarize information from various sources on different aspects of the investigational product(s). Where appropriate, the published reports on related products should be discussed. The overall aim of this section is to provide the investigator with a clear understanding of the possible risks and adverse reactions and of the specific tests, observations, and precautions that may be needed for a clinical trial. Guidance should be provided on the recognition and treatment of possible overdose and adverse drug reactions that is based on previous human experience and on the pharmacology of the investigational product.
References, reports, and appendices	

purpose is to provide the investigators and others involved in the trial with the information to facilitate their understanding of the rationale for, and their compliance with, many key features of the protocol, such as the dose, dose frequency/interval, methods of administration: and safety monitoring procedures. The IB also provides insight to support the clinical management of the study subjects during the clinical trial. The information should be presented in a concise, simple, objective, balanced, and non-promotional form that enables a clinician, or potential investigator, to understand it and make his/her own unbiased risk-benefit assessment of the appropriateness of the proposed trial. For this reason, a medically qualified person should generally participate in the editing of an IB, but the contents of the IB should be approved by the disciplines that generated the described data.

If the investigational product is marketed and its pharmacology is widely understood by medical practitioners, an extensive IB may not be necessary. Where permitted by regulatory authorities, a basic product information brochure, package leaflet, or labeling may be an appropriate alternative if it includes current, comprehensive, and detailed information on all aspects of the investigational product that might be of importance to the investigator. If a marketed product is being studied for new use (i.e. a new indication), an IB specific to that new users should be prepared. The IB should be reviewed at least annually and revised as necessary in compliance with a sponsor's written procedures. More frequent revision may be appropriate depending on the stage of development and the generation of relevant new information. However, in accordance with Good Clinical Practice, relevant new information may be so important that it should be communicated to the investigators, and possibly to the Institutional Review Boards (IRBs)/Independent Ethics Committees (IECs) and/or regulatory authorities before it is included in a revised IB. Generally, the sponsor is responsible for ensuring that an up-to-date IB is made available to the investigator(s), and the investigators are responsible for providing the up-to-date IB to the responsible IRBs/IECs. In the case of an investigator-sponsored trial, the sponsor-investigator should determine whether a brochure is available from the commercial manufacturer. If the investigational product is provided by the sponsor-investigator, then he or she should provide the necessary information to the trial personnel. In cases where preparation of a formal IB is impractical, the sponsor-investigator should provide, as a substitute, an expanded background information section in the trial protocol that contains the minimum current information described in this guideline.

Regulatory authorities, of course, are responsible for protecting the populations over which they have jurisdiction. Likewise, the review of the IND application is a critical component of the process that needs time for critical evaluation and an opportunity for dialogue between reviewers and pharmaceutical sponsors.

Phase 1 Studies

Many different studies are conducted during Phase 1 and beyond that would fall in the category of Phase 1 type studies. Table 7.5 illustrates the major Phase 1 study types, prototypical design elements, and the purpose of each study type.

In most circumstances, subjects in first in man studies will be healthy young men, usually between the ages of 18 and 45 years. The rationale for this age group

Table 7.5 Phase 1 study types and features.

Study type	Design/features	Purpose
Single ascending dose (SAD)	• Randomized, placebo-controlled, healthy volunteers (or patients, in certain cases) • Starting dose determined by preclinical toxicology studies	• Safety/tolerability, identify maximum tolerated dose (MTD) • General PK characteristics, variability, linearity, dose proportionality • Exploration of drug elimination (urine PK, metabolite identification)
Multiple ascending dose (MAD)	• Randomized, placebo-controlled, healthy volunteers (or patients, in certain cases) • Doses guided by SAD results	• Same as SAD under multiple-dose conditions • Assess steady-state parameters (accumulation, time-dependency)
Bioavailability (BA)	• Typically, crossover, single-dose (if linear PK) study in healthy subjects • Measure blood/plasma conc. of parent drug and major active metabolites for ≥ 3 t½	• Relative and /or absolute BA of drug from a formulation and/or route of administration
Bioequivalence (BE)	• Crossover study in fasted, healthy subjects given single doses of test and reference products administered at same molar doses • Measure blood/plasma conc. of parent drug • "Pivotal" BE study required to bridge the to-be-marketed formulation (test) to that used in Phase 3 trials (reference)	• BE acceptance criteria: 90% CI of the geometric mean ratios of C_{max} and AUC between test and reference fall within 80–125%

Table 7.5 (Continued)

Study type	Design/features	Purpose
Mass balance	• Typically, single-dose, healthy males (n = 4–6), from intended route of administration • Radio-labeled (C^{14}) drug molecule • Measure concentrations of parent and metabolite(s) and determine radioactivity in plasma, urine, feces	• Determine primary mechanism(s) of elimination and excretion from the body • Determine the proportion of parent drug converted to metabolite(s)
Food effect	• Single-dose, crossover, two-treatment (fed vs. fasted), two-period, two-sequence study in healthy subjects (n ≥ 12 with data) • Use highest strength of drug product; fed: FDA high-fat high-calorie meal • PK assessments similar to BA study	• Evaluate effect of food on rate and extent of drug absorption from a given formulation • No food-effect if 90% CI of fed/fasted C_{max} and AUC ratios within 80–125%. • Labeling instructions on administration of drug on empty stomach or without regard to meals
Drug interaction	• Typically, crossover design (parallel – if long $t^{1/2}$ drug); healthy subjects (patients for safety considerations or to evaluate PD endpoints) • Choice of doses/dosing intervals/dosage forms of substrate and inhibitor and/or inducer, routes and timing of co-administration, number of doses should maximize possibility of detecting interaction and mimic the clinical setting.	• Evaluate potential of investigational drug as an inhibitor/inducer (I) and substrate (S) of certain metabolizing enzymes and/or transporters
Organ impairment	• Single-dose (if linear and time-independent PK), parallel groups, males, and females with varying degrees of organ impairment (≥6 per group) • Reduced designs and population PK approaches are complementary and/or alternatives	• Effect of organ impairment on drug clearance; dosage recommendations for various stages of organ impairment • Effect of hemodialysis (renal impairment) on drug exposure; info on whether dialysis could be used as treatment for drug overdose

(Continued)

Table 7.5 (Continued)

Study type	Design/features	Purpose
Special populations	• Single-dose, parallel groups, males, and females across strata (e.g. age, trimester; ≥6 per group) • Reduced designs and population PK approaches are complementary and/or alternatives	• Effect of sub-strata on drug clearance; dosage recommendations for various stages (age, pregnancy state, etc.)
Thorough QT (TQT)	• Usually, single-dose study in healthy subjects; evaluate therapeutic and "supratherapeutic" doses of drug versus positive control (e.g. moxifloxacin) • ICH Guidelines, E14: recommendations for design, conduct, analysis, and interpretation	• Identify drugs that prolong QT (95% CI upper bound ≥ 10 ms) that need a more thorough ECG monitoring in pivotal trials • Label instructions regarding QT prolongation risk

is based on the desire to have a homogeneous population in which to study the effects of the new drug and also to limit variability in kinetics and dynamics. It is also presumed that this population will be more able to withstand unexpected toxicity caused by the test drug. Healthy subjects in this context are those who have no underlying diseases that could interfere with the conduct of the study or confound the interpretation of the safety or pharmacokinetic data. Criteria for inclusion into the study based upon medical history, physical examination, use of concomitant medications, alcohol, cigarettes, and recreational drugs, as well as the results of blood testing 12-lead ECG, blood pressure heart rate are laid out in the study protocol. Male subjects are generally preferred because at this early stage of development, reproductive toxicology testing in animals will not have been completed, and the risk to the fetus of female subjects who might be pregnant or become pregnant shortly before or after the study has not been characterized.

While there are many variations of the single alternating dose (SAD) and multiple alternating dose (MAD) study designs, there are commonalities that span all design nuances. The essential elements are the necessity to dose escalate from some initial low dose perceived to be a "no-effect" or minimally effective dose with limited or no projected activity or toxicity in human volunteers based on animal pharmacology and toxicology studies (see previous chapter) to higher

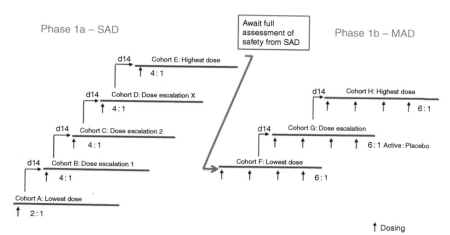

Figure 7.1 Interplay between SAD and MAD FTIM study designs: randomized, double-blind within cohort, placebo-controlled, dose-escalating studies.

doses with both activity, and potentially toxicity may be observed. Each dose cohort will enroll a minimum number of subjects and subjects who receive placebo (no active drug) so that comparative assessments of safety may be made. SAD and MAD study designs are typically coordinated and may, in fact, be included in a single protocol depending on the perceived risk to volunteers. Figure 7.1 illustrates a common design construct for Phase 1 SAD and MAD trials. It would be normally expected that the SAD cohorts are completed first to inform the dose-escalation plans for the MAD cohorts. Variation in Phase 1 study designs can be found in Patat (2000). As mentioned previously, the IND will contain the FTIM protocol(s) so that regulators can evaluate the designs and the assessment of acceptable risk based on the preclinical toxicology evaluation.

Of course, the regulatory review does not substitute for the clinical review from the IRB, who will also weigh on the protocol design and risk assessment as well as the sponsors description in the informed consent that volunteers and/or patients would sign before they would be permitted to participate and enroll in the trial. The process of informed consent occurs when communication between a patient and physician results in the patient's authorization or agreement to undergo a specific medical intervention (and participate in a drug study). In seeking a patient's informed consent (or the consent of the patient's surrogate if the patient lacks decision-making capacity or declines to participate in making decisions), physicians must:

a) Assess the patient's ability to understand relevant medical information and the implications of treatment alternatives and to make an independent, voluntary decision.

b) Present relevant information accurately and sensitively, in keeping with the patient's preferences for receiving medical information. The physician should include information about:

The diagnosis (when known)

The nature and purpose of recommended interventions

The burdens, risks, and expected benefits of all options, including forgoing treatment

c) Document the informed consent conversation and the patient's (or surrogate's) decision in the medical record in some manner. When the patient/surrogate has provided specific written consent, the consent form should be included in the record.

In emergencies, when a decision must be made urgently, the patient is not able to participate in decision making, and the patient's surrogate is not available, physicians may initiate treatment without prior informed consent. In such situations, the physician should inform the patient/surrogate at the earliest opportunity and obtain consent for ongoing treatment in keeping with these guidelines.

Beyond the FTIM studies, other Phase 1 studies are planned and designed, and conducted to evaluate the suitability of the investigational agent for the intended indication based on lifestyle considerations of the patient population (i.e. food effect and drug interaction trials) and the attributes of the investigational drug (QTc and mass balance studies). Figure 7.2 provides a chronology of both the

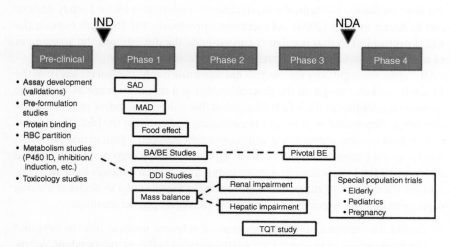

Figure 7.2 Schematic illustrating the timing and relationship of common Phase 1 trials used to support drug development of new chemical entities.

timing and the typical order of these studies. As the figure illustrates, even though we refer to these trials as Phase 1 studies, many of them are actually conducted after Phase 2, and occasionally Phase 3 has been initiated. The rationale for this is based on the fact that these trials are expensive, time-consuming, and often difficult to enroll (see Special Populations chapter). Pharmaceutical sponsors are often conservative with spending the money these trials require before actually knowing they have a potential drug (i.e. compound worth the financial investment these studies demand).

Phase 1 Economics

Relative to later phases of drug development, Phase 1 is less expensive though the range of Phase 1 development programs is quite varied in terms of the number and nature of studies required and likewise the total cost of such studies. As discussed, Phase 1 trials are generally of shorter duration and in fewer subjects as opposed to those required in Phases 2 and 3, from which proof-of-concept, the patient therapeutic window, and the evidence of safety and efficacy are defined, respectively. If one compares cost per patient across the various development phases, we observe that the per patient costs are quite similar: $38 500, 40 000, and 42 000 USD for Phase 1, 2, and 3, respectively (Battelle survey data, 2013, http://www.phrma.org/sites/default/files/pdf/biopharmaceutical-industry-sponsored-clinical-trials-impact-on-state-economies.pdf). Hence, the primary driver differentiating the cost of the various development phases is the number of patients studied.

Focusing on the major costs associated with Phase 1 trials (http://medrio.com/partners/the-top-5-cost-drivers-in-phase-i-clinical-trials), we can easily identify staffing as the biggest expenditure – typically identified as consuming 40–45% of the total cost of a Phase 1 trial. Staffing costs are usually defined based on a cost per patient basis as well. Phase 1 staffing costs have risen over the past decades as companies have hired more clinical research associates (CRAs) per Phase 1 trial. Staff increase is supported primarily by the increased complexity of early phase trials and an increase in the number of procedures per patient. The research site itself is the next biggest drive accounting for 15–35% of the total study budget. Usually, the CRO or sponsor pays the research site on a per bed per day basis. Additional costs include services like staff surveillance, catering, overhead, and administration expenses.

Rounding out the primary Phase 1 cost drivers are subject recruitment (15–20%), the cost of diagnostic equipment (12–20%), and data management tools (8–10%). Challenges with incentivizing healthy volunteers to participate in studies and high dropout rates make patient recruitment a large Phase 1 cost-driver. Likewise, regulation on washout periods (the length of time a study subject must wait before participating in another trial). Advertising and screening costs also factor into recruitment costs. Diagnostic equipment can include fees for ECG, PET scan machines, and cardiac telemetry monitoring devices, as well as lab supplies. Electronic data management tools are used in approximately 70% of studies, and costs can vary widely depending on the system's software delivery model, payment model, and ease-of-use.

In addition to clinical costs, Phase 1 formulation development must establish the physicochemical properties of the early drug candidates: chemical makeup, stability, and solubility. Process chemists and manufacturing scientists must optimize the process they use to make the chemical so they can scale up from a medicinal chemist producing milligrams to manufacturing on the kilogram and ton scale. They further examine the product for suitability to package as capsules, tablets, aerosol, intramuscular injectable, subcutaneous injectable, or intravenous formulations. Together, these CMC activities also contribute to the overall costs of Phase 1. Overall, the cost of product development can account for as much as 30–35% of the total cost of bringing a new drug to the market Suresh and Basu (2008). The quality of product development also affects the time to market and the quality of manufacturing, and therefore cost of manufacturing. Looking across all phases of development puts a value of about $25.5 million USD on Phase 1 development based on 2013 dollars (DiMasi et al. 2016). Of this amount, $1–5 million would be spent on early formulation development with the variance affected by the complexity and the proximity to a final market image the sponsor is willing to pursue on Phase 1 efforts.

Toxicology efforts to support Phase 1 testing have been estimated at $6.5 million USD, with an additional $16 million USD required for toxicology trials necessary to support first patient dosing (start of Phase 2) (Mestre-Ferrandiz et al. 2012). Other minor costs during Phase 1 can be associated with commercial activities attempting to create and quantify a value proposition for a new product and can include early market forecasts, payer and healthcare provider surveys to understand reimbursement considerations, and early health economic studies conducted by the sponsor or external partner or CMO and funded through grants. Typically, these costs are spread-out overall development candidates within in therapeutic franchise, and as such costs are difficult to link with individual programs.

Phase 1 Deliverables Defined

The primary goal of a successful Phase 1 development program is the thorough evaluation of safety in healthy volunteers including a projection of the presumed therapeutic window in patients. In this context, the therapeutic window refers to the range of drug exposures (typically represented by plasma pharmacokinetic metrics such as Cmax or AUC) the complimentary active but safe doses in which a patient may be dosed during their pharmacotherapy (see Figure 7.3).

Additional goals for a successful Phase 1 program include the evaluation and dosing guidance for relevant lifestyle factors such as potential food and drug interaction effects. Likewise, special populations in whom dosing guidance are required (elderly, pediatric pregnant, and/or organ impaired) would also be evaluated as part of an extended Phase 1 program.

During the course of Phase 1, the therapeutic window of development compounds will become known and compared to other agents in development through competitive intelligence and already on the market through prescribing information or the scientific literature. It will be obvious then to the sponsor, prescribers, and healthcare payers (insurance companies) how the potential candidate compares with the standard of care and competitors, which becomes an

Figure 7.3 Idealized therapeutic window shown on time (a) and dose (b) scales.

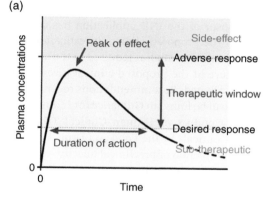

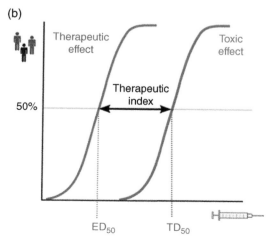

issue hopefully defined in the TPP and the basis for a Go/No-go decision to proceed with patient-level clinical trials. Drugs with a wide therapeutic index may be easier to manage/prescribe, but many drugs with narrow therapeutic indices (NTIs) represent important medicines for a variety of conditions, especially when there are no or few alternatives. NTIs based on FDA designation include Aminophylline, Carbamazepine, Clindamycin Clonidine, Digoxin, Disopyramide, Guanethidine, Isoproterenol, Levoxyine, Valproate, and Sodium Warfarin. While all are important medicines in their target therapeutic areas, they also present challenges to generic competition (see Generics chapter).

Toxicology Milestones

The IND application is expected to contain information about pharmacological and toxicological (laboratory animals or in vitro) studies on the basis of which the sponsor of the IND application has concluded that it is reasonably safe to conduct the proposed clinical investigations. The kind, duration, and scope of animal and other studies required in the application will depend on the duration and nature of the proposed clinical investigations for the investigational agent being evaluated. Recommendations regarding study types and duration for these studies can be found in Guidance for Industry: M3(R2) Nonclinical Safety Studies for the Conduct of Human Clinical Trials and Marketing Authorization for Pharmaceuticals by the Food and Drug Agency.

The completion of preclinical toxicity studies to support the first administration to humans is a time-critical step in the clinical development of medicines and has been complicated by differences in international regulatory requirements. Luckily, ICH has provided some consistency in the requirements, and there is a more consistent expectation of these requirements today. Repeated-dose toxicity studies in two species (one non-rodent) for a minimum duration of two weeks (Table 7.6) would generally support any clinical development trial up to two weeks in duration. Clinical trials of longer duration should be supported by repeated-dose toxicity studies of at least equivalent duration. Six-month rodent and nine-month non-rodent studies generally support dosing for longer than six months in clinical trials (for exceptions, see Table 7.6 footnotes).

Formulation Development

When a compound enters preclinical development for a GLP tox study from the drug discovery stage, we face a question of how to develop a tox and clinical formulation that ensures the success of IND and first dose in humans. Poor

Table 7.6 Recommended duration of repeated-dose toxicity studies to support the conduct of clinical trials: maximum duration of clinical trial recommended minimum duration of repeated-dose toxicity studies to support clinical trials.

Maximum duration of clinical trial	Recommended minimum duration of repeated-dose toxicity studies to support clinical trials	
	Rodents	Non-rodents
Up to 2 wk	2 wk[a]	2 wk[a]
Between 2 wk and 6 mo	Same as clinical trial[b]	Same as clinical trial[b]
> 6 mo	6 mo[b,c]	9 mo[b,c]

[a] In the United States, as an alternative to two-week studies, extended single-dose toxicity studies (see footnote c in Table 3) can support single-dose human trials. Clinical studies of less than 14 days can be supported with toxicity studies of the same duration as the proposed clinical study.

[b] In some circumstances, clinical trials of longer duration than three months can be initiated, provided that the data are available from a three-month rodent and a three-month non-rodent study, and that complete data from the chronic rodent and non-rodent study are made available, consistent with local clinical trial regulatory procedures, before extending dosing beyond three months in the clinical trial. For serious or life-threatening indications or on a case-by-case basis, this extension can be supported by complete chronic rodent data and in-life and necropsy data for the non-rodent study. Complete histopathology data from the non-rodent should be available within an additional three months.

[c] There can be cases where a pediatric population is a primary population, and existing animal studies (toxicology or pharmacology) have identified potential developmental concerns for target organs. In these cases, long-term toxicity testing starting in juvenile animals can be appropriate in some circumstances.

biopharmaceutical properties of compounds are attributed to 39% of the failure of the new drug program under development. A compound with poor biopharmaceutical properties or improper formulation design could lead to a delay in the project or even program termination. The key considerations for successful tox and Phase 1 formulation development consist of the following: pre-formulation studies, biopharmaceutic evaluation, analytical method development, formulation development, and cGMP manufacturing for the FTIM studies.

One of the primary goals of pre-formulation studies is to identify the physico-chemical characteristics of a drug candidate that predict drug product performance in-vitro and in-vivo but also to for the foundation of early formulation development. Pre-formulation studies usually cover the following items: pKa, LogP/LogD, Ph solubility curve, pH stability curve, solvent solubility, particle size distribution, hygroscopicity, API solid-state stability under stressed temperature/humidity conditions, melting point, salts form evaluation, polymorph/hydrate/solvate evaluation, forced degradation under different stressed conditions (light, heat, oxygen, acidic, and alkaline pH, etc.), stability-indicating analytical methods for characterization of active, and impurities. These parameters are important to guide the selection of a potential drug candidate and to determine future formulation strategies.

To accelerate products for early Phase 1 testing, it is not uncommon for a sponsor to utilize a drug reconstitution approach. The procedure requires powder for reconstitution by filling individual doses of active pharmaceutical ingredients, or "API," into glass or plastic bottles, what is generally referred to as a power-in-bottle (PIB) formulation. Drug reconstitution is performed in the clinic, or CRA at the Phase 1 site, by adding water (or some other solvent system) to the drug product and then administering it to the volunteer or patient. This approach offers the advantage of alleviating the need to develop a complex formulation and the analytical testing methodologies required to test a formulation. The PIB approach is also convenient for placebo testing as well, given the simplicity of the approach and the ability to maintain various levels of blinding (to the volunteer/patient and PI as needed).

Impact of Phase 1 Studies on Labeling

A package insert is a document included in the package of a medication that provides information about that drug and its use. For prescription medications, the insert is technical and provides information for medical professionals about how to prescribe the drug. This information is also sometimes referred to as prescribing information, professional labeling, the direction circular, or the package circular. The actual document is authored by the pharmaceutical sponsor with editing and guidance from the FDA. The details of this process and are defined in the general requirements for prescription drug labeling section of the Code of Federal Regulations (21 CFR 201.56). The purpose of the package insert is to provide a summary of the safe and effective use of the drug. The FDA ensures that the summary is informative and accurate, not promotional, false, or misleading and that no implied claims or suggestions for use if evidence of safety or effective is lacking are mentioned, and that the material included is based whenever possible on data derived from human experience. Although patients may obtain useful information from prescription drug labeling, its primary purpose is to give healthcare professionals the information they need to prescribe drugs appropriately.

The actual section headings contained in the package insert are shown in Table 7.7.

The important aspect of the label that is not always clear to the people who read the document is that every sentence contained in the label is referenced to either

Table 7.7 Section headings listed in the package insert document with sections that contain data from Phase 1 trials indicated.

Label section	Phase 1 data referenced?	Details
1. Indications and usage	No	
2. Dosage and administration	Yes	Route of administration decided by drug substance attributes, marketplace, and patient population; dose and route demonstrated in Phase 1 initially and verified in patients during later phases
3. Dosage forms and strengths	No	
4. Contraindications	No	Typically, from patient trials
5. Warnings and precautions	No	Typically, from patient trials
6. Adverse reactions	Can be	Typically, from patient trials
7. Drug interactions	Yes	DDI and/or probe studies conducted in Phase 1
8. Use in specific populations 8.1. Pregnancy 8.2. Labor and delivery 8.3. Nursing mothers 8.4. Pediatric use 8.5 Geriatric use	Yes	All qualify as Phase 1 trials if conducted
9. Drug abuse and dependence 9.1. Controlled substance 9.2. Abuse 9.3. Dependence	No	Typically, from patient trials or preclinical pharmacology studies
10. Overdosage	No	Typically, from patient trials
11. Description	No	
12. Clinical pharmacology 12.1. Mechanism of action 12.2. Pharmacodynamics 12.3. Pharmacokinetics 13. Nonclinical toxicology 13.1. Carcinogenesis, mutagenesis, impairment of fertility 13.2. Animal toxicology and/or pharmacology	Yes	PK and PD described initially from Phase 1 trials but may be augmented with patient data

(Continued)

Table 7.7 (Continued)

Label section	Phase 1 data referenced?	Details
14. Clinical studies	No	
15. References	No	
16. How supplied/storage and handling	No	
17. Patient counseling information	No	

an actual clinical study report (CRS) contained in the NDA submission or a peer-reviewed publication in which the results are deemed relevant. Even though Phase 1 is typically conducted in healthy volunteers for most therapeutic areas, there is still a great deal of information relevant to caregivers and prescribers from this phase of development.

Phase 1 for Oncology

Oncology is often singled out as a therapeutic area with distinct differences in its development paradigm relative to others. The distinction is starting in Phase 1 and is predicated on two main distinctive features: (1) historically, the agents used to combat various forms of cancer were highly toxic based on the nature of the non-specificity of their cell killing and also a presumed greater tolerance for toxicity given the severity of the disease and (2) the usual requirement for combination or multimodal therapy owing to the fact that single-agent therapies were not effective enough. These factors contributed to two distinct features of Phase 1 oncology trials – they are typically conducted in cancer patients, and they are focused on assessing maximum tolerated dose (MTD) as an endpoint of the Phase 1 study design. Regulatory guidance (see references at the end of the chapter) is available, of course, but an additional factor is that even early Phase 1 trials in oncology incorporate combination strategies in the assessment of regimen strategy and the MTD (Wages et al. 2016). Pediatric Oncology is further complicated by the size of the target population and the difficulties in recruiting and enrolling a few patients into a clinical trial with many sites spanning an often global setting. Efficient enrollment strategies in pediatric oncology are likewise essential to conduct such trials at all phases (Skolnik and Barrett 2008; Barrett et al. 2008).

Historically, Phase 1 trials in oncology have been guided by the desire to estimate the MTD in the course of the typical SAD/MAD approach but with constraints around acceptable risk thresholds for toxicity. The traditional 3+3 design remains the prevailing method for conducting Phase 1 cancer clinical trials. It requires no modeling of the dose–toxicity curve beyond the classical assumption for cytotoxic drugs that toxicity increases with dose. This rule-based design proceeds with cohorts of three patients (see Figure 7.4); the first cohort is treated at a starting dose considered to be safe based on extrapolation from animal toxicological data, and the subsequent cohorts are treated at increasing dose levels that have been fixed in advance. Historically, dose-escalation has followed a modified Fibonacci sequence in which the dose increments become smaller as the dose increases (e.g. the dose first increases by 100% of the preceding dose, and thereafter by 67, 50, 40, and 30–35% of the preceding doses). If none of the three patients in a cohort experiences dose-limiting toxicity, another three patients will be treated at the next higher dose level. However, if one of the first three patients experiences dose-limiting toxicity, three more patients will be treated at the same dose level. The dose escalation continues until at least two patients among a cohort of three to six patients experience dose-limiting toxicities (i.e. ≥33% of patients with dose-limiting toxicity at that dose level). The recommended dose for Phase 2 trials is conventionally defined as the dose level just below this toxic dose level.

There are many variations of the 3+3 design, and model-based approaches are also utilized for dose escalation based on the dose-toxicity relationship (Wages et al. 2016). Other designs, more adaptive and real-time methods like the continuous reassessment method (CRM) (Wages et al. 2016) utilize the results of

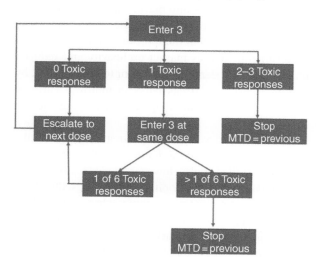

Figure 7.4 Classic 3+3 design criteria and enrollment heuristic based on observing dose-limiting toxicity events and the declaration of maximum tolerated dose (MTD) through dose escalation.

the trial as it progresses to predict the toxicity relationship offers the potential for a more precise estimate of the MTD and a more efficient construct that gets to the MTD with fewer doses. Oncology as a therapeutic area also depends on the identification of more mechanism-specific biomarkers (Park et al. 2004) so that multimodal therapy strategies can be more efficiently chosen and recommended. Phase 1 in oncology likewise accelerates the timing for dose selection so that Phases 2 and 3 can be more focused on confirmation of regimen and dose selection.

Phase 1 for Rare Diseases

In the United States, a rare disease is defined as a condition that affects fewer than 200 000 people. Other countries have their own official definitions of a rare disease. In the European Union, a disease is defined as rare when it affects fewer than 1 in 2000 people. In the EU, as many as 30 million people may be affected by one of over 6000 existing rare diseases. Patients with rare diseases were often without much hope with respect to significant emphasis from pharmaceutical sponsors in the not-so-distant past. The Orphan Drug Act of 1983 was passed in the United States to facilitate the development of orphan drugs – drugs for rare diseases such as Huntington's disease, myoclonus, ALS, Tourette syndrome, and muscular dystrophy which affect small numbers of individuals residing in the United States. Orphan drug designation does not indicate that the therapeutic is either safe and effective or legal to manufacture and market in the United States. That process is handled through other offices in the US Food and Drug Administration. The designation means only that the sponsor qualifies for certain benefits from the federal government, such as market exclusivity and reduced taxes. Only thirty-eight orphan drugs had been approved prior to the 1983 Act; by 2014, 468 indication designations covering 373 drugs had been approved. Partly as a result of the 1983 US Orphan Drug Act, Japan adopted it in 1993, as did the European Union in 2000.

Standards for approval of orphan drugs developed for the treatment of rare diseases are the same as those of common diseases: there must be substantial safety and efficacy evidence from well-controlled trials. However, in some cases it may not be possible to meet these standards when developing orphan drugs; therefore, the FDA applies scientific judgment and regulatory flexibility when making decisions about drug development and approval in rare diseases. Many orphan diseases are serious and/or life-threatening and primarily affect pediatric patients, underscoring both the challenges and urgency of effective drug development. The inappropriateness of administering some therapies to healthy controls and the rarity of orphan diseases also pose logistical challenges for conducting clinical trials. The patient populations are small, limiting the use of extensive dose-ranging Phase 2 studies. In addition, patients are often in relatively poor physical condition, which can limit the number and type of procedures that can be performed.

The implications for Phase 1 are similar to oncology in that healthy volunteers do not really provide a reliable surrogate for defining the therapeutic window for orphan drugs in rare disease populations. In addition, the limited availability of the patient population often necessitates that patients studied in Phase 1 are likewise recruited for Phases 2 and 3 trials. Occasionally, this can be accomplished with a rolling phase design that allows patients to be studied in an acute or multidose Phase 1 trial and then continued on therapy with designs focused on dose-finding and/or clinical endpoint determination. Other nuances with the clinical development of orphan drugs for rare diseases will be discussed in subsequent chapters.

A Successful Phase 1 Program

Desirable properties for Phase 1 candidates include a well-defined PK/PD profile that aligns with the TPP, formulation experimentation that supports further development of a drug product closer to the final market image in time for Phase 3 testing (assuming the candidate gets that far), development of suitable, validated analytical methods that can be transferred to commercial labs supporting later phase testing, and an acceptable toxicology profile supporting future patient trials and complimentary commercial evaluation that suggests there will be a market for the candidate if approved. Getting to this stage requires coordination of the functional groups supporting Phase 1 development (see Chapter 11 for a description of Project Teams) and a great deal of luck.

It is in the best interest of study sponsors to "kill" a drug candidate (terminate development) if it doesn't meet the requirements of the TPP, but this doesn't always happen as sponsors are often determined to recoup their investment by considering promising candidates for other indications or fixing the "warts" of a compound by altering its exposure (PK) or behavior (PD) if possible. There are strategies that do work occasionally, but care must be taken not to be overly optimistic about the potential to do so. This is sometimes referred to as a sunk-cost effect – the tendency to continue investing in something that clearly isn't working. Because human nature tends to make us want to avoid failure, people (in this case, pharmaceutical companies) will often continue spending time, effort, or money to try and fix what isn't working instead of cutting their losses and moving on (e.g. to the backup compound at least). If a compound does fulfill its Phase 1 objectives, its on to Phase 2 and patient trials and dose-finding and proof-of-concept trials to support Phase 3 (see Chapter 8).

References

André, F.E. and Foulkes, M.A. (1998). A phased approach to clinical testing: criteria for progressing from phase I to phase II to phase III studies. *Developments in Biological Standardization* 95: 57–60.

Barrett, J.S., Skolnik, J.M., Jayaraman, B. et al. (2008). Improving study design and conduct efficiency of event-driven clinical trials via discrete event simulation: application to pediatric oncology. *Clinical Pharmacology and Therapeutics* 84 (6): 729–733.

Breder CD, Du W, Tyndall A. What's the regulatory value of a target product profile? *Trends Biotechnology* 2017;35(7):576–579. doi: https://doi.org/10.1016/j.tibtech.2017.02.01.

DiMasi, J.A., Grabowski, H.G., and Hansen, R.W. (2016). Innovation in the pharmaceutical industry: new estimates of R&D costs. *Journal of Health Economics* 47: 20–33.

Mestre-Ferrandiz, J., Sussex, J., and Towse, A. (2012). The R&D Cost of a New Medicine. Office of Health Economics. https://www.ohe.org/system/files/private/publications/380%20-%20R%26D%20Cost%20NME%20Mestre-Ferrandiz%202012.pdf?download=1 (accessed 30 September 2019).

Park, J.W., Kerbel, R.S., Kelloff, G.J. et al. (2004). Rationale for biomarkers and surrogate end points in mechanism-driven oncology drug development. *Clinical Cancer Research* 10: 3885–3896.

Patat, A.A. (2000). Designing and interpreting the results of first-time-to-man studies. *Dialogues in Clinical Neuroscience* 2 (3): 203–212.

Skolnik, J.M. and Barrett, J.S. (2008). Refining the phase 1 pediatric trial. *Pediatric Health* 2 (2): 105–106.

Suresh, P. and Basu, P.K. (2008). Improving pharmaceutical product development and manufacturing: impact on cost of drug development and cost of goods sold of pharmaceuticals. *Journal of Pharmaceutical Innovation* 3 (3): 175–187.

U. S. Food and Drug Administration. Center for Drug Evaluation and Research (1995). Content and Format of Investigational New Drug Applications (INDs) for Phase 1 Studies of Drugs, Including Well-Characterized, Therapeutic, Biotechnology-derived Products, Guidance for Industry. https://www.fda.gov/media/71203/download.

U. S. Food and Drug Administration. Center for Drug Evaluation and Research (2015). Brief Summary and Adequate Directions for Use: Disclosing Risk: Information in Consumer-Directed Print Advertisements and Promotional Labeling for Prescription Drugs, Guidance for Industry. https://www.fda.gov/media/70768/download.

U. S. Food and Drug Administration. Center for Drug Evaluation and Research (2016). Submission of Quality Metrics Data, Guidance for Industry. https://www.fda.gov/media/93012/download.

U. S. Food and Drug Administration. Center for Drug Evaluation and Research (2019a). Rare Diseases: Common Issues in Drug Development, Guidance for Industry Draft. https://www.fda.gov/media/120091/download.

U. S. Food and Drug Administration. Center for Drug Evaluation and Research (2019b). Considerations for the Inclusion of Adolescent Patients in Adult Oncology Clinical Trials, Guidance for Industry Draft. https://www.fda.gov/media/113499/download.

Wages NA, Ivanova A and Marchenko O. Practical designs for phase I combination studies in oncology. *Journal of Biopharmaceutical Statistics* 2016; 26(1): 150–166. doi:https://doi.org/10.1080/10543406.2015.1092029.

Chapter Self-Assessments: Check Your Knowledge

Questions:

- How does the assessment of the pharmacokinetics (PK) and pharmacodynamics (PD) in Phase 1 impact the Target Product Profile (TPP) and impact the evaluation of the drug candidate?
- What is the purpose of the Investigator's Brochure in the IND?
- Why are phase studies typically conducted in healthy, young volunteers?
- Why and how is Phase 1 treated differently for oncology as opposed to other therapeutic areas?

Answers:

- The PK/PD derived from Phase 1 helps define the therapeutic window and dosing frequency of a drug candidate as well as assist in the interpretation of any AEs or ADRs. All of these assessments are required for the TPP and eventually communicated to FDA.
- The purpose of the Investigator's Brochure (IB) is to provide the investigators and others involved in a trial [with the investigation drug] with information to facilitate their understanding of the rationale for, and their compliance with, many key features of the protocol, such as the dose, dose frequency/interval, methods of administration: and safety monitoring procedures. The IB also provides insight to support the clinical management of the study subjects during the clinical trial.
- Healthy volunteers provide researchers with crucial data because their health information can be used as a reference for comparison to patient response. Young, healthy volunteers are also chosen to minimize variability of both the body's actions to eliminate the drug and the drug response again in comparison to patients.
- Phase 1 is treated differently for oncology due to the severity of the disease necessitating MTD determination in patients as opposed to healthy volunteers and the historically toxic nature of the agents used to treat cancer, especially cell-killing agents, which are sometimes indiscriminate in action and toxicity.

Quiz:

1 True or false. Phase 1 is always conducted in healthy young male volunteers without exception.

2 Which of the following is NOT considered a Phase 1 study? ("a" is a Phase 2 trial)
A Patient dose-finding trial
B Drug interaction trial
C SAD trial
D Food-effect trial
E MAD trial

3 True or false. A rare disease is any disease that affects a small percentage of the population. In some parts of the world, an orphan disease is a rare disease whose rarity means there is a lack of a market large enough to gain support and resources for discovering treatments for it, except by the government granting economically advantageous conditions to creating and selling such treatments. (true)

4 The following is an essential component of the IND submission (choose the best answer):
A Chemistry, Manufacturing, and Control Information
B Pharmacology and Toxicology Information
C Previous Human Experience with the Investigational Drug
D Detailed clinical development plan
E a, b, and c
F c and d only
G c only

8

Phase 2

Jeffrey S. Barrett

Aridhia Digital Research Environment

Phase 2 Objectives, TPP Alignment and Deliverables to Phase 3

Phase 2 is the second phase of clinical trials or studies for an experimental new drug, in which the focus of the drug is on its effectiveness. The Center for Drug Evaluation and Research (CDER) or Center for Biologics Evaluation and Research (CBER), in the case of a therapeutic protein or biologic, divisions of the US Food and Drug Administration (USFDA), oversees these clinical trials. Phase 2 trials typically involve hundreds of patients who have the disease or condition that the drug candidate seeks to treat. The main objective of Phase 2 trials is to obtain data on whether the drug actually works in treating a disease or indication, which is generally achieved through controlled trials that are closely monitored, while safety and side effects also continue to be studied. Phase 2 studies also aim to establish the most effective dosage for the drug and the optimum delivery method in the target population. Phase 2 trials usually form the biggest stumbling block in the development of a new drug (Patel et al. 2017).

Phase 2 trials are typically constructed as double-blind, randomized, placebo-controlled studies. This means that some of the patients enrolled in the study will receive the drug candidate, while others will receive a placebo or a different drug. The assignment is done on a random basis, and neither the participant nor the clinical investigator knows whether the participant will be receiving the drug or the placebo. Randomness and anonymity are rigorously enforced to prevent bias in the studies.

Phase 2 trials are considered successful when analysis of the data from enrolled participants indicates that the experimental drug works in treating the disease or indication. Patients who have received the experimental drug should have better clinical outcomes on a statistically significant basis than those who received the placebo or the alternative drug. If Phase 2 trials are successful, the drug proceeds

Fundamentals of Drug Development, First Edition. Edited by Jeffrey S. Barrett.
© 2022 John Wiley & Sons, Inc. Published 2022 by John Wiley & Sons, Inc.
Companion website: www.wiley.com/go/Barrett/FundamentalsDrugDevelopment

to Phase 3 studies. Phase 2 studies only commence if Phase 1 studies do not reveal unduly high toxicity or other safety risks of the experimental drug. While up to a third of drugs in Phase 1 studies do not progress to the Phase 2 stage because they are not safe enough, the odds of a drug progressing from Phase 2 to Phase 3 trials are even lower, about 32–39% (DiMasi et al. 2016).

Because of the relatively low rate of success at the Phase 2 stage, market reaction to a successful Phase 2 outcome is generally rewarded with significant stock price appreciation for the company developing the drug. The degree of stock appreciation depends on a number of factors including the prevailing environment for equities in general and healthcare stocks in particular, the disease or indication that the drug aims to treat, the strength of the Phase 2 results, and price movement in the stock prior to the release of Phase 2 results. Beyond Phase 2 trials, there is a plethora of activity happening from all parts of an organization, and regulatory strategy is being extensively discussed, anticipating positive outcomes from the clinical trials. A milestone in this progression is the End-of-Phase 2 (EOP2) meeting with the FDA and/or equivalent global regulatory agency. The details of a company's aspirations for Phase 2 outcomes are captured in the TPP along with other phases of development. Typical milestones for Phase 2 articulated in the TPP include successful demonstration of proof-of-concept (POC) with a clearly defined threshold for efficacy and safety comparable to or better than the minimally acceptable criteria (usually established as the standard of care), successful dose-ranging study(ies) from which dosing recommendation for Phase 3 can be supported and a suitable formulation (at or close to the final market image) available for Phase 3 trials.

In this chapter, we will discuss the essential elements of Phase 2, defining the terminology for key milestones that are the hallmark of the phase. We will exposure other complementary activities being pursued by non-clinical team members that also represent key Phase 2 deliverables. Common and novel study designs utilized in Phase 2 will be explained, along with a discussion about the necessity of biomarkers for dose justification and clinical endpoint declaration. The End of Phase 2 meeting at the FDA will be described along with an assessment of attrition rates for Phase 2 and discuss under what circumstances it may be reasonable for sponsors to skip Phase 2 entirely.

POC, POP, and POM and the Necessity of Patient and Indication-Specific Biomarkers

Demonstrating POC is an important milestone for Phase 2, as mentioned previously. The goal of proof of concept studies, typically involving a small number of subjects and more latitude in statistical requirements, is to provide evidence that a drug is likely to be successful in later stages of drug development. Although often not published, such studies allow drug developers to make "Go/No Go" decisions about proceeding with larger, more expensive studies (Preskorn 2014). Related concepts include proof-of-principal (POP), which refers to the demonstration of pharmacological impact on the disease in question and proof-of-mechanism (POM), which refers to the engagement of the active drug entities at the intended site(s) of action.

While these concepts are certainly related and factor into the various Phase 2 study designs, an important corresponding feature of these studies is their dependence on the availability of validated biomarkers. Such biomarkers serve as the mechanism to establish these concepts and the data source from which statistical summarization ensues to test the various study designs based on the established statistical analysis plans. Not coincidentally, in the field of oncology, these concepts (POC, POP, and POM) also define the biomarker categories investigated in early phase development to support compound progression (Bradley 2012). In Table 8.1, the various biomarker categories and concepts related to their specific purpose and goal for oncology drug development are illustrated.

The use of POM and POP pharmacodynamic biomarkers allows an early assessment of the pharmacological activity of a new drug. Traditionally, dose-finding first-in-human for oncology trials, in particular, relied on escalating the dose of the drug up to a maximum tolerated dose (MTD; the highest dose of a drug or treatment that does not cause unacceptable side effects), which is then declared the recommended dose for further development. For many emerging oncology therapies, dosing to MTD is either impractical or unwarranted. In the absence of desirable "off-target" pharmacology, dosing beyond a relevant pharmacodynamic plateau is likely to offer little benefit but instead risks increasing toxicity or even producing confounding effects. Rather than blindly escalating the dose to the MTD, applying appropriate POM pharmacodynamic biomarkers during the

Table 8.1   POC, POP, and POM biomarker categories applied to Phase 2 oncology development reproduced from Bradley 2012.

Biomarker category	Goal/purpose
POM	Show that the candidate drug engages at a reliable and quantifiable level in humans, indicating a functional effect.
POP	Show that the candidate drug results in a biological and/or clinical change associated with the disease and the mechanism of action.
POC	Show that the candidate drug results in a clinical change on an accepted endpoint or surrogate, in patients with the disease, plus evidence of a high degree of confidence of success in Phase 3.

dose-escalation stage of a clinical trial can provide an estimated optimum dose without having to expose patients to unnecessary toxicities. POP biomarkers can then be examined in patient expansion cohorts treated with the identified dose in order to confirm the pharmacological impact on the clinical manifestation of the disease (e.g. a tumor).

An additional consideration in the development of any biomarker strategy is the confirmation that the biomarkers proposed actually track relevant actions in the target patient population(s). It is common for drug sponsors to perform early biomarker studies even in Phase 1 to initiate the establishment of exposure-response relationships and to refine and validate the underlying analytical methods. Examples of biomarkers include everything from pulse and blood pressure through basic chemistries to more complex laboratory tests of blood and other tissues. Regardless of source, type of biomarker, or intended use, the sponsor also must clearly demonstrate that the biomarker proposed for Phase 2 studies is relevant in the target population and responds in a manner that is directionally consistent with the mechanism of action and can be used to discriminate dose cohorts based on the variability of the response. There are plenty of examples where biomarker performance looked reasonable in Phase 1 but not Phase 2.

Clinical Endpoint Declaration and Modeling and Simulation

A requirement for every Phase 2 protocol is the development of clear, achievable study objectives that can be evaluated in the population of interest based on pre-specified endpoints. Selecting these endpoints in the context of an appropriate study design provides some confidence that the drug candidate can achieve the best possible chance of success in a confirmatory Phase 3 trial (LeBlanc and Tangen 2012). Endpoints thus refer broadly to an event or outcome that can be measured objectively, allowing the investigator and study sponsor to determine whether the intervention being studied is beneficial. Early thoughts about potential endpoints are described in the TPP, and so the sponsor is assumed to have had regular communication with the FDA regarding the choice of endpoints available and recommended to evaluate the proposed disease indication. The endpoints proposed for a particular clinical trial are likewise included and specified in both the study objectives and statistical analysis plan of the protocol. Some examples of endpoints include assessments of clinical events (e.g. stroke, pulmonary exacerbation, venous thromboembolism), outcomes (e.g. what a final result of treatment long term, mortality, tumor resolution, cure of disease, remission), patient symptoms (e.g. pain, dyspnea, depression), measures of function (e.g. ability to walk or exercise), adverse events or surrogates of these effects or symptoms; clinical outcomes are considered the most reliable endpoint.

Given the importance of defendable endpoints to the success of the overall development program, it is prudent for drugmakers to consider the relationship between endpoint selection and the probability of success. One mechanism to do this has been the incorporation of various modeling and simulation (M&S) techniques to evaluate the impact of endpoint selection and other design elements on the likelihood that a drug candidate successfully completes Phase 2. Simulating from existing clinical trial data sets, and developing clinical trial simulation models, can be useful tools for such evaluation. Examples of M&S application supporting early phase drug development with the intention of informing endpoints selection include blood pressure evaluation for hypertension patients (Karmali et al. 2018), blood pressure reduction to support stroke evaluation (Lassere et al. 2012), and time to virologic failure evaluation in HIV patients (DiRienzo and DeGruttola 2003).

The use of M&S in the design and interpretation of clinical trials has had a dramatic impact on FDA approval and labeling decisions (DellaPasqua 2016). A 2011 review conducted by the FDA found a dramatic increase in both the number of reviews with pharmacometric analysis and the impact of those analyses on drug approval and labeling decisions. Pharmacometric analysis was found to have made an important contribution to 126 drug approval decisions (64%) between the years 2000 and 2008 (Lee et al. 2011). Additionally, pharmacometric analysis was found to impact labeling decisions in 133 applications (67%) during this time period. Recent studies using comparable compounds with high and low M&S usage show the reduction in the number of patients for specific trials and trial completion time (first and last patient visits). As an example, a new drug targeting schizophrenia yielded a 95% reduction in the number of subjects needed for Phase 3, and even though the time saved in Phase 3 was only 12% (about four months in this case), the drug developer was able to avoid certain intermediate trials, bringing the total time savings to almost two years when compared with comparable drugs going through the same process without using M&S. For a recent drug targeting multiple cancers, the benefits are even more dramatic: 90% reduction in the number of patients for Phase 3 trials and a 75% reduction in trial completion time, a savings of over three years (Glass et al. 2016).

Phase 2 Study Designs: Common and Novel

The main goal of Phase 2 clinical trials is to identify the therapeutic efficacy of new treatments. They are usually single-arm studies but may take the form of multiple-arm trials. Multiple-arm trials can be randomized or non-randomized with or without control arms. Phase 2 trials decide whether the new treatment is promising and warrants further investigation in a large-scale randomized Phase 3 clinical trial based on an observed response rate that appears to be an improvement over the standard treatment or other experimental treatments. Because the sample size is small (generally less than 50 patients), Phase 2 clinical trials are only able to detect a large treatment improvement, e.g. greater than 10%. To detect a small difference in treatment, e.g. less than 5%, one would require a much larger sample size, which is not always possible in Phase 2 studies. Phase 2 studies are prominent in cancer therapeutics because new treatments frequently arise from combinations of existing therapies or by varying dose or radiation schedules. In addition to testing whether the therapeutic intervention benefits the patient, other goals of Phase 2 trials are to screen the experimental treatment for the response activity and extending knowledge of the toxicology and pharmacology of the treatment.

A randomized controlled trial (or randomized control trial; RCT) is a type of scientific experiment that aims to reduce certain sources of bias when testing the effectiveness of new treatments (see Figure 8.1); this is accomplished by randomly allocating subjects to two or more groups, treating them differently, and then comparing them with respect to a measured response. One group – the experimental or test group – has the intervention being assessed, while the other – usually called the control group – has an alternative condition, such as a placebo or no intervention. The groups are followed under conditions of the trial design to see how effective the experimental intervention is. Treatment efficacy is assessed in comparison to the control. There may be more than one treatment group or more than one control group. The trial may be blinded, in which information that may influence the participants is withheld until after the experiment is complete. A blind can be imposed on any participant of an experiment, including subjects,

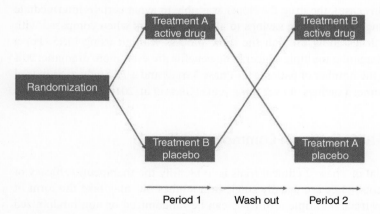

Figure 8.1 Typical double-blind, randomized, placebo-controlled study design often employed in Phase 2.

researchers, technicians, data analysts, and evaluators. Good blinding may reduce or eliminate some sources of experimental bias.

The randomness in the assignment of subjects to groups reduces selection bias and allocation bias, balancing both known and unknown prognostic factors, in the assignment of treatments. Blinding reduces other forms of experimenter and subject biases, as mentioned previously. Another important characteristic of some Phase 2 trial designs is the use of early stopping rules. If there is sufficient evidence that one of the treatments under study has a positive treatment effect, then patient accrual is terminated, and this treatment is declared promising. Also, if treatment is sufficiently shown not to have a desirable effect, then patient accrual is terminated, and this treatment is declared not promising.

While standard Phase 2 trial design has some well-established statistical properties, it is not without flaws. Selection bias is common because these trials are often carried out in a single institution or a small group of academic institutions where the patient population differs significantly from the at-large Phase 3 target population, and multicenter community-based studies only partially overcome this issue. In March 2004, the FDA issued a report recognizing that the approval of innovative medical therapies had slowed over the preceding years. The estimated Phase 2 failure rate in 2006 was 50% vs. 20% in 10 years earlier. A result of the critical pathway initiative of the FDA was increased interest in innovative trial designs. In December 2016, the US Congress passed the Twenty-first Century Cures Act, allotting $500 million to the FDA to establish an "innovation account" for National Institutes of Health funding to speed regulatory approval of medical therapies. Since then, the FDA has devoted efforts to exploring modern trial design and evidence development, including the use of adaptive trial designs (ADs) and real-world evidence.

A number of different designs fall under the category of innovative trial design, all of which allow interim data analysis and modification of the trial. Examples of such designs include enrichment trials, biomarker-stratified trials, and adaptive trials. Enrichment trials allow patient enrollment by clinical criteria, and each is then assayed for a pre-specified drug target. After that, several different trial strategies can be pursued: (1) randomize all enrolled patients and analyze the patients carrying the target in a subgroup analysis; (2) continue the trial with patients who only express the target; or (3) split the trial into two groups (those with the target and those without) and randomize and analyze each group separately. Enrichment trials (see Figure 8.2) may hasten to market therapeutics that benefit a specific patient subpopulation (identified by the presence or absence of a biomarker) rather than a more heterogeneous population with a broad disease designation, but they depend in part on knowing in advance what factors may contribute to disease progression, and then constructing trial populations that contain the various factors. A downside of enrichment trials is that they identify agents that work in enriched populations but may show less efficacy in unselected populations. Such trials may also inadvertently exclude patient subpopulations that are responsive to the drug because a characteristic common to that subpopulation was not recognized in trial design and patient selection. A therapy that might be effective in an untested patient subpopulation would then be inadvertently discarded from further development for lack of efficacy.

A biomarker-stratified design is a commonly used all-comer design for evaluating treatment effects in various biomarker subgroups and the predictive value of the biomarker for optimal treatments. As illustrated in Figure 8.2, this design, all screened patients are randomized to one of two treatments (Test T or Control C) with biomarker as a stratification factor. In a BSD design, the selection probabilities are equal to one so that the expected proportion of biomarker positives in the randomized cohort is equal to the prevalence rate of biomarker positives in the underlying patient population.

An adaptive design (AD) is a clinical trial design that allows for prospectively planned modifications to multiple aspects of the design based on accumulating data from subjects in the trial. In ADs, the goal is to learn from accumulating data in the trial and apply what is learned as quickly as possible in a prospectively specified way during the trial itself to hone flexible aspects of the study while it is still ongoing (Wang et al. 2018). ADs can be classified as prospective, continuously adjusted or concurrent (ad hoc), and retrospective. In prospective ADs, there is a pre-specified protocol to alter aspects of the study, such as size, follow-up period, and clinical endpoints following interim data analysis. This might lead to early

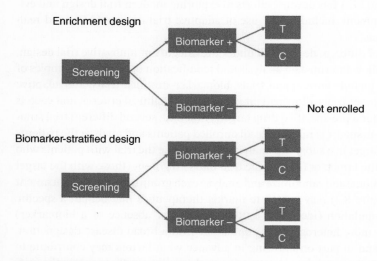

Figure 8.2 More novel Phase 2 trial designs employed to improve patient selection, design efficiency, and treatment response (T = test; C = control).

termination of a study based on futility or unacceptable toxicity, or, alternatively, might require a change in sample size. A platform study or master protocol design is a type of adaptive trial in which multiple treatment arms are simultaneous studies, and interim analysis allows early termination of various arms due to futility or lack of efficacy. Concurrent or ad hoc study designs allow flexibility to alter multiple parameters in a study in a pre-specified way based on interim results. In ad hoc design, investigators are allowed to hone their hypothesis based on interim results and re-steer the study accordingly. Both retrospective and prospective data following changes are used in the analysis. Retrospective ADs allow the investigators to change the primary study endpoint or analysis methodology in a pre-specified way after a study is closed.

To date, analysis of novel Phase 2 trial designs gives mixed and sometimes conflicting results with regard to their effects on study size and duration though there is great encouragement from regulators for drug makers to continue to invest in and refine these approaches in an effort to improve Phase 2 efficiency and decision making and reduce overall costs of drug development.

The End of Phase 2A Meeting at FDA

The purpose of an EOP2 meeting is to facilitate interaction between FDA and sponsors who seek guidance related to clinical trial design employing a variety of tools including clinical trial simulation and quantitative modeling of prior knowledge (e.g. drug, placebo group responses, disease), designing trials for better dose-response estimation and dose selection, and other related issues. With respect to timing EOP2 meetings should be held before Phase 3 trials begin, and topics include determination of the safety of proceeding to Phase 3, evaluation of the Phase 3 plan and protocols for adequacy and to assess adult and pediatric safety and effectiveness, identification of information necessary to support a marketing application. Recall that the sponsor has provided a TPP at the IND stage with some description of the target attributes the candidate is expected to have in order for development to proceed along with some detail regarding the clinical development plan, target populations of interest, proposed indications, and details of Phase 2 and 3 trials including designs and endpoints under consideration. The EOP2 meeting provides a more candid checkpoint for revisiting the plans initially laid out in the TPP with a more detailed discussion. This meeting is so highly regarded by the FDA that they created guidance for the industry on this meeting (see references - Guidance for Industry (2009)). This guidance is intended to further FDA initiatives directed at identifying opportunities to facilitate the development of innovative medical products and improve the quality of drug applications through early meetings with sponsors.

Prior to the actual EOP2 meeting, sponsors are required to provide FDA with a "briefing document" that summarizes the proposed meeting flow and essentially provides the supporting data to be presented and discussed at the meeting. Table 8.2 provides an example of a briefing document contents with some interpretation of why these topics might be discussed at the EOP2 meeting.

Table 8.2 Example of a typical EOP2 meeting briefing document contents.

Topic	Purpose for EOP2 meeting discussion
Summaries of Phase 1 and Phase 2 investigations	Summary of human experience documenting the therapeutic window in patients and the MTD established in healthy volunteers along with any established relationships between exposure and toxicity (AEs or ADRs). This data serves as a baseline for discussion regarding Phase 2 designs including proposed biomarkers.
Summary information on plans for Phase 3 trials	Expansion/refinement of concepts outlined in the TPP. Serves as the basis for discussion of dose recommendation, study population(s) and event schedules and sampling schemes.
Choice of comparator	Review and discussion of what constitutes the standard of care along with any suitable active comparator that could be used in Phase 3 trials.
Definition and time point for assessment of primary endpoint	Review of thought leader opinions on acceptable endpoints and clinical performance in recent trials; necessary to get consensus with FDA if possible.
Statistical analysis approach and criterion for success and failures of the primary efficacy and secondary endpoints	Review of proposed SAP defined in the Phase 3 protocol. Seek to get confirmation with FDA on proposed plan specifically the trial enrollment criteria, sample size and study population along with proposed endpoints; review of regulatory precedents.
Discussion of pediatric study (ies)	Study design proposals reviewed to seek agreement as a consequence of PDUFA and BPCA legislation.
Size of the safety database	Confirmation that the sponsor has (or will have) accumulated enough patient exposures to support an NDA filing.
Plans for pediatric studies to address PREA	As above, review of proposed studies, designs, and proposed timing relative to adult development.
Plans for additional non-clinical studies (if required)	Discussion of NDA submission completeness relative to proposed Phase 3 plans; discussion of what would be acceptable as Phase 4 commitments.

Exposure Response and the Opportunity for Phase 2 Trials to Support Registration

The cost of drug development is substantial, and sponsors try very hard to streamline drug development to deliver safe and efficacious medicines to the patients that need them in the shortest time frame possible. Of course, as a business, there are also financial incentives to maximize the patent life, reduce the cost of development for the particular candidate and recoup the R&D expenditure in as short a time frame as possible. One of the biggest line items in the R&D budget is the cost of the Phase 3 trials (estimated to range between $11.5 and $52.9 M USD based on the complexity of the trial and the duration of therapy [Sertkaya et al. 2016]). One opportunity to lower the Phase 3 study cost allocation is by reducing the number of Phase 3 trials required for submission and market access. Regulatory authorities provide a mechanism to do just that, assuming that the sponsor can establish adequately defined and clinically relevant exposure-response relationships that allow regulators to understand the conditions upon which the proposed relationships are dependent and the ability to both interpolate this relationship to doses unstudied but within the range of evaluation and extrapolate outside the range of clinical experience with reasonable uncertainty. The circumstances under which sponsors may propose such an approach and development plan are outlined in several guidelines by the FDA and EMA. Examples exist where market access has indeed been granted based on a single Phase 3 trial and well-defined and clinically relevant exposure-response relationships (see Table 8.3).

In terms of basic principles that govern the establishment of a good exposure-response relationship, the approach has been around for a long time and grounded in the basic understanding of pharmacology and PK/PD relationships. The main principle is based on the concept of a target effect and use of a PD model to predict the target concentration needed to achieve that effect and a PK model to predict the dose required to achieve the target concentration – Target Effect → Target Concentration → Dose relationship.

From a simplistic approach, we can define the target concentration as a function of the C50 (the concentration that yields 50% of the maximum effect achievable), Emax (maximum effect achievable or observed), and the target effect desired (see below).

$$\text{Target Concentration} = \text{C50} \times \text{Target Effect} / \left(\text{Emax} - \text{Target Effect} \right)$$

If rapid attainment of the target concentration is essential, a combined loading and maintenance dose can also be derived from simple PK relationships (see below).

$$\text{Loading Dose} = \text{Volume of Distribution} \times \text{Target Concentration}$$

$$\text{Maintenance Dose Rate} = \text{Clearance} \times \text{Target Concentration}$$

Table 8.3 Examples of drug candidates utilizing exposure-response relationships as part of their registration strategy.

Drug	Indication	E-R Relationship	Outcome
PD 0348292, an oral factor-Xa inhibitor (Cohen et al. 2013)	Thromboprophylaxis after total knee replacement surgery.	Concentration and risk of either VTE or bleeding	Characterization of the dose-response relationship using an adaptive Phase 2 study design would have informed Phase 3 dose selection if not discontinued.
Cariporide (Weber et al. 2002]	Acute coronary syndrome (ACS)	PK exposure/time-to-event, survival model	Model-based assessment lead to design and conduct of a second Phase 3 trial on modified dose regimen in coronary artery bypass graft surgery patients only.
Abatacept (Roy et al. 2007)	Rheumatoid Arthritis	Exposure and serum interleukin (IL)-6 concentration	Justified that the body weight-tiered abatacept doses approximating 10mg/kg ensure optimal exposure and IL-6 suppression.
Busulfan [Booth et al. 2007]	Immunosuppression prior to hematopoietic stem cell transplantation	Exposure – toxicity (VOD)	Dosing regimen recommended; achieved adequate target exposure in pediatric patients. Results used for labeling busulfan in the United States.

This simple example shows that only four parameters are required to find the right dose. These are maximum drug effect (E-max) and potency (C50) to predict the target concentration and volume of distribution (V) and clearance (CL) to predict the dose. In actual drug development programs and certainly the examples referred to in Table 8.3, the PD and PK models are usually more complex, but the principle remains the same.

Skipping Phase 2

As previously stated, Phase 2 studies are intended to explore the effectiveness of the product for a particular indication over a range of doses and to assess short-term side effects. Studies typically involve a few hundred patients who have the target condition but do not generally have other diseases that might obscure the effect of the drug on the target condition. Phase 2 trials may be randomized and/or controlled but often measure laboratory values or other biomarkers rather than clinical outcomes (i.e. effects on how a patient feels, functions, or survives). When a Phase 2 study does assess clinical outcomes, it is usually for relatively short periods of time and in a relatively small number of people. Sponsors assess Phase 2 results to determine if the preliminary results are sufficiently promising to justify a Phase 3 study. While that is the general framework for Phase 2, there are situations in which the sponsor feels an alternative path may be acceptable; either (1) outcomes in a broader, representative patient population can be evaluated directly in a Phase 3 study with less risk of misinterpretation or (2) the preclinical data is so compelling in terms of the dose selection that the dose-finding aspect of the traditional Phase 2 paradigm is unwarranted. In these situations, pharmaceutical companies have actually skipped Phase 2 entirely and gone straight ahead with Phase 3 testing. We will consider each of these situations.

When a sponsor skips Phase 2 because they feel the value (of doing studies in this phase) gained is low along with the risk of a failed Phase 3 program, the decision is typically based on several beliefs. Primarily, this decision is based on the sponsor's feeling that POC has either been established preclinically or based on similar characteristics (e.g. PK/PD) of the development candidate to approved therapies or another compound in development for which POC has been established. Another rationale is the commitment to study more than one dose in Phase 3. Hence, dose-finding is essentially established in a greater number of patients representing the target population – risky and more expensive. Oncology and infectious disease are often the two therapeutic areas where this situation happens. An additional consideration is when the study population in so small that effectively the entire population is studied in either phase (2 or 3) as is often the case for rare diseases, and effective outcomes are evaluated in all studies (see Rare Disease FDA guidance).

In the other situation, there is great confidence in the preclinical assessment of POC and/or a well-understood rationale for the dosing requirements independent of a Phase 2 trial. These conditions have been met in the past for antibiotics and anti-infectives. For both classes of agents, clinical trials are challenging. In the case of antibiotics, patients are urgently started on empiric therapy to reduce mortality and morbidity, which may obscure the effect of an antibiotic under investigation. Imprecise diagnosis of infection under study can weaken the conclusion of a new antibiotic's effectiveness. Uncertainty of bacterial pathogen can lead to additional antibiotic coverage with overlapping spectrum with the antibiotic under study. With regards to the study initiation, the severity of the acute illness can make obtaining informed consent and completing enrollment procedures challenging, and hospital policies often encourage early discharge, increasing the operational challenges of studying IV only antibiotics for an in-patient trial. Whereas ~70% of Phase 2 POC trials in other therapeutic areas are unsuccessful, this doesn't apply to antibiotic trials. As a successful antibiotic POC combines informative preclinical work, demonstrating a thorough understanding of PK/PD in Phase 1, showing that target exposures for key pathogens can be reliably and safely achieved (good safety margin). This, coupled with the fact that small Phase 2 trials often don't reveal safety risks, is enough motivation for some sponsors to skip Phase 2.

Phase 2 Attrition

In a recent publication (Wong et al. 2019), clinical trial success rates and durations by indication were estimated based on a sample of 406 038 entries of clinical trial data for more than 21 143 compounds from 1 January 2000 to 31 October 2015. The highest three success rates were 32.6% for clinical studies of ophthalmology drug candidates, 25.5% for cardiovascular drug candidates, and 25.2% for infectious disease products. The lowest percentage came from oncology trials, at just 3.4%. Interestingly, Wong and colleagues recorded a 15% success rate for CNS candidates – even though the category includes investigational drugs for Alzheimer's disease – which according to a 2014 Cleveland Clinic study (Cummings et al. 2014), showed a 99.6% failure rate between 2002 and 2012. That study found high attrition rates for Alzheimer's treatments, with 72% of agents failing in Phase 1, 92% failing in Phase 2, and 98% failing in Phase 3.

Given the risky nature of drug development, the summary is not new, unfortunately, but drilling down further into where and why drug development fails offers an additional perspective (DiMasi et al. 2016). Figure 8.3 looks at success

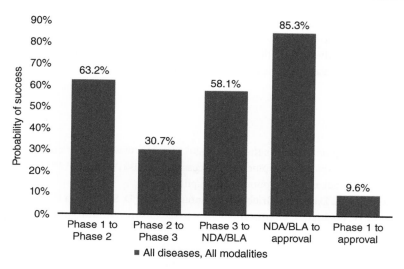

Figure 8.3 The difference in success rates of new molecular entities (NME) by development phase (BIO, Clinical Development Success Rates 2006–2015).

rates by phase of development and shows the high attrition for Phase 2 in particular. Despite the new tools and innovations, Phase 2 stands out as the low point in the development phase suggesting that there are conditions or poor decisions made in this phase that are either not present in the other phases or more problematic in Phase 2. Commonly held opinions on why this is the case include the following: poor understanding of the target biology and target/pathway-associated biomarkers, unknown or known interpatient differences in the disease not properly controlled and absence of a predictive biomarker, studied dose range that is too low or too high, treatment duration too short for efficacy, medications that suppress or mask response to active treatment, imbalance in confounding patient factors between active and control arms, geographic differences in treatment standards and clinical endpoints that are not consistent or reproducible (Patel et al. 2017).

A recent evaluation of clinical trial failures (Philippidis 2019) reviewed the most prominent failures of 2018 and examined the root cause behind many of the candidates across development phases. Perhaps not surprisingly, many of the reasons cited were consistent with the observations about Phase 2 attrition rates (Patel et al. 2017). Specifically, Philippidis mentioned inadequate study design, improper dose selection, non-optimal assessment schedules, inappropriate efficacy metrics/markers, and issues with how the data were analyzed as the high-level reasons behind the most recent failures. As this chapter reinforces, Phase 2 is a pivotal component of the current overall development paradigm. While there may be a rationale to skip Phase 2 for some sponsors, it would seem more prudent to do it better. Phase 2 and overall drug development attrition are obviously linked, but it our quest to bring new, important medicines to patients who need them, the fundamentals are still valid – give the right dose to the right patients at the right time. This would seem to be the overall objective of Phase 2 and the mission of drug development.

References

BIO, Clinical Development Success Rates 2006–2015. https://www.bio.org/sites/default/files/Clinical%20Development%20Success%20Rates%202006-2015%20-%20BIO,%20Biomedtracker,%20Amplion%202016.pdf.

Booth BP, Rahman A, Dagher R, Griebel D, Lennon S, Fuller D, Sahajwalla C, Mehta M, Gobburu JV. Population pharmacokinetic – based dosing of intravenous Busulfan in pediatric patients. *The Journal of Clinical Pharmacology* 2007; 47:101–111 DOI: https://doi.org/10.1177/0091270006295789.

Bradley, E. (2012). Incorporating biomarkers into clinical trial designs: points to consider. *Nature Biotechnology* 30: 596–599.

Cohen, A.T., Boyd, R.A., Mandema, J.W. et al. (2013). An adaptive-design dose-ranging study of PD 0348292, an oral factor Xa inhibitor, for thromboprophylaxis after total knee replacement surgery. *Journal of Thrombosis and Haemostasis* 11 (8): 1503–1510. https://doi.org/10.1111/jth.12328. 23782955 http://dx.doi.org/10.1111/jth.12328 30.

Cummings, J.L., Morstorf, T., and Zhong, K. (2014). Alzheimer's disease drug-development pipeline: few candidates, frequent failures. *Alzheimer's Research and Therapy* 6: 37.

DiMasi, J.A., Grabowski, H.G., and Hansen, R.A. (2016). Innovation in the pharmaceutical industry: new estimates of R&D costs. *Journal of Health Economics* 47: 20–33.

DiRienzo, A.G. and DeGruttola, V. (2003). Design and analysis of clinical trials with a bivariate failure time endpoint, with application to AIDS clinical trials group study A5142. *Controlled Clinical Trials* 24 (2): 122–134.

Glass, H.E., Glass, L.M., Tran, P., and Alghamdi, H. (2016). Pharmaceutical organizational size and phase 3 clinical trial completion times. *DIA Therapeutic Innovation and Regulatory Science* 50: 1–7.

Guidance for Industry (2009). End-of-Phase 2A Meetings, U.S. Department of Health and Human Services, Food and Drug Administration Center for Drug Evaluation and Research (CDER). https://www.fda.gov/media/72211/download.

Hansen AR, Cook N, Amir E, Siu LL, Abdul Razak AR. Determinants of the recommended phase 2 dose of molecular targeted agents. *Cancer* 2017;123(8):1409–1415. doi: https://doi.org/10.1002/cncr.30579.

Karmali, K.N., Lloyd-Jones, D.M., van der Leeuw, J. et al. (2018). Blood pressure-lowering treatment strategies based on cardiovascular risk versus blood pressure: A meta-analysis of individual participant data. *PLoS Medicine* 15 (3): e1002538. https://doi.org/10.1371/journal.pmed.1002538.

Lassere, M.N., Johnson, K.R., Schiff, M., and Rees, D. (2012). Is blood pressure reduction a valid surrogate endpoint for stroke prevention? An analysis incorporating a systematic review of randomised controlled trials, a by-trial weighted errors-invariables regression, the Surrogate Threshold Effect (STE) and the Biomarker-Surrogacy (BioSurrogate) Evaluation Schema (BSES). *BMC Medical Research Methodology* 12: 27.

LeBlanc, M. and Tangen, C. (2012). Choosing phase II endpoints and designs: evaluating the possibilities. *Clinical Cancer Research* 18 (8): 2130–2132. https://doi.org/10.1158/1078-0432.CCR-12-0454. (accessed 8 March 2012). PMID: 22407830; PMCID: PMC4820349.

Lee JY, Garnett CE, Gobburu JV, Bhattaram VA, Brar S, Earp JC, Jadhav PR, Krudys K, Lesko LJ, Li F, Liu J, Madabushi R, Marathe A, Mehrotra N, Tornoe C, Wang Y, Zhu H. Impact of pharmacometric analyses on new drug approval and labelling decisions: a review of 198 submissions between 2000 and 2008. *Clinical Pharmacokinetics* 2011;50(10):627–635. doi: https://doi.org/10.2165/11593210-000000000-00000.

Patel, D.D., Antoni, C., Freedman, S.J. et al. (2017). Phase 2 to phase 3 clinical trial transitions: reasons for success and failure in immunologic diseases. *The Journal of Allergy and Clinical Immunology* 140: 685–687.

Philippidis, A. (2019). Unlucky 13: Top Clinical Trial Failures of 2018: Biopharma's pursue costly studies despite data showing low success rates. *Genetic Engineering and Biotechnology News Magazine* 39 (3): 14–16. https://www.genengnews.com/a-lists/unlucky-13-top-clinical-trial-failures-of-2018.

Preskorn, S.H. (2014). The role of proof of concept (POC) studies in drug development using the EVP-6124 POC study as an example. *Journal of Psychiatric Practice* 20 (1): 59–60.

Roy, A., Mould, D.R., Wang, X.F., Tay, L., Raymond, R., Pfister. M. (2007). Modeling and Simulation of Abatacept Exposure and Interleukin-6 Response in Support of Recommended Doses for Rheumatoid Arthritis. *J Clin Pharmacol.* Nov; 47 (11): 1408–20.

Sertkaya, A., Wong, H.-H., Jessup, A., and Beleche, T. (2016). Key cost drivers of pharmaceutical clinical trials in the United States. *Clinical Trials* 13 (2): 117–126.

Wang X, Zhou J, Wang T and George SL. On enrichment strategies for biomarker stratified clinical trials. *Journal of Biopharmaceutical Statistics* 2018; 28(2): 292–308. doi:https://doi.org/10.1080/10543406.2017.1379532.

Weber, W., Harnisch, L., and Jessel, A. (2002). Lessons learned from a phase III population pharmacokinetic study of cariporide in coronary artery bypass graft surgery. *Clinical Pharmacology and Therapeutics* 71 (6): 457–467.

Wong CH, Siah KW, Lo AW. Estimation of clinical trial success rates and related parameters. *Biostatistics* (2019) 20(2): 273–286, doi:https://doi.org/10.1093/biostatistics/kxx069.

Chapter Self-Assessments: Check Your Knowledge

Questions:
- What are the main objectives of Phase 2 drug development?
- Under what conditions are Phase 2 perceived as successful?
- What does MTD refer to in drug development and how is it useful for Phase2?
- What is the purpose of the end-of-Phase 2 meeting at FDA?

Answers:
- The main objective of Phase 2 trials is to obtain data on whether the drug actually works in treating a disease or indication, which is generally achieved through controlled trials that are closely monitored, while safety and side effects also continue to be studied. Phase 2 studies also aim to establish the most effective dosage for the drug and the optimum delivery method in the target population (Hansen et. al., 2017).
- Phase 2 trials are considered successful when analysis of the data from enrolled participants indicates that the experimental drug works in treating the disease or indication. Patients who have received the experimental drug should have better clinical outcomes on a statistically significant basis than those who received the placebo or the alternative drug. If Phase 2 trials are successful, the drug proceeds to Phase 3 studies.
- MTD (maximum tolerated dose) refers to the highest dose of a drug or treatment that does not cause unacceptable side effects, which is then declared the recommended dose for further development. For many emerging oncology therapies, dosing to MTD is either impractical or unwarranted. In the absence of desirable "off-target" pharmacology, dosing beyond a relevant pharmacodynamic plateau is likely to offer little benefit but instead risks increasing toxicity or even producing confounding effects. Rather than blindly escalating the dose to the MTD, applying appropriate proof-of-mechanism (POM) pharmacodynamic biomarkers during the dose-escalation stage of a typical Phase 2 clinical trial can provide an estimated optimum dose without having to expose patients to unnecessary toxicities.
- The purpose of an End of Phase 2 (EOP2) meeting is to facilitate interaction between FDA and sponsors who seek guidance related to clinical trial design

employing a variety of tools including clinical trial simulation and quantitative modeling of prior knowledge (e.g. drug, placebo group responses, disease), designing trials for better dose response estimation and dose selection, and other related issues. With respect to timing, EOP2 meetings should be held before Phase 3 trials begin, and topics include determination of the safety of proceeding to Phase 3, evaluation of the Phase 3 plan and protocols for adequacy and to assess adult and pediatric safety and effectiveness, identification of information necessary to support a marketing application.

Quiz:

1 Choose the best answer to fill in the blanks." Phase 2 trials are typically constructed as _____, _____, _____ studies."
   A open-label, placebo, controlled, crossover
   B parallel-group, randomized, dose-finding
   C double blind, randomized, dose-finding
   D double blind, randomized, placebo-controlled
   E none of the above are correct

2 Some examples of endpoints include all but which of the following?
   A assessments of clinical events (e.g. stroke, pulmonary exacerbation, venous thromboembolism),
   B outcomes (e.g. result of treatment long term, mortality, tumor resolution, cure of disease, remission), patient symptoms (e.g. pain, dyspnea, depression), measures of function (e.g. ability to walk or exercise),
   C adverse events or surrogates of measured effects or symptoms.
   D all are examples of endpoints

3 Choose the best answer to fill in the blanks. A number of different designs fall under the category of innovative trial design, all of which allow interim data analysis and modification of the trial. Examples of such designs include _____, _____ and _____.
   A enrichment trials, biomarker-stratified trials, and adaptive trials
   B POC trials, POP trials, and POM trials
   C enrichment trials, biomarker-stratified trials, and POC trials
   D parallel-group, crossover, and enrichment trials
   E None are correct

4 True or False. One opportunity to lower the Phase 3 study cost allocation is by reducing the number of Phase 3 trials required for submission and market

access. Regulatory authorities provide a mechanism to do just that, assuming that the sponsor can establish adequately defined and clinically relevant exposure-response relationships that allow regulators to understand the conditions upon which the proposed relationships are dependent and the ability to both interpolate this relationship to doses unstudied but within the range of evaluation and extrapolate outside the range of clinical experience with reasonable uncertainty.

9

Phase 3

Zhaoling Meng, PhD

Data and Data Sciences, Sanofi R&D USA

After a successful Phase 2 program, Phase 3 of the drug development is officially initiated. Phase 3 is considered as the confirmation phase in a drug development journey, where the candidate drug's efficacy is confirmed, the drug safety is further demonstrated, and the benefit-risk profile is defined. Planning of the Phase 3 program usually starts early in the clinical development process, often as early as when a compound is considered to enter clinical development. The targeted drug label and desired benefit-risk profile are the ultimate end goals for a compound under the evaluation and drivers for many decisions in clinical development planning. As a compound progresses through earlier development phases, it paves the road for Phase 3 planning and success. To achieve desired efficacy, safety outcomes, and final drug labels in a targeted patient population, adapting to early phases' learning and designing Phase 3 accordingly play a key role. In addition, indication-specific requirements and historically accepted standards (a set of minimal requirements) frequently exist and need to be satisfied for many therapeutic areas to gain regulatory approval. Beyond satisfying these "minimal requirements" for approval, drug differentiations are also desired and further decide many additional considerations in the Phase 3 planning. Therefore, strategic Phase 3 planning goes beyond the approval and includes post-approval patient, payer, and reimbursement considerations as well.

This chapter summarizes the key considerations for Phase 3 planning, which include the dose proposal for Phase 3, design considerations for safety and efficacy confirmation, statical analysis plan, pre-specification, patient population, efficacy and safety endpoint planning, various factors impacting Phase 3 success. Developmental prioritization and innovations in attempt to enhance Phase 3 success and drug's benefit-risk profile for patients will also be discussed.

Fundamentals of Drug Development, First Edition. Edited by Jeffrey S. Barrett.
© 2022 John Wiley & Sons, Inc. Published 2022 by John Wiley & Sons, Inc.
Companion website: www.wiley.com/go/Barrett/FundamentalsDrugDevelopment

Efficacy vs. Safety at Phase 3 Dose and Design for the Confirmation

Phase 3 is generally regarded as the confirmation phase of drug development (Sheiner 1997). In practice, it carries much more than a straightforward confirmation. An ideal drug development scenario would be advancing one single dose to Phase 3 based on the most promising dose selected from a Phase 2 multiple-dose dose-ranging study – see Figure 9.1. Then Phase 3 would be carried out to confirm the selected dose's efficacy and safety effects. The reality is that the dose selected from Phase 2 is frequently further "tuned" in Phase 3, to say the least. Lyauk (Lyauk et al. 2019) reviewed the doses that appeared in 60 approved indications' labels between February 2015 to February 2017 and summarized doses explored and confirmed in these approvals. Out of 56 development indications with both early phases and Phase 3 information, 24 programs out of 56 (12 + 10 + 2, 42.8%) advanced one dose after first-in-patient (FIP) Phase directly to the confirmative Phase (or Phase 3). In comparison, 32 programs (12 + 1 + 16 + 3, 57.2%) embarked on exploratory Phase 2 with multiple doses. Among 24 FIP-direct-to-confirmatory programs, not surprisingly, 15/24 (4 + 9 + 2, 62.5%) studied multiple doses in Phase 3. Interestingly, 14 (3 + 8 + 1 + 2) out of the rest 32 indications (43.8%) with

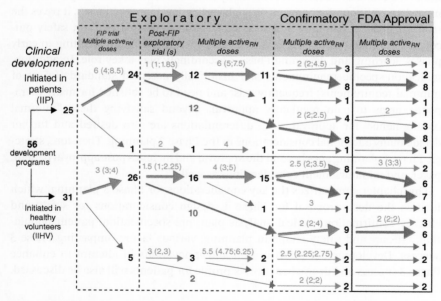

Figure 9.1 Representative clinical development paths to label-dose identification and approval for 56 US FDA-approved drugs (Lyauk et al. 2019). Green arrow indicates "Yes" to the multiple-dose investigation, and red arrow indicates "No".

Post FIP/Phase 2 multiple-dose exploration still studied multiple doses in Phase 3. At the end, less than 13% (7 out of 56) programs received more than one dose approval by Food and Drug Administration (FDA) per indication. These numbers illustrated additional dose explorations that happened in Phase 3. Multiple-dose explorations in Phase 3 are relatively easy to understand for the drug development programs advancing from FIP Phase directly to Phase 3, which was discussed in the previous chapter. For other cases, the authors cited inadequate "exploration" in Phase 2 as one reason and advocated additional efforts needed at the Phase 2 stage. However, the characteristic differences between Phase 2 and Phase 3 warrant some of these further multiple-dose "explorations" in Phase 3, which will ultimately confirm the drug's efficacy and illustrate the drug's safety.

Compared to a Phase 2 program, Phase 3 studies are usually larger and often with longer treatment exposure per patient. One apparent rationale is to have a sufficiently large study size and ensure the study power in confirming the drug efficacy. (Study power: the probability of detecting a difference between the study drug vs. the control when a true difference exists.) Phase 3 studies are frequently powered at 90–95%, which are higher than the usual 80–90% power requirements of Phase 2 studies. A larger Phase 3 study variability is also expected, which attributes to a combination of influential factors including diversifying study eligibility criteria, embracing global enrollment, requiring a longer per patient treatment duration, etc. Jointly anticipating larger study variability and larger study power, a Phase 3 study usually results in a larger study size even designed to confirm the same targeted treatment effect size as that of Phase 2. For many disease indications, to gain the final approval, a longer than Phase 2 treatment duration per patient is often required. For example, a pivotal Phase 3 study in chronic obstructive pulmonary disease (COPD) is usually 52 weeks by agency's requirement compared to a common 24-week Phase 2 study. This requirement is to ensure the treatment efficacy observed in Phase 2 can be confirmed as a long-term benefit in Phase 3 (Donohue et al. 2019). These indication-specific requirements dictate the per patient on-treatment exposure duration differences between Phase 2 and Phase 3.

Having sufficiently large studies and long treatment exposure also provides a good opportunity to better investigate a drug's safety profile and potentially discover less frequent adverse events that were missed in early phases. For many indications, a certain sized safety database with disease-population-appropriate on-treatment exposure are required prior to the approval. That is, the required treatment exposure needs to be sufficient long to expose any potential safety signal. The goal is to adequality assess and characterize a drug's risk profile in case signals were missed in previous developmental phases jointly consider targeted patient population, drug mechanism of action and intended usage length (Determining the Extent of Safety Data Collection Needed in Late-Stage Premarket a Post-approval Clinical Investigations. FDA 2016a). For example, to get a new drug approved in Type 2 diabetes indication, "Type 2 Diabetes Mellitus: Evaluating the Safety of New Drugs for Improving Glycemic Control Guidance for Industry" guideline requires "at least 4,000 patient-years of exposure to the new drug in Phase 3 clinical trials; at least 1,500 patients exposed to the new drug for at least 1 year, at least 500 patients exposed to the new drug for at least 2 years" (FDA 2020). There are also number of patients exposed requirements for special population such as patients with chronic kidney disease, patients with cardiovascular risk, and elderly patients. Phase 3 program planning accommodates these patient exposure requirements and the associated uncertainty.

As a drug's safety profile might not be established adequately by Phase 2 due to its limited study sizes and treatment duration, a Phase 3 program with only one dose could lead to an increased risk of failure in identifying the dose with the right benefit-risk balance. Therefore, many drug developers choose to design a multiple-dose Phase 3 program, where the optimal dose is further identified or confirmed (Lisovskaja and Burman 2013). A multiple-dose Phase 3 study provides flexibility in assessing and balancing the drug risk-benefit profile across different dose levels to increase the program's probability of success (POS). The drug developer weights a higher development cost burden to increase the program's POS. Of course, post-marketing surveillance and post-approval commitment can be required by an agency to follow up on relatively rare events beyond the pre-approval safety database and further test the winner dose. The development requirements, when to do what, between Phase 3 and post-marketing rely on a benefit-risk balance, which is based on the medical and mechanistic understanding of the new drug, patient needs and historical experiences.

Patient Population Diversity and Variability

The driver of the Phase 3 planning is the target drug label, which often determines the types of patient populations enrolled in the development. To satisfy the usual minimal two studies in providing substantial efficacy evidence for approval (see the later section for details), two Phase 3 studies can be conducted in the same targeted patient population. As pivotal studies supporting the approval, the patient population needs to be clearly defined and carefully selected via a list of patient inclusion and exclusion criteria. The enrolled patients should include sufficient diversity and be representative of the targeted patient population including disease severity and population demographic characteristic diversity such as sex, age, race, ethnicity, region distributions (Enhancing the Diversity of Clinical Trial Populations – Eligibility Criteria, Enrollment Practices, and Trial Designs. US FDA 2020). On one hand, the concern is the drug is approved based on a narrower patient population enrolled in clinical studies and then, after approval, given to a much broader population without studying its efficacy or safety in these patients (Nazha et al. 2019). On the other hand, excessive patient heterogeneity could reduce the study result consistency and lead to a lower study probability of success. To strike a balance, the drug developer can plan multiple studies in patients with different disease severity or characteristics in each instead of "packing" all sources of diversity in a single study. The increased relative homogeneity within each study improves the chance of demonstrating the drug efficacy and safety in each subpopulation. The approval support for diverse disease severity groups is then achieved by synthesized evidence across studies. For example, type II diabetes drug development programs are usually consistent of studies with patients on different diabetic background therapy such as prior metformin treatment failure study, prior insulin treatment failure study, prior sulfonylurea treatment failure study, etc. (Gourgari et al. 2017).

Another aspect in diversity includes sex, age, racial, ethnicity, special patient population diversity, and various country or region representation. In clinical trials, minority populations and women being underrepresented is well-recognized issue. Khan et al. conducted a study on a 10-year trend of women and minorities' participation in pivotal trials for cardiovascular new drug approval from 2008 to 2017 (Khan et al. 2020) – see Figure 9.2. For example, the authors found the woman representation rates ranged from 30% to below 50% with an average 36% participation, and blacks were under-represented at 4%, even with a 13% composition of US population. FDA guideline on "Enhancing the Diversity of Clinical Trial Populations – Eligibility Criteria, Enrollment Practices, and Trial Designs

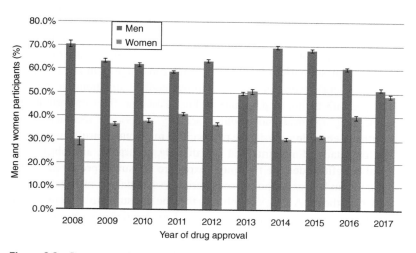

Figure 9.2 Recent regulatory assessment of underrepresentation of minority populations and women in Phase 3 clinical trials. Percentage of men and women participating overall in cardiovascular and diabetes mellitus pivotal drug trials according to the year of drug approval (Khan et al. 2020).

Guidance for Industry" (US FDA 2020) advocated patient diversity representation in clinical trials by boarding trial eligibility criteria and warned an approval review issue when the sponsors failed to do so. In line with these considerations, adolescent patients are encouraged to be included in the adult clinical development program, as long as similarity in disease histology, and biologic responses can be justified (Considerations for the Inclusion of Adolescent Patients in Adult Oncology Clinical Trials. US FDA 2019a). The intention is to speed up the current pediatric development timeline and mitigate the delay and lack of completion in pediatric trials, which resulted in pediatric patients' access lag to efficacious treatments (Hwang et al. 2018). To satisfy regional and country requirements for the drug approval and fulfill another source of diversity, global and multiple regional clinical trials become a frequent solution (ICH E17, US FDA 2018). The approach is well received and supported by agencies from various countries such as the United States, Europe, China, Japan, etc. to reduce the lag time of their patients' access to new treatments (Shenoy 2016). Much research and commentary have been presented to address operation, regulatory, statistical, and ethical issues on these trials including assessing treatment efficacy and safety consistency across counties (Chen and Quan 2016).

Efficacy and Safety Endpoints in Phase 3

With Phase 3 clinical trial patient populations determined, specifying appropriate efficacy and safety endpoints is then the next critical step to support new drug's benefit and risk assessment in the target disease and patient population. Clinical efficacy endpoints can be objective such as clinical events or lab measurements, or subjective such as questionaries (Evans 2010). They can range from solid survival endpoints (e.g. patient's overall survival), improvement of functions, to disease symptoms or patient's feeling (e.g. quality of life). Based on their importance in supporting the drug approval and labeling, primary, supportive secondary, and exploratory efficacy endpoints need to be prespecified in pivotal Phase 3 studies (see later section for details). Endpoints that can directly, accurately, and objectively reflect clinical benefit under the treatment exposure is usually preferred as the primary endpoint, and other endpoints considered as secondary to provide support or exploratory hypotheses. A drug's benefit can then be measured to demonstrate if it brings a clinically meaningful improvement based on specified efficacy endpoints. However, when a direct clinical benefit measurement is not possible or not practical in a given clinical trial setting, such as it takes too long to obtain, a surrogate endpoint can substitute instead (Aronson 2005). Some surrogate endpoints are extensively validated to represent a surrogacy for clinical benefits and can directly support a drug's final approval (Table of Surrogate Endpoints That Were the Basis of Drug

Approval or Licensure. US FDA 2021). Such examples include lowering LDL for cardiovascular drug approval and lowering HbA1c for diabetes drug approval. Some surrogate endpoints are considered "reasonably likely to predict a clinical benefit" and used to support accelerated approval with a need for later final approval assessment for efficacy (Surrogate Endpoint Resources for Drug and Biologic Development. FDA 2018a). An example is objective response rate (ORR) in oncology as a "surrogate" endpoint for progression-free survival (PFS) or overall survival (OS) or PFS as a "surrogate" endpoint for OS (Mulkey et al. 2020). The acceptance and usage of an endpoint as surrogate in the approval is not stationary and can evolve over time. Coming back to the oncology example, PFS was initially considered as a "reasonable likely" surrogate endpoint for OS. However, as new and more efficacious treatments were introduced in some tumor areas, patients' overall survival was extended. Patient staying free of progression becomes a clinically meaningful benefit over time. PFS was then considered as a valid clinical endpoint for some tumor areas (Clinical trial endpoints for the approval of cancer drugs and biologics guidance for industry. FDA 2018b).

For some disease areas, more than one efficacy endpoint is needed to indicate different aspects of the clinical benefit (Multiple Endpoints in Clinical Trials. FDA 2017). Multiple efficacy endpoints could represent the need for improvement in different disease attributes, which are equally important and individually essential to demonstrate a drug's clinical benefit. These multiple endpoints can either be considered as co-primary endpoints in a clinical trial. That is, a study is considered statistically significant only if each of these co-primary efficacy endpoints is statistically significant. When combining multiple endpoints into a single score can better demonstrate a drug's clinical benefit, the combined endpoint is referred as a composite endpoint (Brown-Tuttle 2018). Composite endpoints are used when a drug benefits different disease attributes that are more related or sometimes competing to occur, or each component is relatively infrequent. Using a composite endpoint can improve the feasibility of trial conduct in a reasonable timeframe for the primary endpoint accumulation. When a composite endpoint is used and shows a statistically significant effect, each of the individual components is desired to show an effect, at least an overall consistent trend. Major adverse cardio event (MACE) with components of cardiovascular death, non-fatal MI, and non-fatal stroke is one composite endpoint example in cardiovascular indication, where the endpoint is achieved whichever event occurs.

Patient reported outcomes and quality of life measures become more and more important, which illustrates a more patient-centric emphasis in current Phase 3 studies. This will be discussed further in a later section. At the end, it is worth to mention endpoints in a vaccine trial (Design of vaccine efficacy trials to be used during public health emergencies; WHO n.d.). Vaccine efficacy (VE) is usually used to support efficacy assessment in a vaccine trial, which is a similar concept as that in a drug trial. However, there is frequently a minimal efficacy lower bound requirement for approval in a vaccine trial, e.g. 95% confidence interval lower bound to be > 30% instead of 0. That is regulatory agency sets a more rigorous requirement in a minimal efficacy demonstration before approving a preventive intervention. In addition to a drug's efficacy, safety assessment becomes more important in Phase 3 studies as enrolled patient number and treatment exposure increase. Besides spontaneous reporting, some adverse events of special interest can be prespecified and proactively collected based on disease-specific needs or the compound class prior knowledge. Safety monitoring will be further discussed below.

Additional Phase 3 Design Considerations

For Phase 3 studies, some design features employed in Phase 2 are also recommended by researchers and regulators, for similar considerations, to achieve actionable safety and efficacy assessments of a new drug. These design considerations become more rigorously prespecified for the drug approval consideration compared to the early exploratory stages.

Randomized controlled trial (RCT) design, regarded as the gold standard for drug development trials, is preferably employed in Phase 3. Trial control arm can be either a current standard treatment (standard of care or SOC) or a placebo treatment. A placebo-controlled study compared to an investigational drug given on its own is a desired design in measuring "absolute" drug efficacy and safety (ICH E10 2001). It is the so-called placebo-controlled monotherapy study. A drug can be developed as a monotherapy treatment when it is intended to be given to patients on its own. The safety profile of monotherapy is relatively easier to demonstrate as no interference from other treatments. When a development objective is to assess the improvement of a new drug on top of SOC compared to SOC alone on either efficacy, safety, or both, the drug is developed as a combination therapy. When both the new drug and control arms are simultaneously treated with SOC as patient's background therapy, it refers to as an add-on study design. It can be applied to enroll a patient population that previously failed the SOC. In this case, the control SOC arm is often given placebo treatment for blinding treatment assignment purpose. In contrast, combination therapy can also be developed when two or more drugs, often at a prefixed dosage of each component, are intended to be given together to a naïve patient population. It is common in oncology or infection diseases, where a single drug is likely insufficient to provide adequate efficacy (Humphrey et al. 2011). When only one component is new, the new combination therapy is often compared to the other component (or rest components) in illustrating improvements in either efficacy or safety. There is also a fixed-dose combination development scenario, where both components (A and B) were approved before, and the combination (A+B) at the fixed dose of each is expected to provide additional benefit either on efficacy or safety. A factorial design (a design with arms treated with each component and all possible combinations of components) with A+B, A, B, and sometimes placebo arm could be required to illustrate A+B being superior to both A and B with placebo as a calibrator arm. The combination drug development is a complex topic with many agency guidelines to reference (FDA 2017; EMA 2017)

Depending on the choice of the control arm and the study objective, a Phase 3 study can be designed as a superiority study or a non-inferiority (NI) study. In a superiority study, the primary objective is to demonstrate a treatment difference and show the new drug being superior to the control. In a confirmative study, a prerequisite for success is the primary comparison passing a prespecified type one error control threshold (more details see below section). However, the decision on claiming a new drug's benefit could also depend on the estimated treatment difference meeting a clinical meaningful margin per disease area. The superiority design can be applied for the control arm being either a placebo or active control. For NI studies, the primary objective is to show a new drug not being less effective than an existing drug. The new drug could provide other benefits to patients such as being safer, more convenient to in-take (e.g. oral vs. subcutaneous), or administered less frequently (bi-weekly vs. once monthly). The NI design is usually applied to active-controlled studies. To assess "not being less effective," the requirement is not to demonstrate a positive difference (if positive means benefit) but to illustrate the treatment difference not being worse than a prespecified amount (Non-Inferiority Clinical Trials to Establish Effectiveness. FDA 2016b). The prespecified amount of effect loss that can be tolerated is referred to as a NI boundary. Determining a NI boundary is not trivial and depends on disease area, historical experience, and clinical and statistical justifications. It has been a highly interesting research topic for both researchers and regulators (Choice of a non-inferiority margin. EMA 2006). There is also an equivalence trial design where the primary objective is to demonstrate equivalence of two drugs on an endpoint, and it is more commonly applied in Phase 2 setting or generic drug development in Phase 3. Figure 9.3 below illustrates the interlink among the above three types of designs (Points to consider on switching between superiority and non-inferiority. EMA 2000).

Where M is a prespecified non-inferiority margin.

In Phase 3, treatment double-blind (treatment assignment is blinded for both participants and experimenters) is applied whenever possible to reduce both participants' and experimenters' potential subjective bias towards either arm. Exceptions include when the feasibility of blinding is problematic such as intravenous doing of a new drug formulation verse a subcutaneous control formulation. The type of

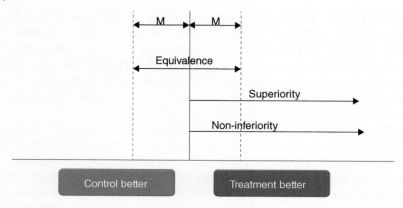

Figure 9.3 Comparison of various Phase 3 study designs with respect to performance criteria and the prespecified study constructs.

studies and number of studies required for approval will vary case by case (for details, see the later section). Although randomized, controlled, double-blind studies are especially preferred in Phase 3 confirmative setting, there are exceptions including open-label studies, single-arm studies for various practical considerations (Tenhunen et al. 2020). In additional, registry, post-marketing surveillance study, observational study, and real-world study can be carried out as supplements after the drug is on the market. The drug effectiveness and safety can be further understood, especially for rare diseases.

Statistical Analysis Plan and Pre-specification

For Phase 3 studies, especially ones playing a critical role in the regulatory approval, planning, communicating, and committing with regulatory agencies are essential. As described in the previous chapter, an end-of-Phase 2 meeting or equivalent will usually be held prior to Phase 3 start to seek agreements and inputs from regulatory agencies. As these submission critical trials are considered as confirmatory and pre-specification prior to data unblinding is essential to unbiased regulatory decision making (ICH E9, US FDA 1998). The most noticeable difference between Phase 3 and early phases is more rigorous pre-specification. In addition to submitting a protocol prior to the study starts as other phases, a statistical analysis plan (SAP) is highly encouraged to be included as part of the protocol (Good Review Practice. US FDA 2013a). SAP should include sufficient details describing the primary analysis, key supportive analyses, and any analytical aspects impacting the study integrity and regulatory decision making. Gamble et al. (2017) published a recommendation paper on a minimal set of items to be included in SAP, which is summarized below.

- Administrative information: title and trial registration, SAP version, Protocol version, roles and responsibility, and signatures
- Introduction: background and rationale, and objectives
- Study methods: trial design, randomization, sample size, framework, statistical interim analyses and stopping guidance, the timing of final analysis, and timing of outcome assessments
- Statistical principles: confidence intervals and P values, adherence and protocol deviations, analysis populations
- Trial population: screening data, eligibility, recruitment, withdrawal/follow-up, baseline patient characteristics
- Analysis: outcome definitions, analysis methods, missing data, additional analyses, harms, statistical software, and references.

A part from randomization at baseline and treatment assignment blinding, pre-specifying critical data collection and analysis methods play a key role in reducing bias in post-randomization behavior such as dropout, loss to follow up, treatment rescue, on-treatment concomitant medication adjustment, etc. (ICH E9(R1) US FDA 2017). Although a separate SAP document can be submitted later to include additional analysis-related details, the current trend in recommendation is to pre-specific these procedures, intended approaches in data collection, and analysis related to a clear intended objective of treatment effect estimation (i.e. estimand). These recommendations not only reduce suspicion of bias and data stretching in a post-hoc missing data imputation approach but also handle missing data from its roots: preventing (see Figure 9.4 for workflow considerations). Nevertheless,

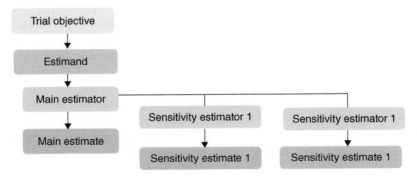

Figure 9.4 Relationship and workflow of key statistical design protocol elements for Phase 3 trials. *ICH E9(R1) US FDA 2017 addendum: estimands and sensitivity analysis.

details or some changes could be added or updated post the trial start as researchers accumulating knowledge via either protocol amendment or a separate SAP update. SAP is required to be finalized and submitted before the study unblinding, and any important changes should be made prior to unblinded data access to avoid suspected biases.

A related concept approach is the pre-specification of the analysis population, intended-to-treat or per-protocol. Intended-to-treat analysis includes all enrolled or randomized patients in the primary analysis to maintain the integrity of the randomization. The per-protocol analysis excludes patients with pre-identified major protocol violations to keep the integrity of treatment effects when patients adhered to planned treatments (Eric McCoy 2017). Debates and regulatory acceptance differences exist on impacts of these analyses, especially potential differential impacts on superiority vs. non-inferiority studies (Ranganathan et al. 2016). Frequently, both analyses are conducted in the study with one prespecified as the primary analysis and the other as a sensitivity analysis to aim for result consistency and study robustness.

Type One Error Control, Study Power, and the Amount of Evidence for Approval

The evidence required for the regulatory confirmation, i.e. approval, varies. Generally, two confirmative Phase 3 trials are needed (ICH E9, US FDA 1998). The rationale is to control the so-called program-wide type one error rate, i.e. falsely approving a new drug for its efficacy or benefit when the efficacy or benefit is not real. If each Phase 3 trial's type one error is controlled at 5% level under a 2-sided hypothesis test, two positive Phase 3 trials simultaneously will control the program-wide type one error rate at $0.05*0.05 = 0.25\%$. Although the type one error rate of 5% (i.e. P-value = 5%) is an "arbitrarily" pre-determined number and drew some criticisms on its universal justification (Wasserstein et al. 2019), the 5% rate comes with a long establishing history and is well-accepted in most drug development regulatory approval process to substantiate the evidence (Di Leo and Sardanelli 2020). When there are multiple hypotheses to be tested and established, either from multiple clinical endpoints, doses, or sub-populations, a statistical procedure can be applied to control the Phase 3 study's type I error rate at 5%. The type I error control methodologies range from more rigorous family-wide method to false discover rate and to simulation-based etc., with more rigorous procedure considered better acceptable for the confirmative stage and by the regulatory agency (Dmitrienko et al. 2009). The multiple testing control procedure also needs to be prespecified. If an interim analysis is planned, additional type one error control needs to be considered, and more details are described in the innovation section below.

To achieve a positive study (i.e. P-value < 0.05), the study sample size planning to ensure an adequate study power (probability of a study to demonstrate the drug effect given a controlled type one error rate) is the next critical consideration. As mentioned above, Phase 3 studies are frequently powered at 90–95%. A sufficiently large Phase 3 sample size also provides a more precise estimate of the treatment effect (i.e. narrower confidence interval for the estimated effect). It provides researchers, patients, and regulators a better understanding of the treatment benefit. Besides a statistical significance (P-value is less than 5%), estimated treatment benefit indicating a clinically relevant and meaningful improvement is also frequently required for approval. As a larger clinical study is more costly and time-consuming, a sensible Phase 3 study is usually powered to detect a meaningful treatment effect but not too large to detect a small and less meaningful effect (Faber and Fonseca 2014).

There are exceptions of the minimal two trials for approval. Gaining the approval based on one large outcome trial is an example (Schnell et al. 2020). As the approval evidence is only based on one study, a more stringent type one error control on the study is usually required. The significant level of 1% is frequently used for one single study approval, which is more stringent at 5% level for one of two studies and less stringent than requiring two studies at 0.25%. The single study approval case is usually for less frequent event type of efficacy (or safety) endpoint, where one large trial with a sufficiently long follow-up makes better sense than two separate smaller trials. Another exception could be for rare diseases and unmet medical need where two-studies requirement is considered unrealistic. However, increasing approval based on a single study recently, such as in the oncology area where unmet medical needs and disease rareness is justified, is still questioned for its ability of providing "substantial evidence" for approval (Ladanie et al. 2019).

As mentioned previously, the amount of evidence accumulated to support safety and then benefit/risk ratio assessment is equally important for the regulatory approval decision. However, the safety assessment is usually conducted in a "signal-detection" rather than a formal hypothesis pre-specification and then conformation framework. At the submission stage, the efficacy and safety information collected across multiple clinical studies could be pooled together to conduct the analyses of integrated summary of safety (ISS) and integrated summary of efficacy (ISE) (Integrated Summaries of Effectiveness and Safety. US FDA 2009). ISS is frequently required and reviewed by the regulatory agency to better assess a new drug safety profile and identify relatedly infrequent safety signals. ISE provides more data to generate more precise estimates of the treatment effects of interests.

Trial Operation Consideration, Enrollment, Monitoring, and Trial Integrity

Harrer (Harrer et al. 2019) presented an astonishing picture of the drug development cycle including the cost, time needed, and success rate across different stages of the development phases. From the drug discovery to regulatory submission and post-marketing, a 15-year period was expected on average with Phase 3 development spanning 4 out of 15 years. A compound selection started from 5000 to 10 000 candidates, which were reduced to ~5 drug candidates just before the start of the clinical trials. After entering the clinical phase, the failure rate from Phase 1 to Phase 3 was more than 85% and reduced to one-third from Phase 3 to FDA approval – see Figure 9.5. On the cost side, Phase 3 cost was estimated around 30% of the total average R&D cost of US$ 1.5 to 2 billons. It illustrated Phase 3 as an efficient development phase to manage the failure risk and ensure the investment return. Besides scientific and design factors already mentioned above, there are many trial operation aspects can be carefully carried out to manage the risks associated with Phase 3.

As Phase 3 trials are relatively large and frequently carried out globally in multiple counties, patient enrollment and retention become critical factors for trial success. Fogel (2018) reviewed factors associated with clinical trial failures and cited multiple cases including 25% of cancer trials failed to enroll enough patients (Feller 2015), and one-third investigated public-funded trials that required a time-extension due to slow enrollment (Campbell et al. 2007). On the other hand, patient retention ensures sufficient and high-quality data to be captured to support the research questions.

As the multiregional trails add to the complexity and variability of patient enrollment and retention, fierce competitions among drug developers require even more patient-friendly, efficient data and tech-savvy approaches. Real-world

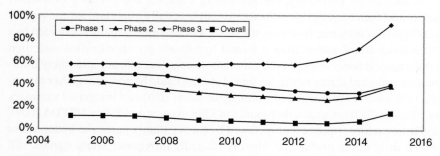

Figure 9.5 Historical regulatory data examining the probability of success by development phase over time (1 January 2005, to 31 October 2015) (Wong et al. 2019).

data is more and more advocated to assist in patient needs and enrollment feasibility assessment, provide feedbacks on patient inclusion and exclusion criteria selections and measure practicality of the protocol setups (Sturges 2019). In addition, new wearable and wireless devices (Apple watch, venturebeat.com 2018) provide continuous data collection and patient performance monitoring in trials, which could reduce patient burden in entering the data. Another example is to use current data science advances and simplify patient assessment. Hematoxylin and eosin (H&E) is a gold standard stain approach for evaluating tumor samples and is routinely available. However, Multiplexed imaging (mIHC) allows deep assessment of the tumor microenvironment and is costly and resource-intensive. Burlingame et al. (2018) developed an artificial intelligence (AI) method to predict the need of mIHC assessment from readily available H&E images, simply the process and reduce the cost.

While the clinical trials are ongoing and patients/sites are enrolled and followed up, site monitoring is essential to ensure the protocol process implementation, participant risk management, and high-quality data collection. Although trials in all phases can benefit from trial monitoring, Phase 3 trials benefit particularly due to their size, study length, patient diversity, and complexity. To avoid introducing bias via the clinical trial conduct, the site monitoring is usually conducted under treatment blinded conditions. One of the main tasks for the patient/site monitoring is to minimize trial missing data and increase patient retention. High levels of missing data or patient dropout raise a red flag for the trial conduct as well as the interpretability of the study outcomes (Molloy and Henle 2016). Efficient and early fraud detection can promote the drug developer to mitigate the issue early on and reduce its impact on the study integrity. To strike a balance between resources and quality of the monitoring, FDA promoted the risk-based monitoring approach and leveraging the modern clinical trials to achieve efficiency and effect by focusing on the critical aspects of the trials (A risk-baseline approach to monitoring. US FDA 2013b).

To ensure the clinical trial smooth progression and integrity, another layer of trial monitoring is the data monitoring committee (DMC, or independent data monitoring committee, IDMC, or data and safety monitoring board, DSMB). All clinical trials need closely monitor patient safety, but late phase studies frequently require a dedicated committee to take on this important task considering the trial complicity and high regulatory scrutiny on the trial integrity (Establishment and Operation of Clinical Trial Data Monitoring Committees. US FDA 2006). In addition, some safety signals, especially less frequent events, only become apparent in large Phase 3 studies. This also raises the need for treatment unblind review, where a committee independent from the study team is critical to avoid any operational bias. Besides monitoring safety, DMC can take on an important interim analysis decision-making role for efficacy (more details see later section on Phase 3 innovation).

Special Protocol Assessment, Fast Track Approval, Success, and Failure

In this section, selected processes related to specific scientific contents of Phase 3 development are described. If the efficacy evidence collected in a Phase 3 trial is the primary support for approval, the Phase 3 trial's protocol can be submitted for Special Protocol Assessment (SPA) with FDA's agreement for review (ref). Note SPA process can also be granted to other trial phases, such as Phase 2 with critical data for a subpart H accelerate the approach. The main objective of the assessment is to agree on critical study design elements in supporting the drug approval, such as sample size, dose selection, endpoint selection, population selection, etc. Even though the review result is not legally binding for either sponsor or FDA, reaching a written agreement and implementing the agreement in the development process are much likely to reduce the drug application review time or increase the drug approval probability. In 2014, Maher (Maher et al. 2014) summarized all special protocol assessment cases in a 10-year period from January 2003 to January 2013 in the office of hematology and oncology products (OHOP) at FDA. 132 out of 532 total SPA submissions (25%) reached an agreement, and 30 of these 132 submissions (23%) were submitted or planned to submit a New Drug Application (NDA) at the time of assessment. The 25% rate (132/532) in reaching the agreement during a SPA is worth noticing, as critical aspects of the protocol need to be detailed and agreed. Nevertheless, drug developers frequently announce such achievements in press release to indicate an important positive progression in the development process. European Medicines Agency (EMA) has a similar procedure called Scientific Advice and Protocol Assistance (SA/PA), although it is less utilized by sponsors.

Starting in 2018, FDA started a Complex Innovative Trial Design (CID) pilot meeting program to fulfill a performance goal agreed under PDUFA VI. The meeting program goal is to facilitate and advance the use of complex and innovative trial designs such as adaptive designs and Bayesian analysis in the late stage of the development. The program is to span from 2019 to 2022, provide opportunities for sponsors and FDA to publicly discuss innovative study designs, gain some transparency on the regulatory acceptance of these new approaches, and ultimately promote innovations in clinical development. Many innovations discussed in the later section of this chapter will fall under the CID category including historical trial information borrowing and application of the Bayesian method in pivotal studies. To increase sponsor's uptake of this process, FDA later in 2019 clarified a main sticking point: public disclosure of trial-related information. The clarification listed some exclusions of the required disclosure including sponsor name, product name, subject level data, recruitment strategies, etc.

After the results from Phase 3 trials (or Phase 2 trials for some cases) demonstrated a preferable benefit and risk profile of a new drug compared to current therapies or meeting unmet medical needs, an NDA or biologic license application (BLA) will be filed with FDA for the drug approval. For serious or life-threatening conditions with unmet medical needs, FDA established four fast-track programs (accelerated approval, priority review destination, fast-track destination, and breakthrough therapy destination) to expedite the development and shorten the time for an efficacious treatment to reach patients (FDA 2017). Accelerated approval is usually granted to a drug demonstrated the effect on an endpoint reasonable likely to predict the clinical benefit, where the clinical benefit will take a long time to achieve. After the accelerated approval is granted based on the intermediate or potential surrogate endpoint, the sponsor has a post-approval commitment to demonstrate the final clinical benefit in a predefined timeframe. The accelerated approval pathway will have a significant impact on the drug development planning, especially Phase 3 planning. Priority Review is requested at the submission and requires FDA to shorten the submission review time to 6 months, compared to a 10-month standard review time. Priority review is usually granted when a new drug or indication is deemed effective or safer for these conditions. Both Fast-Track and breakthrough therapy designations allow a priority and rolling review of drugs targeting serious or life-threatening conditions with unmet medical needs, and both can expedite the development program and impact Phase 3 planning. The fast-track designation requires either nonclinical or clinical data to support a potential benefit for the request. FDA CDER lists each year's fast track approvals, and 36 occurred in 2020 (https://www.fda.gov/drugs/nda-and-bla-approvals/fast-track-approvals   2020a). Breakthrough therapy designation has been established more recently in July 2012 and requires clinical data to support a potential substantial treatment benefit over current therapies. From 2013 to end of 2020, there were 190 drug and biologics approvals based on the breakthrough designation with 34 approvals in 2020 (https://www.fda.gov/drugs/nda-and-bla-approvals/breakthrough-therapy-approvals 2020b). An example is FDA's priority review, breakthrough therapy designation, and approval of Osimertinib as adjuvant therapy in December 2020 for non-small cell lung cancer patients with EGFR mutations.

As a new drug gets developed for an indication and moves through each clinical development phase, the risk of failure persists. As mentioned earlier in this chapter, the cost associated with Phase 3 to approval is usually highest among the clinical development phases with the corresponding risk being immediate or low. However, there are certainly disease and therapeutic area differences. Wong et al. (2019) estimated clinical trial probability of success (POS, success in starting the next phase) for each phase (Phase 1 to phase 2, Phase 2 to Phase 3, and Phase 3 to approval) and overall by various disease areas using historical clinical trial outcomes. The authors summarized 406 038 trials (185, 994 unique trials over 21 143 compounds) from 1 January 2000 to 31 October 2015. The overall POS for all drugs and vaccines is 13.8%, with Phase 3 to approval POS being 59%. Compared to other disease areas, a lower overall POS rate of 3.4% and a lower Phase 3 to an approval rate of 35.5% in oncology are especially striking. The authors also estimated the clinical trial POS evolving trend over time. Interestingly, the overall success rates increased over time and were apparently driven by higher Phase 3 successes over time. In this paper and a subsequent paper (Lo et al. 2019), machine learning methods were applied to identify factors influencing these success rates. For example, they discovered clinical trials designed based on biomarkers had higher POS compared to those independent of biomarkers in oncology. These could reflect higher successes of targeted therapies, precision medicine advances, or even more successes in the utilization of patient enrichment trials. This historical information is utilized to help sponsors making decisions, prioritizing the development plan, or optimizing the development plan.

Patient-centric Consideration, Target Product Profile, and Target Value Profile

As direct beneficiary, fundamental focus, and critical success factor of drug development, patients should always be the center of the clinical trials. With raised requirements from patients, patient advocates and regulators, integrating patient's experience and feedback on the disease and from the clinical trial become increasingly important in the drug development cycle. First, these patient-centric considerations improve the regulator and drug developer's understanding of the unmet medical needs and assessment of the drug' benefit-risk profile. Secondly, implementing these patient-centric considerations can lessen patient's burden in the trial, boost trial enrollment, reduce study dropout, and increase trial success. Patient-centric consideration should be implemented across clinical trial phases (Timpe et al. 2020). Phase 3 trial, with its relatively large size, diverse patient population and long study follow up, provides a good opportunity in collecting patient's feedback data and demonstrating patient-assessed benefit. It also presents a great need in integrating these considerations for the trial and for the regulatory approval success.

Patient-reported outcomes (PRO) have a long history being measured in clinical trials mostly using questionnaires. It often includes patient symptoms and quality of life measurements and could be used as primary, secondary, or exploratory endpoints (Rivera et al. 2019). Merciera-Bebber (Mercieca-Bebber et al. 2018) summarized PRO's contributions to clinical trials, which include supporting clinical data interpretation and regulatory decisions. In the approval of mitoxantrone in men with metastatic prostate cancer, the primary endpoint of a PRO, palliation of pain, supported the approve even though the overall survival and another important clinical endpoint serum prostate-specific antigen level did not show a difference. Gnanasakthy (Gnanasakthy et al. 2012) reported 24% (28 out of 116) of FDA approvals between 2006 and 2010 and a decreased 16.5% (30 out of 182) of FDA approvals between 2011 and 2015 had received PRO-based labeling. Only 7.5% (3 out of 40) of FDA approvals between 2010 and 2014 in the hematology and oncology area (Gnanasakthy et al. 2016) had received PRO-based labeling. The authors discussed barriers and challenges of PROs uses and impacts in clinical trials, which includes not sufficient understanding and implementing across stages of the trials, insufficient reflecting patient's need evolving over time, inappropriate choice of PRO, poor reporting and missing value, high variability in PRO to support conclusive decision, etc. New data collection, patient engagement, and innovation adaptations are anticipated to bring in the resolution. To compliant with the Twenty-first Century Cures Act, FDA, in 2017, started a 5-year effort on developing patient-focused drug development guidance (Patient-Focused Drug Development. FDA 2020). The scope covers (1) collection comprehensive and representative inputs, (2) methods to identify what is important for patients, (3) selecting, development or modifying fit-for-purpose clinical outcomes assessments, and (4) incorporating clinical outcome assessment into endpoints for regulatory decision making. The plan focuses on methodology development to fulfill the promises of a patient-centric clinical trial.

Although the success of the drug development is often measured by the approval, its true success and impacts go beyond the regulatory approval event. A drug's target product profile (TPP) captures minimal and also desired efficacy, safety, formulation, etc., features to serve the strategic development planning for a drug's differentiation. It was initiated as a sponsor internal document but recognized its usage in regulatory communications recently (Breder et al. 2017; Tyndall et al. 2017). Target value profile (TVP) evolves from TPP and emphasizes on values to patients and unmet medical needs. It indicates the recognition of insights and differentiations brought by patient-centric trial considerations in the drug development and promotes an early integration of health economics and outcomes research (HEOR) supports and robust evidence generation in pivotal Phase 3 trials.

Innovation in Phase 3

Many drug development innovations occurred at the early stage of development, such as compound selection, drug indication selection, drug-drug combination selection, patient responder population identification, and dose selection. Phase 3 could be viewed as a less exciting stage to confirm the efficacy and safety signals identified in previous phases with much fewer "surprises". However, considering

Phase 3's features of large sample size, diverse patient population, and long study follow-up time, additional innovative approaches are needed to either ensure its success or optimize the drug's benefit and risk profile. As the development Phase 1/2/3 dividing lines become increasingly blurry in recent years, especially in oncology and rare disease areas, innovations are often carried across stages.

Adaptative clinical trial with interim analysis and decision points prior to the final analysis is a well-recognized innovative approach. Although with a long history of being introduced by FDA in a 2004 strategic path initiative (Mahajan and Gupta 2010) and well-acknowledged contributions in increasing trial probability of success and/or efficiency, adaptive trials were applied more frequently in the early stages of the drug development with primary objectives being exploratory and signal identification (Lai et al. 2015). The group sequential design trials are one type of adaptive trials that were better accepted by regulators and therefore more applied in Phase 3 confirmative setting (Mazumdar and Bang 2007). It allows one or multiple preplanned interim analyses to stop the trial early for sufficient efficacy (trial success) or unlikely efficacy evidence (futility) with preplanned stopping rules. It was regarded as "better-understood" type of adaptive trial and better accepted by regulators with the availability of theoretical supports on its capability in controlling erroneous claims (type I error). There was a lack of broader adaptation of other adaptive approaches in Phase 3 confirmative trials with reasons including operation complexity, burden, increased resources and expertise needs, etc. Among them, regulatory concerns in introducing bias, making wrong claims, and maintaining trial integrity were higher hurdles to overcome. Previously, FDA referred to some of these adaptive designs as being "less understood." However, in recent years, with gained experiences in applications and advancements in methodology assessing and controlled these concerns, FDA updated its guidance (Adaptive Designs for Clinical Trials of Drugs and Biologics, US FDA 2019b) and opened doors for more adaptations in the confirmative setting. These adaptive approaches include sample size re-estimation to increase the study sample size at interim, seamless Phase 2/3 trials with treatment dose selection and even hypothesis selection at interim, patient population adaptation from narrow to a broader for safety consideration or from board to narrower for efficacy enrichment (Enrichment Strategies for Clinical Trials to Support Determination of Effectiveness of Human Drugs and Biological Products, US FDA 2019c), etc. With recent genetic/ genomic, biomarker, and biotechnology advancement, the last type of adaptive designs, specifically referred as enrichment designs, hold great promises in precision medicine era for oncology and other complex diseases (Polley et al. 2019).

While there is still a debate on whether current Phase 3 clinical studies become increasingly complex (Glass et al. 2015), large simple trials, at the other extreme of Phase 3 complexity, gain much attraction and application. A large simple trial is developed mainly for chronic diseases where a new drug's treatment effect is expected to be moderate and could take a relatively long time to establish on top of the existing standard of care. As a large study sample size is needed to detect the moderate treatment improvement, simplified protocol design and trial objectives are needed to keep the trial cost in check and increase patient retention (Eapen et al. 2014). In addition, careful endpoint selection and design consideration are also essential to ensure the evidence generation and the support of a drug's long-term benefit-risk assessment. At the same time, large simple trials provide a good opportunity to increase trial-studied patient population diversity and study patients' responses closer to "real-world" settings compared to conventional clinical trials (Large Simple Trials and Knowledge Generation in a Learning Health System: Workshop Summary 2013). As pragmatic trials are conventionally conducted to understand a drug's effectiveness in a real-world setting, pragmatic randomized controlled trials (RCT) become a recent transition trial design between Phase 3 studies and post-marketing studies (Dal-Ré et al. 2018). The new trend is welcomed under the same notion of weakening the drug development phase dividing line and extending Phase 3 drug development focusing from TPP transition to TVP (ENCePP Guide on Methodological Standards in Pharmacoepidemiology. Section 5.6 n.d.).

As discussed earlier in the chapter, one major hurdle in conducting and completing Phase 3 trials is patient recruitment. For rare diseases and sub-patient populations considering a specific biomarker or genetic mutation, locating, and enrolling enough patients in the trial becomes even more challenge. Furthermore, for severe or life-threatening diseases, enrolling patients, or a large number of patients to a placebo arm is considered unethical. Even though a sufficiently size placebo arm is needed to power a placebo-controlled study. For unmet medical needs, there could be no other available treatment (SOC) to be used as the control arm in a study. To address these issues, increase Phase 3 trial efficiency, and accelerate unmet-medical-need drug development, synthetic and external controls can be considered in Phase 3 trials in regulatory settings (Ghadessi et al. 2020, Framework for FDA's real-world evidence program. US FDA 2018). Synthetic control refers to a clinical trial control arm consisted of both study-enrolled patients on the control arm and mathematically borrowed control arm information from an external data source (see Figure 9.6). External control refers to a control arm completely consisted of external data. The external data sources could include historical clinical trials, patient registries, electronic medical records, claim data, etc. (van Rosmalen et al. 2018). Recognizing the use of non-randomized control data in these approaches, there are evolving views on pre-specification requirements and checklist, Bayesian borrowing methods, patient population matching, and weighting methods to avoid bias and potential misleading results via inappropriate leveraging the external data (Andre et al. 2020; Thorlund et al. 2020). Also, recognizing some of these data sources are not from convention clinical trials, data fit-for-purpose assessment also becomes essential (Chen et al. 2021).

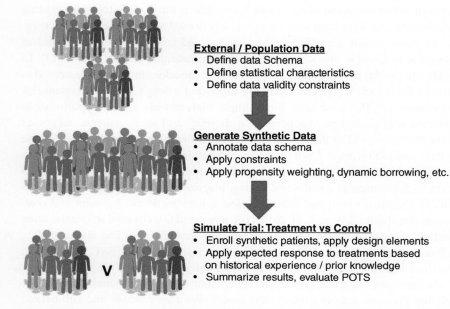

Figure 9.6 Schematic illustrating the use of a controlled clinical trial with a synthetic control arm (Kristian Thorlund et al. 2020).

With the advancement of digital technologies, more innovations are expected. Modeling and simulation approaches can be applied to optimize, reduce, and even replace the need for some clinical trials. New digital endpoints can be developed via devices closely monitoring patient treatment responses, which can increase Phase 3 development efficiency and better address patient's need. New digital data capturing methods can reduce patients' physical site visits, reduce patients' burden and reduce cost (Inan et al. 2020).

References

Andre, E.B., Reynolds, R., Caubel, P. et al. (2020). Trial designs using real-world data: the changing landscape of the regulatory approval process. *Pharmacoepidemiology Drug Safety* 29: 1201–1212.

Apple watch, venturebeat.com (2018). Apple Watch Series 4 can detect falls, take ECGs, and lead you through breathing exercises. https://venturebeat.com/2018/09/12/apple-watch-series-4-can-detect-falls-take-ecgs-and-lead-you-through-breathing-exercises.

Aronson, J.K. (2005). Biomarkers and surrogate endpoints. *British Journal of Clinical Pharmacology* 59 (5): 491–494.

Breder, C.D., Du, W., and Tyndall, A. (2017). What's the regulatory value of a target product profile? *Trends in Biotechnology* 35 (7): 576–579.

Brown-Tuttle, M. (2018). The Regulatory Strategist Toolbox: Clinical Endpoint Analysis Tools. Regulatory Focus. Regulatory Affairs Professionals Society.

Burlingame, E.A., Margolin, A.A., Gray, J.W., and Chang, Y.H. (2018). SHIFT: speedy histopathological-to-immunofluorescent translation of whole slide images using conditional generative adversarial networks. *Proceedings of SPIE The International Society fot Optical Engineering* 10581: 1058105.

Campbell, M.K., Snowdon, C., Francis, D. et al. (2007). Recruitment to randomised trials: strategies for trial enrollment and participation study: the STEPS study. *Health Technology Assessment* 11: 105.

Chen, J. and Quan, H. (2016). *Multiregional Clinical Trials for Simultaneous Global New Drug Development*. ISBN 9781498701464. Chapman and Hall/CRC.

Chen, J., Ho, M., Lee, K. et al. (ed.) (2021). *The Current Landscape in Biostatistics of Real-World Data and Evidence: Clinical Study Design and Analysis*. Statistics in Biopharmaceutical Research.

Dal-Ré, R., Janiaud, P., and Ioannidis, J.P.A. (2018). Real-world evidence: how pragmatic are randomized controlled trials labeled as pragmatic? *BMC Medicine* 16 (1): 49.

Di Leo, G. and Sardanelli, F. (2020). Statistical significance: p value, 0.05 threshold, and applications to radiomics – reasons for a conservative approach. *European Radiology Experimental* 4: 18. https://doi.org/10.1186/s41747-020-0145-y.

Dmitrienko, A., Tamhane, A.C., and Bretz, F. (ed.) (2009). *Multiple Testing Problems in Pharmaceutical Statistics*, 1e. Chapman and Hall/CRC.

Donohue, J.F., Kerwin, E., Sethi, S. et al. (2019). Maintained therapeutic effect of revefenacin over 52 weeks in moderate to very severe Chronic Obstructive Pulmonary Disease (COPD). *Respiratory Research* 20: 241. https://doi.org/10.1186/s12931-019-1187-7.

Eapen, Z.J., Lauer, M.S., and Temple, R.J. (2014). The imperative of overcoming barriers to the conduct of large, simple trials. *JAMA* 311 (14): 1397–1398.

EMA (2000). *Points to consider on switching between superiority and non-inferiority*. EMA.

EMA (2006). *Choice of a Non-Inferiority Margin*. EMA.

EMA (2017). *Guideline on Clinical Development of Fixed Combination Medicinal Products*. EMA.

ENCePP Guide on Methodological Standards in Pharmacoepidemiology (n.d.). 5.6. Pragmatic trials and large simple trials. http://www.encepp.eu/standards_and_guidances/methodologicalGuide5_6.shtml.

Eric McCoy, C. (2017). Understanding the intention-to-treat principle in randomized controlled trials. *Western Journal of Emergency Medicine* 18 (6): 1075–1078.

Evans, S.R. (2010). Fundamentals of clinical trial design. *Journal of Experimental Stroke Translational Medicine* 3 (1): 19–27.

Faber, J. and Fonseca, L.M. (2014). How sample size influences research outcomes. *Dental Press J Orthod.* 19 (4): 27–29.

FDA (2016a). *Determining the Extent of Safety Data Collection Needed in Late-Stage Premarket a Post-approval Clinical Investigations. Guidance for Industry.* U.S. Department of Health and Human Services Food and Drug Administration.

FDA *Non-Inferiority Clinical Trials to Establish Effectiveness.* Guidance for Industry. FDA 2016b.

FDA. *Multiple Endpoints in Clinical Trials.* Guidance for Industry. FDA 2017. Office of Biostatistics in the Office of Translational Sciences in the Center for Drug Evaluation and Research at the Food and Drug Administration. https://www.fda. gov/media/102657/download.

FDA (2018a). *Surrogate Endpoint Resources for Drug and Biologic Development.* FDA https://www.fda.gov/drugs/development-resources/ surrogate-endpoint-resources-drug-and-biologic-development.

FDA (2018b). Clinical trial endpoints for the approval of cancer drugs and biologics guidance for industry.

FDA (2020). *Type 2 Diabetes Mellitus: Evaluating the Safety of New Drugs for Improving Glycemic Control Guidance for Industry.* Guidance for Industry. U.S. Department of Health and Human Services Food and Drug Administration.

FDA CDER (2020a). FDA CDER fast track approval list. https://www.fda.gov/drugs/ nda-and-bla-approvals/fast-track-approvals.

FDA CDER (2020b). FDA CDER Breakthrough approvals list. https://www.fda.gov/ drugs/nda-and-bla-approvals/breakthrough-therapy-approvals.

Feller, S. (2015). One in Four Cancer Trials Fails to Enroll Enough Participants. https://www.upi.com/Health_News/2015/12/30/One-in-four-cancer-trials-fails-to-enroll-enough-participants/2611451485504.

Fogel, D.B. (2018). Factors associated with clinical trials that fail and opportunities for improving the likelihood of success: a review. *Contemporary Clinical Trials Communications* 11: 156–164.

Carrol Gamble A. Krishan, D. Stocken et al. Guidelines for the content of statistical analysis plans in clinical trials. *JAMA* 2017;318(23):2337–2343. doi:https://doi. org/10.1001/jama.2017.18556.

Ghadessi, M., Tang, R., Zhou, J. et al. (2020). A roadmap to using historical controls in clinical trials – by Drug Information Association Adaptive Design Scientific Working Group (DIA-ADSWG). *Orphanet Journal of Rare Diseases* 15: 69.

Glass, H.E., DiFrancesco, J.J., Glass, L.M., and Tran, P. (2015). Are phase 3 clinical trials really becoming more complex? *Therapeutic Innovation and Regulatory Science* 49: 852–860.

Gnanasakthy, A., Mordin, M., Clark, M. et al. (2012). A review of patient-reported outcome labels in the United States: 2006–2010. *Value in Health* 15: 437–442.

Gnanasakthy, A., DeMuro, C., Clark, M. et al. (2016). Approved by the office of hematology and oncology products of the US Food and Drug Administration (2010–2014). *Journal of Clinical Oncology* 34 (16): 1928–1934.

Gourgari, E., Wilhelm, E.E., Hassanzadeh, H. et al. (2017). A comprehensive review of the FDA-approved labels of diabetes drugs: Indications, safety, and emerging cardiovascular safety data. *Journal of Diabetes and its Complications* 31 (12): 1719–1727.

Harrer, S., Shah, P., Antony, B., and Hu, J. (2019). Artificial Intelligence for Clinical trial design. *Trends in Pharmacological Sciences* 40 (8): 577–591.

Humphrey, R.W., Brockway-Lunardi, L.M., corresponding author et al. (2011). Opportunities and challenges in the development of experimental drug combinations for cancer. *Journal of the National Cancer Institute* 103 (16): 1222–1226.

Thomas J Hwang, Paolo A Tomasi, Florence T Bourgeois. Delays in completion and results reporting of clinical trials under the paediatric regulation in the European Union: a cohort study. *PLoS Medicine* 2018;15(3) doi: https://doi.org/10.1371/journal.pmed.1002520.

ICH E9, US FDA *Statistical Principles for Clinical Trials*. Guidance for Industry. US FDA 1998.

ICH E9(R1), US FDA (2017). *Statistical Principles for Clinical Trials: Addendum: Estimands and Sensitivity Analysis in Clinical Trials*. US FDA.

ICH Topic E 10 (2001). *Choice of Control Group in Clinical Trials*. European Medicines Agency.

ICH, US FDA E17 *General Principles for Planning and Design of Multiregional Clinical Trials*. Guidance for Industry. US FDA. 2018.

Inan, O.T., Tenaerts, P., Prindiville, S.A. et al. (2020). Digitizing clinical trials. *NPJ Digital Medicine* 3: 101.

Muhammad Shahzeb Khan, Izza Shahid, Tariq Jamal Siddiqi, Safi U. Khan, Haider J. Warraich, Stephen J. Greene, Javed Butler, and Erin D. Michos. Ten-year trends in enrollment of women and minorities in pivotal trials supporting recent US Food and Drug Administration approval of novel cardiometabolic drugs. *Journal of American Heart Association* 2020;9: e015594. DOI: https://doi.org/10.1161/JAHA.119.015594.

Ladanie, A., Speich, B., Briel, M. et al. (2019). Single pivotal trials with few corroborating characteristics were used for FDA approval of cancer therapies. *Journal of Clinical Epidemiology* 114: 49–59.

Lai, T.L., Lavori, P.W., and Tsang, K.W. (2015). Adaptive design of confirmatory trials: advances and challenges. *Contemporary Clinical Trials* 45, Part A: 93–102.

Large Simple Trials and Knowledge Generation in a Learning Health System: Workshop Summary (2013). *Roundtable on Value and Science-Driven Health Care; Forum on Drug Discovery, Development, and Translation; Board on Health Sciences Policy*; Institute of Medicine. Washington (DC): National Academies Press (US).

Lisovskaja, V. and Burman, C.-F. (2013). On the choice of doses for phase III clinical trials. *Statistics in Medicine* 32: 1661–1676.

Lo, A.W., Siah, K.W., and Wong, C.H. (2019). Machine learning with statistical imputation for predicting drug approvals. *Harvard Data Science Review* 1 (1): https://doi.org/10.1162/99608f92.5c5f0525.

Lyauk, Y.K., Jonker, D.M., and Lund, T.M. (2019). Dose finding in the clinical development of 60 US Food and Drug Administration – Approved drugs compared with learning vs. confirming recommendations. *Clinical Translational Science* 12: 481–489.

Mahajan, R. and Gupta, K. (2010). Adaptive design clinical trials: methodology, challenges and prospect. *Indian Journal of Pharmacology* 42 (4): 201–207.

Virginia Ellen Maher, Alice Kacuba, Yangmin M. Ning, Anthony J. Murgo, Amna Ibrahim, Ann T. Farrell, Patricia Keegan, Robert L. Justice, Richard Pazdur. Special protocol assessments: 10 years of experience in FDA's Office of Hematology and Oncology Products. *Journal of Clinical Oncology* 32, 2014 (suppl; abstr e17511).

Mazumdar, M. and Bang, H. (2007). Sequential and group sequential designs in clinical trials: guidelines for practitioners. *Handbook of Statistics* 27: 491–512.

Mercieca-Bebber R, King MT, Calvert MJ, Stockler MR, Friedlander M. The importance of patient-reported outcomes in clinical trials and strategies for future optimization. *Patient Related Outcome Measures* 2018;9:353–367. doi: https://doi.org/10.2147/PROM.S156279.

Sile F. Molloy and Patricia Henle. Monitoring clinical trials: a practical guide. *Tropical Medicine and International Health.* volume 21 no 12 pp 1602–1611 2016.

Mulkey F, Theoret MR, Keegan P, et al. Comparison of iRECIST versus RECIST V.1.1 in patients treated with an anti-PD-1 or PD-L1 antibody: pooled FDA analysis. *Journal for ImmunoTherapy of Cancer* 2020;8:e000146. doi:https://doi.org/10.1136/jitc-2019-000146.

Bassel Nazha, Manoj Mishra, Rebecca Pentz, and Taofeek K. Owonikoko. Enrollment of racial minorities in clinical trials: old problem assumes new urgency in the age of immunotherapy. *American Society of Clinical Oncology Educational Book* 39 (2019) 3–10. DOI: https://doi.org/10.1200/EDBK_100021.

Polley, M.-Y.C., Korn, E.L., and Freidlin, B. (2019). Phase III precision medicine clinical trial designs that integrate treatment and biomarker evaluation. *JCO Precision Oncology* 3: 1–9.

Ranganathan, P., Pramesh, C.S., and Aggarwal, R. (2016). Common pitfalls in statistical analysis: Intention-to-treat versus per-protocol analysis. *Perspectives in Clinical Research* 7 (3): 144–146.

Rivera, S.C., Kyte, D.G., Aiyegbusi, O.L. et al. (2019). The impact of patient-reported outcome (PRO) data from clinical trials: a systematic review and critical analysis. *Health Quality of Life Outcomes* 17 (1): 156.

van Rosmalen, J., Dejardin, D., van Norden, Y. et al. (2018). Including historical data in the analysis of clinical trials: Is it worth the effort? *Statistical Methods in Medical Research* 27 (10): 3167–3182.

Schnell, O., Standl, E., Cos, X. et al. (2020). Report from the fifth cardiovascular outcome trial (CVOT) summit. *Cardiovascular Diabetology* 19: 47. https://doi.org/10.1186/s12933-020-01022-7.

Sheiner, L.B. (1997). Learning versus confirming in clinical drug development. *Clinical Pharmacology Therapeutics* 61: 275–291.

Shenoy, P. (2016). Multi-regional clinical trials and global drug development. *Perspective in Clinical Research* 7 (2): 62–76.

Sturges, P. (2019). Utilization of real-world data to enhance recruitment and retention of clinical research participants. *Clinical Researcher* 33 (7). https://acrpnet.org/2019/08/13/utilization-of-real-world-data-to-enhancerecruitment-and-retention-of-clinical-research-participants/

Tenhunen, O., Lasch, F., Schiel, A., and Turpeinen, M. (2020). Single-arm clinical trials as pivotal evidence for cancer drug approval: a retrospective cohort study of centralized european marketing authorizations between 2010 and 2019. *Therapeutic Innovations in Oncology* 108 (3): 653–660.

Thorlund, K., Dron, L., Park, J.J.H., and Mills, E.J. (2020). Synthetic and external controls in clinical trials – A primer for researchers. *Clinical Epidemiology* 12: 457–467.

Timpe, C., Stegemann, S., Barrett, A., and Mujumdar, S. (2020). Challenges and opportunities to include patient-centric product design in industrial medicines development to improve therapeutic goals. *British Journal of Clinical Pharmacology* 86 (10): 2020–2027.

Tyndall, A., Du, W., and Breder, C. (2017). The target product profile as a tool for regulatory communication: advantageous but underused. *Nature Reviews Drug Discovery* 16: 156.

US FDA. *Establishment and Operation of Clinical Trial Data Monitoring Committees.* Guidance for Clinical Trial Sponsors. US FDA 2006.

US FDA. *Integrated Summaries of Effectiveness and Safety: Location Within the Common Technical Document.* Guidance for Industry. US FDA 2009.

US FDA (2013a). *Good Review Practice: Clinical Review of Investigational New Drug Applications.* US FDA.

US FDA. *Oversight of Clinical Investigations — A Risk-Based Approach to Monitoring.* Guidance for Industry. US FDA 2013b.

US FDA (2018). *Framework for FDA's Real-World Evidence Program.* US FDA.

US FDA. *Considerations for the Inclusion of Adolescent Patients in Adult Oncology Clinical Trials.* Guidance for Industry. US FDA. 2019a. https://www.fda.gov/media/113499/download.

US FDA. *Adaptive Designs for Clinical Trials of Drugs and Biologics.* Guidance for Industry. US FDA 2019b. https://www.fda.gov/media/78495/download.

US FDA. *Enrichment Strategies for Clinical Trials to Support Determination of Effectiveness of Human Drugs and Biological Products.* Guidance for Industry. US FDA 2019c. https://www.fda.gov/media/121320/download.

US FDA. *Enhancing the Diversity of Clinical Trial Populations — Eligibility Criteria, Enrollment Practices, and Trial Designs.* Guidance for Industry. US FDA. 2020.

US FDA (2021). *Table of Surrogate Endpoints That Were the Basis of Drug Approval or Licensure.* US FDA https://www.fda.gov/drugs/development-resources/table-surrogate-endpoints-were-basis-drug-approval-or-licensure.

Wasserstein, R.L., Schirm, A.L., and Lazar, N.A. (2019). Moving to a world beyond "p<0.05". *The American Statistician* 73: 1–19. https://doi.org/10.1080/0003130 5.2019.1583913.

WHO (n.d.). *Design of Vaccine Efficacy Trails to be used During Public Health Emergencies – Points of Considerations and Key Principles.* WHO.

Chi Heem Wong, Kien Wei Siah, Andrew W Lo. Estimation of clinical trial success rates and related parameters. *Biostatistics* 2019;20(2):273–286. doi: https://doi.org/10.1093/biostatistics/kxx069.

Chapter Self-Assessments: Check Your Knowledge

Questions:
- Discuss the focus and timing of Phase 3 drug development?
- Under what conditions might a drug developer choose to study more than one dose in a Phase 3 program?
- What is the primary driver for planning of the Phase 3 program? Describe in detail.
- Discuss situations where more than one endpoint are needed to support a Phase 3 clinical trial?

Answers:
- Phase 3 is considered as the confirmation phase in a drug development journey, where the candidate drug's efficacy is confirmed, the drug safety is further demonstrated, and the benefit-risk profile is defined. Planning of the Phase 3 program usually starts early in the clinical development process, often as early as when a compound is considered to enter clinical development. The targeted drug label and desired benefit-risk profile are the ultimate end goals for a compound under the evaluation and drivers for many decisions in clinical development planning.

- As a drug's safety profile might not be established adequately by Phase 2 due to its limited study sizes and treatment duration, a Phase 3 program with only one dose could lead to an increased risk of failure in identifying the dose with the right benefit-risk balance. Therefore, many drug developers choose to design a multiple-dose Phase 3 program, where the optimal dose is further identified or confirmed. A multiple-dose Phase 3 study provides flexibility in assessing and balancing the drug risk-benefit profile across different dose levels to increase the program's probability of success (POS). The drug developer weights a higher development cost burden to increase the program's POS.
- The driver of the Phase 3 planning is the target drug label, which often determines the types of patient populations enrolled in the development. To satisfy the usual minimal two studies in providing substantial efficacy evidence for approval, two Phase 3 studies can be conducted in the same targeted patient population. As pivotal studies supporting the approval, the patient population needs to be clearly defined and carefully selected via a list of patient inclusion and exclusion criteria. The enrolled patients should include sufficient diversity and be representative of the targeted patient population including disease severity and population demographic characteristic diversity such as sex, age, race, ethnicity, region distributions.
- For some disease areas, more than one efficacy endpoint is needed to indicate different aspects of the clinical benefit. Multiple efficacy endpoints could represent the needs of improvement in different disease attributes, which are equally important and individually essential to demonstrate a drug's clinical benefit. These multiple endpoints can either be considered as co-primary endpoints in a clinical trial. That is, a study is considered statistically significant only if each of these co-primary efficacy endpoints is statistically significant. When combining multiple endpoints into a single score can better demonstrate a drug's clinical benefit, the combined endpoint is referred as a composite endpoint. Composite endpoints are used when a drug benefits different disease attributes that are more related or sometimes competing to occur, or each component is relatively infrequent.

Quiz:

1 Fill in the blanks with the answer that best completes the sentence. Compared to a Phase 2 program, Phase 3 studies are usually _____ and often with _____ treatment exposure per patient.

   A smaller, greater

   B smaller, longer

   C larger, greater

   D larger, longer

   E none are correct

2 True or False. Phase 3 studies are frequently powered at 90–95%, which are higher than the usual 80–90% power requirements of Phase 2 studies.

3 Fill in the blanks with the answer that best completes the sentence. Endpoints that can directly, accurately, and objectively reflect _____ benefit under the treatment exposure is usually preferred as the primary endpoint and other endpoints considered as _____ to provide support.

   A positive, complimentary
   B optimal, complimentary
   C clinical, secondary
   D meaningful, secondary

4 What is commonly referred to as the "gold standard" for Phase 3 clinical study designs?

   A Superiority trial design
   B Randomized controlled trial (RCT) design
   C Non-inferiority (NI) trial design
   D Equivalence trial design

10

Phase 4, Special Populations and Post-marketing
Jeffrey S. Barrett

Aridhia Digital Research Environment

Phase 4 loosely refers to the phase of development that occurs post-approval. The scope of activity during this phase is typically dictated by regulators in the guise of post-approval activities that the sponsor would have to agree to begin market access based on the NDA submission. Phase 4 trials are often conducted as post-marketing efforts to further evaluate the characteristics of the new drug regarding safety, efficacy, new indications for additional patient populations, and new formulations. While life cycle management may be at the core of the new formulations and indications efforts, Phase 4 clinical trials in special populations are often conducted as expected fulfillment of regulatory requirements and a condition of approval and market access. Other post-marketing trials, including drug surveillance or real-world evidence trials, are conducted to confirm safety and effectiveness in real-world patients that likely fall outside of the range of patients considered for Phase 3 trials (e.g. outside perhaps narrow inclusion-exclusion criteria) and may be conducted with guidance from the payer community (e.g. healthcare providers and insurance companies). As the healthcare industry moves more to a "value-based" economy, these will likely represent a more commonplace component of Phase 4 efforts.

The term "Special Populations" originated as a means to define patient subpopulations outside the "mainstream" patient population – typically defined as the targeted (or "to be enrolled") Phase 3 population. Special populations were loosely defined as those within which drug administration may be contraindicated or in which dosing adjustments may be warranted. The most considered special populations include geriatrics, pediatrics, pregnant women, and organ impairment (typically, renal and hepatic impairment). The implication of this designation has an impact on both screening criteria and patient exclusions (e.g. severe hepatic or renal functions and patients less than 18 years of age). Regulatory concerns regarding the "extremes" of a population within which little

Fundamentals of Drug Development, First Edition. Edited by Jeffrey S. Barrett.
© 2022 John Wiley & Sons, Inc. Published 2022 by John Wiley & Sons, Inc.
Companion website: www.wiley.com/go/Barrett/FundamentalsDrugDevelopment

data was collected also drove the need for separate clinical investigations of such patients. The actual clinical investigation of special populations typically focuses on safety with the intention to assess the extent to which a subpopulation deviates from the mainstream patient population. The actual evidence for differences (if they exist) tends to be pharmacokinetic (PK)-centric with dosing adjustments, if warranted, guided by PK metrics. PK targets follow equivalence criteria with mainstream "normal" reference population as the comparator. Pharmacometric strategies can facilitate optimal designs, propose meaningful labeling, and encourage the engagement of scientific-driven dialogue with regulators. As such, these approaches are tremendously beneficial for decision-making and have a well-established return on investment. It should also be appreciated that the conventional special population groups as listed above do not always constitute the full range of special populations of interest. Recent emphasis on personalized medicine approaches also suggests that genetic polymorphisms in drug-metabolizing enzymes, as well as other genetic differences, can also reflect populations in whom additional dosing guidance is warranted (e.g. coumadin and prasugrel) and likewise can become "special populations" warranting separate clinical evaluation in Phase 4, post-approval.

Definitions – Why are they "Special?"

Following registration approval, most drugs are used by people from a broad spectrum of ages, races, and ethnic groups, as well as both sexes. Yet historically, most early phase clinical studies have been almost exclusively conducted in adults (mostly white males) aged 18–65 years. Special populations typically refer to categories of patients excluded from early-phase testing for a variety of reasons, mostly related to the risk of early exposure of an untested agent. Special populations typically include women, children (pediatrics), the elderly (geriatrics), pregnancy (obstetrics), and patients with concurrent disease states. Each of these populations has specific considerations that must be considered in relation to study design and regulations. Given the vulnerability of some subpopulations and the challenges and cost of performing clinical studies in these populations, cutting-edge approaches are needed to effectively develop evidence-based and individualized drug dosing regimens. Solutions to address some of these concerns and a proposal for a path forward using more quantitative approaches has recently been put forth (Krekels et al. 2017). The authors propose 5 key issues to support and expedite the development of drug dosing regimens in these populations using model-based approaches: (1) model development combined with proper validation procedures to extract as much valid information from available study data as possible, with limited burden to patients and costs; (2) integration of existing data and the use of prior pharmacological and physiological knowledge in study design and data analysis, to further develop knowledge and avoid unnecessary or unrealistic (large) studies in vulnerable populations; (3) clinical proof-of-principle in a prospective evaluation of a developed drug dosing regimen, to confirm that a newly proposed regimen indeed results in the desired outcomes in terms of drug concentrations, efficacy, and/or safety; (4) pharmacodynamics studies in addition to pharmacokinetics studies for drugs for which a difference in disease progression and/or in exposure-response relation is anticipated compared to the reference population; and (5) additional efforts to implement developed dosing regimens in clinical practice once drug pharmacokinetics and pharmacodynamics have been characterized in special patient populations.

In the end, these populations are considered special because they are associated with additional risk with respect to conventional dosing recommended for mainstream patients considered as the primary or target indication for the new drug. This additional risk is typically linked to a presumed shift in the therapeutic window of the drug in this population. The challenge to the drug developer is to decide if this subpopulation is still likely to benefit from the drug if dosing adjustments can be made to again achieve exposures that are both safe and efficacious. If the answer to this question is yes, it likely implies that the sponsor will proceed with a separate investigation, often post-approval, as part of a Phase 4 commitment.

Physiology and Regulatory Implications

FDA regulations require sponsors to present a summary of safety and effectiveness data by demographic subgroups for age, gender, and race/ethnicity (OMB Policy Directive 15 standard). This assessment provides the basis of analysis of whether modifications of dose or dose intervals are needed to ensure safe and effective use (Federal Register, 2016). Part of the challenge for sponsors is to adjust their tolerance for risk as the drug candidate moves through stages of development. Inclusion/exclusion criteria define boundaries in many relevant demographic categories that define these "special populations." For instance, age requirements during Phase 1 are typically bounded between 18 years at the lower end (hence excluding pediatrics) with an upper bound often at 55–65 years. Also, women who could possibly become pregnant are typically excluded, and BMI is typically capped at 25–30 kg/m^2 (hence obese subjects are excluded). Table 10.1 provides a simple comparison of common Phase 1 vs. Phase 3 demographic boundaries. It is reasonable then to ask what percentage of the population is excluded from drug development clinical research simply based on these criteria. If one uses the US population as an

Table 10.1 Phase 1 vs. Phase 3 demographics.

Demographic factor	Phase 1 subjects	Phase 3 patients
Age	Usually 18–40	Higher % of older patients
Pregnancy	Excluded	Excluded
Race	Predominantly Caucasian	All races
Organ (hepatic/ renal) function	Normal	Often have at least minor degrees of impairment
Weight	BMI usually lower than 25 kg/m^2	Greater % of obese

example and the census demographic statistics, it concludes that approximately 38% of the population is excluded from drug development research simply based on inclusion/exclusion criteria (Owen 2010).

The integrated summary of safety (ISS) and integrated summary of efficacy (ISE) are vital components of a successful submission for regulatory approval in the pharmaceutical industry. ISS and ISE allow reviewers to easily compare individual outcomes, tracking subjects' results across the entire clinical development lifespan of the investigational product. ISS and ISE are crucial aspects of New Drug Applications (NDAs), uniquely required by FDA (United States) regulation. These integrated analyses are not strictly required for NDA submissions to the MHLW (Japan) and the EMA (EU); however, MHLW and EMA submissions must follow the common technical document (CTD) format, which contains sections in line with ISS and ISE. Whether for regulatory submission or not, integrated summaries are useful for discovering rare and/or unexpected trends in patient subgroups, improving the precision of results with a larger population size by integrating studies, comparing variation in study results to assess the risk and benefits, and reaching a strong and defendable statistical conclusion. With respect to special populations and Phase 4 commitments in general, the ISS and ISE are often used by regulators as the impetus behind requesting that a sponsor conduct such a trial based on the identification of such groups as having differing safety or efficacy from mainstream patients.

Organ (Hepatic and Renal) Impairment

Kidney and liver are the main organs involved in the elimination of drugs. Both have a metabolic and a direct excretory capacity, although the first is predominant for drugs eliminated by the liver, while the most frequent mechanism of renal

Table 10.2 Regulatory-based categories defining stages of renal function and renal impairment.

Group	Description	Estimated creatinine clearance (ml/min)
1	Normal renal function	>80
2	Mild renal impairment	50–80
3	Moderate renal impairment	30–50
4	Severe renal impairment	<30
5	ESRD	Requires dialysis

ESRD = End-stage renal disease.

clearance is direct excretion of the unchanged drug or its circulating metabolites. To optimize drug dosing, it is critical to understand how various intrinsic and extrinsic factors affect systemic exposure of the drug and the response. In general, the elimination capacity of the kidney is lower than that of the liver because of the smaller size and associated blood flow. Renal excretion can be limited by the glomerular filtration rate (GFR) in case of passive excretion or by the transporter capacity of the total renal blood flow in case of active secretion (see Table 10.2 for regulatory-based renal function categories). Chronic kidney disease (CKD) or renal impairment can affect the PK characteristics of a therapeutic drug and its metabolites and therefore is one of the most important intrinsic factors that can affect a patient's response to drugs. During drug development, it is critical to understand how renal impairment can affect a drug's pharmacokinetics so that appropriate dosing recommendations can be included in the label. The liver is involved in the clearance of many drugs through a variety of oxidative and conjugative metabolic pathways and/or through biliary excretion of unchanged drugs or metabolites. Alterations of these excretory and metabolic activities by hepatic impairment can lead to drug accumulation or, less often, failure to form an active metabolite. Likewise, hepatic impairment often becomes a critical factor in the designation of appropriate dosing adjustments during regulatory submission, in the proposed drug label and in the prescription to patients suffering from this condition.

With age, there is a decline in total nephron size and number, tubulointerstitial changes, glomerular basement membrane thickening, and increased glomerulosclerosis (see Figure 10.1). This age-related histologic appearance is frequently described as nephrosclerosis, and it describes a combination of two or more histologic features: any global glomerulosclerosis, tubular atrophy, interstitial fibrosis >5%, and any arteriosclerosis (O'Sullivan et al. 2017). The effect of age on change in the level of renal function appears to be more complicated than the effect on

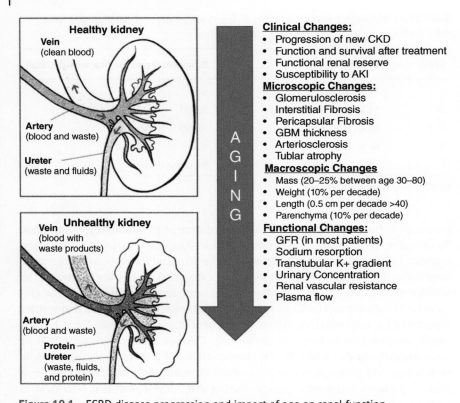

Figure 10.1 ESRD disease progression and impact of age on renal function.

progression to end-stage renal disease (ESRD). Older age is a risk factor for the development of CKD, most likely reflecting both lower mean levels of eGFR and higher rates of renal function loss in older compared with younger patients with an estimated glomerular filtration rate (eGFR) ≥60 ml/min/1.73 m² (Prakash and O'Hare 2009).

The severity of hepatic dysfunction can be defined using a number of validated scales. The Child-Pugh classification has been widely used in clinical practice to categorize chronic cirrhotic patients based on the severity of liver function impairment and is recommended by FDA (FDA Guidance 2003). This classification system consists of five components that assess the degree of impairment, including laboratory parameters: serum bilirubin, serum albumin, and prothrombin time, and clinical symptoms: the presence of encephalopathy and presence of ascites (see Table 10.3). Based on disease severity, patients are categorized into groups defined as mild (class A), moderate (class B), or severe (class C), corresponding to 5–6, 7–9, and 10–15 scores, respectively.

Table 10.3 Regulatory-based categories defining stages of hepatic function and hepatic impairment with 1 and 2-year survival rates.

Child-Pugh score	Grade of dysfunction	1-year survival	2-year survival
5–6	Grade A or well compensated	100%	85%
7–9	Grade B or significant functional compromise	80%	60%
10–15	Grade C or decompensated liver disease	45%	35%

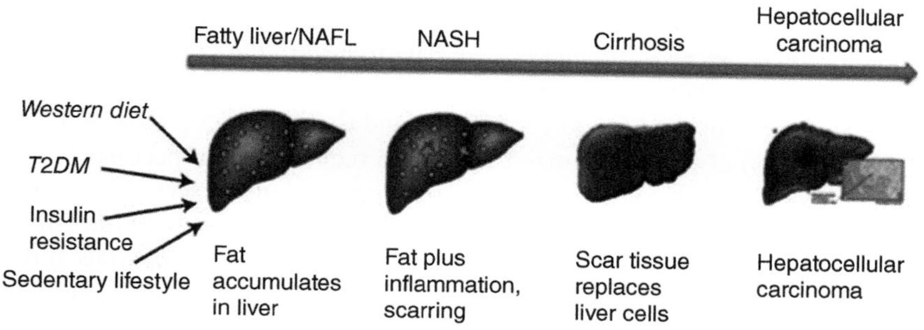

Figure 10.2 Common comorbidities associated with hepatic impairment and typical disease progression associated with fatty-liver disease (Thrasher and Abdelmalek 2016).

Comorbidity affects the prognosis of cirrhosis patients. Measures of a patient's total burden of comorbidity are important for epidemiologic studies and for clinical use (Jepsen 2014). Nonalcoholic fatty liver disease (NAFLD) is now the most common chronic liver disease in the developed world and affects about 25–30% of adults in the United States and 30% of veterans who receive care in the VHA system. Comprised of a spectrum of disease severity, NAFLD ranges from simple steatosis to nonalcoholic steatohepatitis ([NASH] steatosis with hepatocyte inflammation, necrosis, and fibrosis) (see Figure 10.2). Patients with NAFLD have significantly increased mortality because of both hepatic (such as cirrhosis and hepatocellular carcinoma [HCC]) and extrahepatic complications (such as metabolic syndrome [MetS], cardiovascular disease [CVD], and malignancy) (Glass et al. 2019).

Pregnant Women

Pregnant women represent an important segment of the population, with over 6 million pregnancies occurring per year, based on recent national vital statistics (Curtin et al. 2015). Pregnant women may have chronic conditions, such as diabetes, seizure disorders, or asthma, that need to be treated during pregnancy, or pregnant women may develop acute or serious medical conditions during pregnancy that require treatment. As nearly half of all pregnancies in the United States may be unintended, potential inadvertent exposure to drugs and biological products in pregnancy is possible if a woman is exposed to a drug when she is not aware, she is pregnant. Therefore, there is an important need for safety information on product exposure during pregnancy. During clinical development of most drugs and biological products, pregnant women are actively excluded from trials, and if pregnancy does occur during a trial, the usual procedure is to discontinue treatment and monitor the women to assess pregnancy outcomes. Consequently, at the time of a drug or biological product's initial marketing, except for drugs and biological products developed to treat conditions unique to pregnancy, there are no or limited human data to inform the safety of a drug or biological product taken during pregnancy.

Several sections of the Federal Food, Drug, and Cosmetic Act (FD&C Act) (21 U.S.C. 355(o)(3)) of 2007 authorize the FDA to require certain post-marketing studies or clinical trials for prescription drugs approved under section 505(b) of the FD&C Act and biological products approved under the section of the Public Health Service Act (42 U.S.C. 262). FDA can require such studies or trials at the time of approval to assess a known serious risk related to the use of the drug, to assess a signal of serious risk related to the use of the drug, or to identify an unexpected serious risk when available data indicates the potential for a serious risk. FDA can also require such studies or trials after approval if it becomes aware of new safety information. Post-approval studies using data collected in pregnancy registries may be required to assess potential serious risks to the pregnancy that may affect the health of the fetus or the woman due to drug or biological product use during pregnancy. However, gaps in safety data in pregnant women still exist. Specific critical factors in evaluating the effects of product exposure in human pregnancies may include, but are not limited to, the following (FDA Guidance 2019a): a detailed description of the adverse pregnancy outcome, a detailed description of the exposure including the specific medication, the dose, frequency, route of administration, and duration, the timing of the exposure in relation to the gestational age, the maternal age, medical and pregnancy history, and use of concomitant medications, supplements, and other substances, and exposures to known or suspected environmental teratogens. Quite often, the exact nature of these clinical trials in pregnancy commitments is discussed at the sponsor's End-of-Phase 2 meeting with FDA and equivalent meetings with global regulatory authorities to ensure that the nature of the final submission is agreed upon in advance with no surprises at the time of filing.

Pregnancy-induced maternal physiological changes may affect gastrointestinal function and hence drug absorption rates (see Table 10.4 and Figure 10.3 for a comparison of prescribing vulnerability and pharmacokinetic factors impacting

Table 10.4 Vulnerability of pediatrics and pregnancy for prescription of new agents without general dosing guidance considerations – Pharmacokinetic factors.

PK factor(s)	Pregnancy	Children[a]
Distribution	• Increased body water volume and fat • Decreased serum protein levels • Increased volume of distribution of drugs and tissue distribution of fat-soluble drugs • Increased free fraction due to decreased protein binding	• At birth, a neonate is about 80% water. Water-soluble drug dosing in a neonate requires a higher mg/kg amount for comparable plasma concentration (e.g. for drugs like morphine, gentamicin, and vancomycin). • Protein concentrations and affinity are decreased in the first year of life; newborns exhibit lowered binding to drugs like penicillin and phenytoin.
Metabolism	• Decreased hepatic blood flow and synthesis of enzymes • Competition for hepatic enzymes slows metabolism of some drugs • Induction of some hepatic enzymes can increase the rate of metabolism for certain drugs • Some metabolic activity takes place in the placenta – implications unclear	• Dealkylation is normal in neonates. • Conjugation with acetyl coenzyme A is reduced in the first month of life. • Glucuronidation is normalized by 3-6 mo. • Oxidation is reduced and normalizes to the adult process by 6-12 mo. • Hydroxylation and esterification activity are reduced. • CYP activity matures over time; levels can fluctuate though eventually reach adult expression and activity.
Elimination	• Renal clearance of drugs increases mainly due to increased GFR • Tubular resorption of substances increases, counteracting GFR increase • Renal clearance difficult to predict likewise	• GFR, tubular secretion, and tubular reabsorption are all decreased in the newborn. • GFR in the newborn is about 40 ml/min/173 m^2. It approaches adult values of 100 ml/min/173 m^2 at about 3 mo of age. Afterwards, it may surpass adult values. • Tubular secretion depends on renal blood flow and increases until age 6–12 mo. This can decrease clearance of penicillin's, aminoglycosides, and cephalosporins.

[a] Children in this context refers to all pediatric populations from neonates to adolescents.

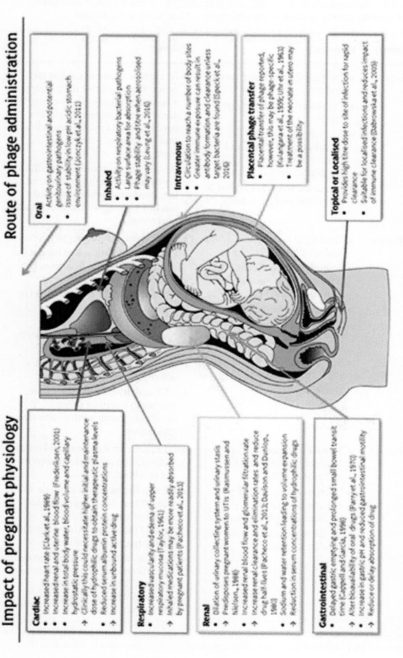

Impact of pregnant physiology

Route of phage administration

Cardiac
- Increased heart rate (Clarke et al., 1989)
- Increase in renal and uterine blood flow (Frederiksen, 2001)
- Increase in total body water, blood volume and capillary hydrostatic pressure
- ↑ Clinically this could necessitate higher initial and maintenance dose of hydrophilic drugs to obtain therapeutic plasma levels
- ↓ Reduced serum albumin protein concentrations
- ↑ Increase in unbound active drug

Respiratory
- Increased vascularity and edema of upper respiratory mucosa (Taylor, 1961)
- ↑ Inhaled medications may be more readily absorbed by pregnant patients (Pacheco et al., 2013)

Renal
- Dilation of urinary collecting system and urinary stasis predisposes pregnant women to UTIs (Rasmussen and Nielsen, 1988)
- Increased renal blood flow and glomerular filtration rate
- ↑ Increase renal clearance and elimination rates and reduce drug half-lives (Pacheco et al., 2013; Davison and Dunlop, 1980)
- Sodium and water retention leading to volume expansion
- ↓ Reduction in serum concentrations of hydrophilic drugs

Gastrointestinal
- Delayed gastric emptying and prolonged small bowel transit time (Cappell and Garcia, 1998)
- ↑ Alter bioavailability of oral drugs (Parry et al., 1970)
- ↑ Increase in gastric pH and reduced gastrointestinal motility
- ↓ Reduce or delay absorption of drug

Oral
- Activity on gastrointestinal and potential genitourinary pathogens
- Issue of stability in low pH acidic stomach environment (Jonczyk et al., 2011)

Inhaled
- Activity on respiratory bacterial pathogens
- Large surface area for absorption
- Phage stability and size when aerosolised may vary (Leung et al., 2016)

Intravenous
- Circulation to reach a number of body sites
- Greater immune exposure can result in antibody formation and clearance unless target bacteria are found (Speck et al., 2016)

Placental phage transfer
- Placental transfer of phage reported, however, this may be phage-specific (Kulangara et al., 1959; Uhr et al., 1962)
- Treatment of the neonate in utero may be a possibility

Topical or Localised
- Provides high titre dose to site of infection for rapid clearance
- Suitable for localised infections and reduces impact of immune clearance (Dabrowska et al., 2005)

Figure 10.3 Physiological changes during pregnancy and their impact on drug pharmacokinetics and consideration of these factors in the application of antenatal phage therapy. This figure includes a licensed image obtained by the authors.

doing requirements for pediatrics and pregnancy). Ventilatory changes may influence the pulmonary absorption of inhaled drugs. As the glomerular filtration rate usually increases during pregnancy, renal drug elimination is generally enhanced, whereas hepatic drug metabolism may increase, decrease, or remain unchanged. A mean increase of 81 in total body water alters drug distribution and results in decreased peak serum concentrations of many drugs. Decreased steady-state concentrations have been documented for many agents because of their increased clearance. Pregnancy-related hypoalbuminemia, leading to decreased protein binding, results in increased free drug fraction. However, as more free drug is available for either hepatic biotransformation or renal excretion, the overall effect is an unaltered free drug concentration. Since the free drug concentration is responsible for drug effects, the above-mentioned changes are probably of no clinical relevance. The placental and fetal capacity to metabolize drugs together with physiological factors, such as differences in acid-base equilibrium of the mother vs. the fetus, determine the fetal exposure to the drugs taken by the mother. As most drugs are excreted into the milk by passive diffusion, the drug concentration in milk is directly proportional to the corresponding concentration in maternal plasma. The milk to plasma (M:P) ratio, which compares milk with maternal plasma drug concentrations, serves as an index of the extent of drug excretion in the milk. For most drugs, the amount ingested by the infant rarely attains therapeutic levels.

Clinical trials in pregnant women differ somewhat from those in other special populations in that the stage of pregnancy is a consideration in the study design as well as the duration of treatment. Another key design factor is whether the agent is intended to treat a condition of pregnancy or another disease indication in which pregnancy is simply a confounder. If the former objective is intended, duration of use is a consideration as well as complimentary biomarker or clinical event observation consistent with the intended action of the drug. If the latter condition is relevant, the trial is likely to be more PK-centric as the other special population trials. In general, the third trimester is often avoided unless that is the clinical window for which the drug is intended.

Pediatrics

Ethical concerns have historically impeded early clinical studies in the pediatric population. Thus, clinical pharmacokinetic and pharmacodynamic studies in the pediatric population did not begin until the 1970s. The FDA Modernization Act (FDAMA 1997) and the Pediatric Rule (1998) have been driving forces for the conduct of pediatric studies in more modern times. These studies have demonstrated the existence of many PK and some pharmacodynamic differences among the pediatric population. Traditional studies demonstrated that PK parameters including half-life, apparent volume of distribution (Vd), and total plasma clearance vary

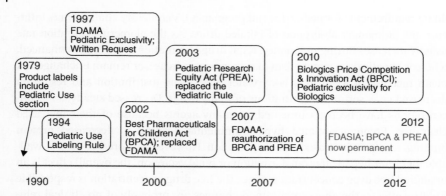

Figure 10.4 Timeline of the changes to the regulations related to pediatric product development (reprinted from Grimsrud et al. 2015).

among different age groups even when normalized by body weight. These findings were supported by population analyses across broad age ranges, which found that age, in addition to body size, is an important determinant of PK parameters in the pediatric population. Age dependency is a function of body composition, organ functions, the ontogeny of drug biotransformation pathways, disease progression, pharmacological receptor functions, and appears to be especially important during the first two years of life.

The result of the regulatory intervention and guidance has been that sponsors embrace the need for pediatric drug development to a much greater extent than in the past and that proactive PIPs (Pediatric Investigation Plans) and PSP (Pediatric Study Plans) are deliverables for early development project teams and not an afterthought prior to NDA submission (see Figure 10.4 for timeline of relevant regulations). In most cases, current NDA filings include a pediatric protocol with the commitment from the sponsor that the pediatric trial will be conducted as part of the post-marketing expectations with specific timelines well defined and agreed upon with the relevant regulatory authorities.

Recommendations for Special Population Trials

In special populations, there are often significant deficits in the quality and efficacy of studies used to determine PK, PD, appropriate dosing, application in different age periods, and many questions surrounding the derivation of doses for neonatal to children from adult studies in addition to patients with organ impairment and pregnant women. Current practice for drug sponsors is to submit these protocols for regulatory review so that agreed-upon metrics for study design, enrollment, and other constructs can be agreed upon prior to study conduct, along with some discussion about how the results will be described in the package insert and drug monograph. Regulatory guidance from global authorities provides expectations with regard to the study population inclusion/exclusion criterion, sample size, and duration of study (Grimsrud et al. 2015; Winter et al. 2018) as well as the necessity of when during drug development (pre or post initial application filing) the various studies are expected to be completed. Many of these trials are the subject of negotiation between the sponsor and regulatory authorities, particularly if the special population in question does not represent a population for whom the drug is likely to be indicated for use.

Regarding the actual trials, one of the more difficult aspects for the sponsor to contend with is the actual availability of the subject populations for these trials. While some CROs specialize in maintaining databases of these special populations, the reality is that they are dynamic based on either developmental (age, disease severity) or condition (pregnant or not, obese or not) factors and hence must be closely monitored, assessed, and likely incentivized to participate in such trials, notwithstanding the additional ethical considerations in the case of pediatrics. Many feel that the best solution is more focused recruitment of special populations into Phase 3 clinical trials obviating the need for special population trials in general (Winter et al. 2018). This, too, is not a perfect solution as data suggest that study participation rates for special populations have fallen to levels that could endanger the successful performance of some types of research. Current census data reveals that demographic trends in the United States continue to shift towards an older and Hispanic population with fewer rural citizens. This situation must likewise be reflected in future study populations along with significant educational efforts to explain the benefit of participating in clinical trials.

The actual study designs of special population trials are remarkably similar. They tend to be small in the total number of subjects enrolled with objectives tied to achieving/determining PK metrics that will be used to judge the adequacy of dosing guidance attained from Phase 3 trials in mainstream patients for generalization to the "special population" of interest. Treatment groups are assigned to the various categories that define the population (e.g. pregnancy state, age cut-points defining neonate, infant child, adolescent groups, or stage of renal or hepatic impairment based on biomarker endpoints, roughly mild, moderate, and severe categories. Enrollment sizes within each category are typically small (e.g. $n = 6\text{--}12$ subjects per group) with some allowance based on the availability of the subgroup or likelihood that they will be indicated for the drug (e.g. smaller sample size for severe hepatic impairment or neonates). As the design is PK-centric, dosing duration is typically limited to single-dose administration with acute safety monitoring only though there are exceptions to this generality. Study results typically include comparative differences among PK metrics such as C_{max} (maximum plasma concentration observed), AUC (area under the plasma concentration-time profile) and half-life among the various subgroups and relative to historical data in mainstream patients. Of course, safety indices (commonly associated AEs and any ADRs) are assessed as well and compared across subgroups. Dosing adjustments and/or contraindication may be recommended if PK and/or safety differences are of a magnitude that cannot be managed safely with conventional (approved) dosing guidance.

Real-World Evidence

The 21st Century Cures Act (Cures Act), signed into law on 13 December 2016, was intended to accelerate medical product development and bring new innovations and advances faster and more efficiently to the patients who need them. Among other provisions, the Cures Act added section 505F to the Federal Food, Drug, and Cosmetic Act (FD&C Act), creating a framework for evaluating the potential use of real-world evidence (RWE) to help support the approval of a new indication for a drug already approved under section 505(c) of the FD&C Act or to help support or satisfy drug post-approval study requirements. Section 505F(b) of the FD&C Act defines RWE as "data regarding the usage, or the potential benefits or risks, of a drug derived from sources other than traditional clinical trials" (21 U.S.C. 355g(b)). In developing its RWE program, FDA believes it is helpful to distinguish between the sources of real-world data (RWD) and the evidence derived from that data. Evaluating RWE in the context of regulatory decision-making depends not only on the evaluation of the methodologies used to generate the evidence but also on the reliability and relevance of the underlying RWD; these constructs may raise different types of considerations. For the purposes of this framework, FDA defines RWD and RWE as follows:

- RWD is data relating to patient health status and/or the delivery of healthcare routinely collected from a variety of sources.
- RWE is the clinical evidence about the usage and potential benefits, or risks of a medical product derived from analysis of RWD.

Examples of RWD include data derived from electronic health records (EHRs); medical claims and billing data; data from product and disease registries; patient-generated data, including from in-home-use settings; and data gathered from other sources that can inform on health status, such as mobile devices. RWD sources (e.g. registries, collections of EHRs, administrative and medical claims databases) can be used for data collection and, in certain cases, to develop analysis infrastructure to support many types of study designs to develop RWE, including, but not limited to, randomized trials (e.g. large simple trials, pragmatic clinical trials) and observational studies (prospective or retrospective).

As it pertains to Phase 4 commitments, the nature of RWE that a sponsor could engage in to fulfill this regulatory expectation is varied and encompasses both hybrid design prospective trials and observational trials. For example, certain elements of a clinical trial could rely on the collection and analysis of RWD extracted from medical claims, EHRs, or laboratory and pharmacy databases. A hybrid trial could use RWD for one clinical outcome (e.g. hospitalization, death), while other elements were more traditional (e.g. specified entry criteria, monitoring, and collection of additional study endpoints by dedicated study personnel). FDA will consider these hybrid trial designs to have the potential to generate RWE. Observational studies (i.e. non-interventional clinical study designs that are not considered clinical trials) can be used to identify the population and determine the exposure/treatment from historical data (i.e. data generated prior to the initiation of the study). The variables and outcomes of interest are determined at the time the study is designed. More commonly, these trials are also conducted in collaboration with the payer community and healthcare providers, at least from the definition of endpoints and outcome measures considered perspective.

It should not be viewed that RWE trials are the solution to discrepancies between Phase 3 results and real-world effectiveness, but they are not without their own issues, and there are some (Suvarna 2018) who feel that, at best, they are hypothesis-generating. There has indeed been the occasion of discordance between the results of an observational, real-world study and an randomized control trial (RCT) (Kosiborod et al. 2017). As happened in the US cohort of CVD real, a large retrospective analysis of outcomes in patients on sodium-glucose cotransporter-2 (SGLT-2) inhibitors vs. in patients on other glucose-lowering drugs; the all-cause mortality was reduced by canagliflozin by 62%, which was statistically significant for superiority, but in the RCT, the integrated analysis of the CVOT, the CANVAS Program, the all-cause mortality risk reduction with canagliflozin was only 13% and not significant. A post-hoc comparison of the two studies dissected out the registry data and pointed out reasons why the results seemed exaggerated including time-lag bias (e.g. for patients to have been prescribed SGLT-2 inhibitors, they would first have had to be other glucose-lowering drugs, and then an SGLT-2 inhibitor was added). This meant that some of the benefits attributed to the gliflozin could have been due to the other glucose-lowering drugs prescribed earlier, suggesting that the comparison was likely unfair. Of course, there are positive examples where RWE trials were complimentary and enhanced the findings from RCTs including early examples such as the Salford Lung Study (New et al. 2014) and the ADAPTABLE study from the Patient-Centered Outcomes Research Institute's PCORnet (National Patient-Centered Clinical Research Network) (Johnston et al. 2016). In the end, RWE trials represent a frontier for the pharmaceutical industry, and understanding the manner and mechanism for how best to implement them is still a work in progress. There is plenty of support for their continued evaluation from both healthcare providers (Xia et al. 2019) and the global regulatory community (FDA Guidance 2019b), but the cost of conduct is not trivial and there likely has to be additional incentives for these to become a mainstay of Phase 4 in the future.

References

Curtin, S. C., Abma, J. C., & Kost, K. (2015). 2010 Pregnancy Rates Among US Women. National Center for Health Statistics. Retrieved 15 January 2021.

Federal Register (2016). Vol. 81, No. 190, Notices, Standards for Maintaining, Collecting, and Presenting Federal Data on Race and Ethnicity. https://www.govinfo.gov/content/pkg/FR-2016-09-30/pdf/2016-23672.pdf.

Glass, L.M., Hunt, C.M., Fuchs, M., and Su, G.L. (2019). Comorbidities and nonalcoholic fatty liver disease: the chicken, the egg, or both? *Federal Practitioner* 36 (2): 64–71. PMID: 30867626; PMCID: PMC6411365.

Grimsrud, K.N., Sherwin, C.M., Constance, J.E. et al. (2015). Special population considerations and regulatory affairs for clinical research. *Clin Res Regul Aff.* 32 (2): 47–56. https://doi.org/10.3109/10601333.2015.1001900. PMID: 26401094; PMCID: PMC4577021.

Guidance for Industry (2003). *Pharmacokinetics in Patients with Impaired Hepatic Function: Study Design, Data Analysis, and Impact on Dosing and Labeling.* U.S. Department of Health and Human Services Food and Drug Administration, Center for Drug Evaluation and Research (CDER), Center for Biologics Evaluation and Research (CBER) https://www.fda.gov/media/71311/download.

Guidance for Industry (2019a). Post-approval Pregnancy Safety Studies. U.S. Department of Health and Human Services Food and Drug Administration, Center for Drug Evaluation and Research (CDER), Center for Biologics Evaluation and Research (CBER). May 2019. https://www.fda.gov/media/124746/download.

Guidance for Industry (2019b). Submitting Documents Using Real-World Data and Real-World Evidence to FDA for Drugs and Biologics, U.S. Department of Health and Human Services, Food and Drug Administration Center for Drug Evaluation and Research (CDER) and Center for Biologics Evaluation and Research (CBER). May 2019. https://www.fda.gov/media/124795/download (accessed 7 August 2020).

Jepsen, P. (2014). Comorbidity in cirrhosis. *World Journal of Gastroenterology* 20 (23): 7223–7230. https://doi.org/10.3748/wjg.v20.i23.7223. PMID: 24966593; PMCID: PMC4064068.

Johnston A, Jones WS, Hernandez AF. The ADAPTABLE trial and aspirin dosing in secondary prevention for patients with coronary artery disease. *Current Cardiology Reports* 2016;18(8):81. doi:https://doi.org/10.1007/s11886-016-0749-2.

Kosiborod, M., Cavender, M.A., Fu, A.Z. et al. (2017). Lower risk of heart failure and death in patients initiated on sodium-glucose cotransporter-2 inhibitors versus other glucose-lowering drugs: The CVD-REAL study (Comparative effectiveness of cardiovascular outcomes in new users of sodium-glucose cotransporter-2 inhibitors). *Circulation* 136: 249–259.

Krekels, E.H.J., van Hasselt, J.G.C., van den Anker, J.N. et al. (2017). Evidence-based drug treatment for special patient populations through model-based approaches. *European Journal of Pharmaceutical Sciences* 109 (Supplement): S22–S26.

New, J.P., Bakerly, N.D., Leather, D., and Woodcock, A. (2014). Obtaining real-world evidence: the SalfordLung Study. *Thorax* 69: 1152–1154.

O'Sullivan ED, Hughes J, Ferenbach DA. Renal aging: causes and consequences. *Journal of the American Society of Nephrology* 28: 407–420, 2017. doi: https://doi.org/10.1681/ASN.2015121308.

Owen RP (2010). Clinical Discussion of Specific Clinical Discussion of Specific Populations. FDA's Clinical Investigator Course. Office of Clinical Pharmacology, Office of Translational Sciences, Center for Drug Evaluation & Research, USFDA. https://www.fda.gov/media/84964/download ().

Prakash, S. and O'Hare, A.M. (2009). Interaction of aging and chronic kidney disease. *Semininors in Nephrology* 29 (5): 497–503. https://doi.org/10.1016/j.semnephrol.2009.06.006. PMID: 19751895; PMCID: PMC2771919.

Suvarna, V.R. (2018). Real world evidence (RWE) – Are we (RWE) ready? *Perspectives in Clinical Research* 9 (2): 61–63. https://doi.org/10.4103/picr.PICR_36_18. PMID: 29862197; PMCID: PMC5950611.

Thrasher T, Abdelmalek MF. Nonalcoholic Fatty Liver Disease. *North Carolina Medical Journal* 2016, 77 (3) 216–219; DOI: https://doi.org/10.18043/ncm.77.3.216.

Winter, S.S., Page-Reeves, J.M., Page, K.A. et al. (2018). Inclusion of special populations in clinical research: important considerations and guidelines. *Journal of Clinical Translational Research* 4 (1): 56–69. PMID: 30873495; PMCID: PMC6410628.

Xia, A.D., Schaefer, C.P., Szende, A. et al. (2019). RWE framework: an interactive visual tool to support a real-world evidence study design. *Drugs Real World Outcomes* 6 (4): 193–203. https://doi.org/10.1007/s40801-019-00167-6. PMID: 31741199; PMCID: PMC6879703.

Chapter Self-Assessments: Check your knowledge

Questions:
- What does the term "special populations" refer to?
- What is the focus of special population studies?
- Why should a pharmaceutical sponsor risk litigation by doing drug trials in pregnant women?
- Has legislation helped the often-cited condition of pediatrics being "therapeutic orphans" with respect to drug research?

Answers:

- The term "Special Populations" defines patient subpopulations outside the "mainstream" patient population (i.e. the targeted or "to be enrolled" Phase 3 population). Special populations are loosely defined as those within which drug administration may be contraindicated or in which dosing adjustments may be warranted. The most considered special populations include geriatrics, pediatrics, pregnant women, and organ impairment (typically, renal and hepatic impairment).

- The actual clinical investigation of special populations typically focuses on safety with the intention to assess the extent to which a subpopulation deviates from the mainstream patient population. The actual evidence for differences (if they exist) tends to be pharmacokinetic (PK)-centric with dosing adjustments, if warranted, guided by PK metrics. PK targets follow equivalence criteria with mainstream "normal" reference population as the comparator.

- Pregnant women may have chronic conditions, such as diabetes, seizure disorders, or asthma, that need to be treated during pregnancy, or pregnant women may develop acute or serious medical conditions during pregnancy that require treatment. It is unethical for study sponsors not to provide guidance for their drug if there is a likelihood that it can be used to treat pregnant women.

- Ethical concerns historically impeded early clinical studies in the pediatric population. Thus, clinical pharmacokinetic and pharmacodynamic studies in the pediatric population did not begin until the 1970s. The FDA Modernization Act (FDAMA 1997) and the Pediatric Rule (1998) have been driving forces for the conduct of pediatric studies in more modern times.

Quiz:

1 Phase 4 refers to . . .. (Choose the best answer)?
   A The phase of development where safety and efficacy in the target population are established
   B It is not a phase at all, its where commercial groups promote, sell, and distribute the drug
   C Refers to the phase of development that occurs post-approval
   D None of these are correct

2 True or False. Phases 1 and 3 populations are (and should be) exactly the same.

3 Choose the best answer to fill in the blanks. The two biomarkers used to grade organ function for renal and hepatic impairment trials are _____ and _____ respectively.
   A Estimated glomerular filtration rate (eGFR) and Child-Pugh classification
   B Renal disease stage and Child-Pugh classification
   C CKD disease grade and NALFD stage
   D All are acceptable

4   Examples of real-world data (RWD) include which sources (choose the best answer):

   A Data derived from electronic health records (EHRs); medical claims and billing data; data from product and disease registries.

   B Patient-generated data, including from in-home-use settings;

   C Data gathered from other sources that can inform on health status, such as mobile devices and social media.

   D All of the above

11

Role and Function of Project Teams

Jeffrey S. Barrett

Aridhia Digital Research Environment

A project team is a team whose members belong to different groups or functions and are assigned to activities for the same project. A team can be divided into sub-teams according to need. Usually, project teams are only utilized for a defined period and are then disbanded after the project is deemed complete. Within pharmaceutical development, the project team is the engine that drives the development of new drugs, devices, or vaccines. While there may be additional decision-making bodies that receive recommendations from the project teams and focus on more of the strategic or business aspects of development, the project teams are typically responsible for the development plans, the achievement of critical project milestones, and the project budget and timelines.

Of course, project teams are formed with the support of the functional groups that provide the representative team members and the senior leadership of the company who support the creation and eventually the termination of project teams. Functional groups refer to the scientific and operational disciplines that support various aspects of drug development including research, development, and commercial parts of the company. The project management group typically has the role of coordinating specific project team activities and ensuring that agendas are generated to track team activities and that communication of relevant timing considerations is both accounted for and properly communicated within the team and to decision-making bodies (see Figure 11.1 for Venn diagram of project team stakeholders). Functional managers would have the role of assigning staff to represent the various scientific, and operational disciplines that complete project team activities and ensure that their project team representative is adequately resourced to fulfill project team requirements.

Upper management secures the portfolio (supported with either internal or acquired project assets – compounds, drugs, vaccines, devices, etc.) and provides resources (financial and staffing) for the functional groups and the project teams

Fundamentals of Drug Development, First Edition. Edited by Jeffrey S. Barrett.
© 2022 John Wiley & Sons, Inc. Published 2022 by John Wiley & Sons, Inc.
Companion website: www.wiley.com/go/Barrett/FundamentalsDrugDevelopment

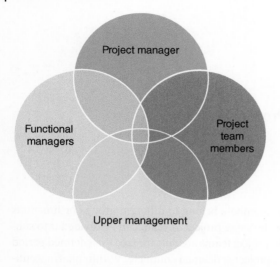

Figure 11.1 Influential stakeholders that drive and support project team creation, dissolution, and project interactions.

either directly or indirectly. On the occasion of a company seeking to acquire a compound, therapeutic area, or an entire company, due diligence teams will often be formed with project team representatives working on related areas. The due diligence team would have the task of evaluating the opportunity within their functional area of expertise and providing feedback and a recommendation to move forward (or not) with the acquisition.

In addition to planning, designing, and completing the various tasks required to advance the development plans and move the project to critical milestones, the project team makes recommendations to decision-making bodies that review the recommendations and ultimately decide the fate of the project. Decision-making bodies will focus their efforts on the early and late-stage portfolios and be comprised of the internal senior leadership supporting the therapeutic areas of interest from either a research or commercial perspective (Jekunen 2014). Figure 11.2 illustrates one type of governance structure by which project teams receive feedback and guidance. In this example, Therapeutic Area Review Boards make decisions on the early-stage portfolio, such as what compounds will move forward into preclinical testing based on high throughput screening criteria. Early-stage project teams would generate such data and rank the compounds based on proposed criteria, ultimately making recommendations to the review board. Ultimately, a more senior decision-making authority such as the Global Research & Development Board shown in the figure would provide guidance on the early stage portfolio taking all therapeutic areas into consideration. This body may be influenced by agreements the company may have made with

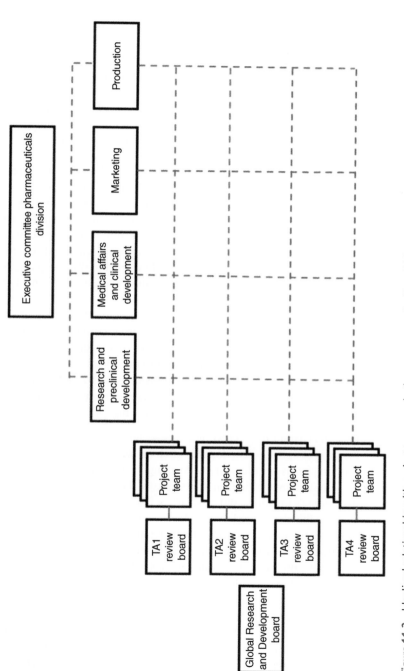

Figure 11.2 Idealized relationship with project teams and other governance bodies within a corporate R&D environment.

certain development partners, the current regulatory climate, the marketplace considerations, etc. (i.e. some information outside the project team's focus) and ultimately have responsibility for the portfolio prioritization.

Within the workings of any project team, there are numerous activities that are advanced by the various project team members under the coordination of the project manager supporting the team and the direction of the project team leader(s). These activities are tracked in the project team's meeting minutes and captured in coordination with project timelines on Gantt charts (a type of bar chart that illustrates a project schedule). Figure 11.3 illustrates a generalized workflow of project team actions driving to a decision point with the typical interplay between governing bodies and R&D hierarchy for early, late-stage, and commercial interests. The influence of these bodies' changes with the stage of development and the various milestones under evaluation at any given time. The emphasis of the team is to drive the process to make informed recommendations for all Go/No go decision milestones (see also Chapter on Compound progression). Subject matter experts (SMEs) both with and external to the company, are often also brought in to help define the target product profile (TPP) and help identify patient populations of interest.

A typical project team meeting will last one to three hours depending on the stage of development and the frequency of the meeting (usually monthly or bi-weekly). The agenda is usually defined by the project manager in consultation with the project team leader, with a draft agenda sent out ahead of finalization to give team members a chance to add or delete items relevant to their functional areas. While the agendas are very project-specific, there are some general characteristics for every meeting to ensure the program is moving forward (see below).

1) Reiteration of the current meeting's objectives – "Today, our goal is to X, etc."
2) Round table project update – The focus is often on what didn't get done, why, and what the impact is on the project. Working as a team, no one will ever collectively complete everything on time always. This is an opportunity to make sure everyone's priorities and tasks are in line.
3) Discussion of roadblocks and risks – Emphasize what roadblocks and risks are projected. Focus on short-term (what might happen in the week) and long-term (what might happen in a month or two months from now). This is an opportunity to solve problems and eliminate obstacles.
4) Discuss deadlines and major milestones – Identify important dates, whether it be an upcoming presentation, a product launch, etc. Make sure everyone in the team knows of the important dates.
5) Assess budget considerations – While this won't be relevant to all team members and every project, it will be relevant to spend some time talking about where the team is in budgeting. Are you over or under budget, and what needs to be prioritized to stay in line?

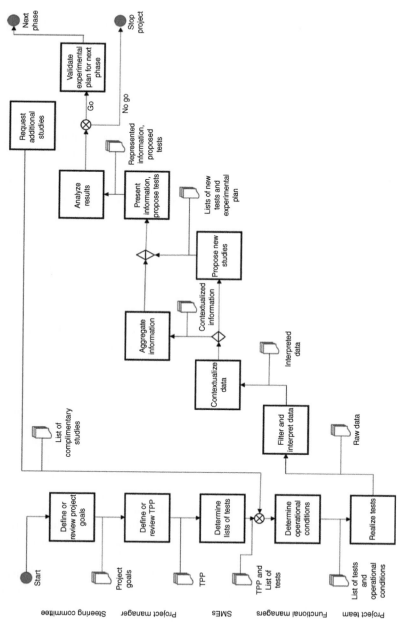

Figure 11.3 Example workflow of project team activities in coordination with stakeholder input and review of recommendations prior to decisions on milestone achievement.

As each project team is required to provide periodic updates to other decision-making bodies, addressing these items with some frequency ensures that the update is current and reflects all key inputs for transparency in decision making.

Project Team Composition and the Influence of Development Phase

Project team membership is dependent upon the stage of development, the unique features of the project being developed, and any specific legal or commercial challenges the project may present (Zeller 2002). There is always a delicate balance between having a small team to efficiently drive the process and having the right skillsets represented to ensure that all relevant areas of expertise are represented.

Regarding the team membership specifically, two roles are central to the team dynamics independent of the development stage – the project team leader and the project manager. The project manager has the overall responsibility for the successful initiation, planning, design, execution, monitoring, controlling, and closure of a project. The project manager helps drive the agenda and facilitates team dynamics. Project managers often have some prior experience from one of the underlying support functions represented on the team. The project team leader leads team members to perform their roles efficiently so that the project goals are achieved. They review and give individual feedback to improve the skills of the team members, so that performance of the project improves overall.

The training and experience of the project team leader are also dictated in part by the stage of development of the project, with early-stage projects often having project leaders from a discovery stage functional group and emphasis (e.g. pharmacology, medicinal chemistry, etc.) while late-stage leaders typically come from a clinically oriented group (e.g. Clinical Pharmacology or therapeutic area physician). Functional team members are recommended by their functional managers based on staffing availability, expertise, and experience with the project therapeutic area. The flower diagram (Figure 11.4) provides an example of project team composition for an early development candidate. Table 11.1 provides a description of the individual functional group member and their role on the project team.

Table 11.1 also describes the typical training/educational background of the various team members. These reflect the most common qualifications and are not prescriptive. In reality, there are no rules on such roles with respect to training; experience is the biggest determinant of the value of the team member. It is also likely that individuals will sit on multiple project teams that may or not even be in the same therapeutic area.

Team composition changes with the stage of development, as mentioned previously. For example, the flower diagram would look slightly different for a

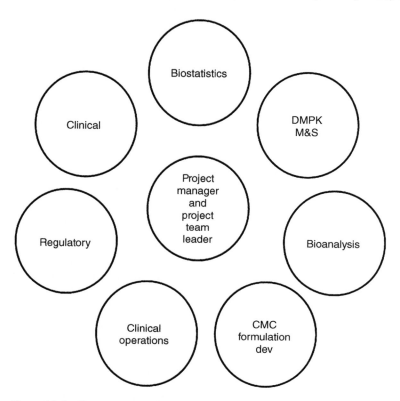

Figure 11.4 Flower diagram showing the typical functional group representation supporting early clinical development of a new molecular entity (drug, biologic, or vaccine).

discovery-stage candidate with the clinical team member and biostatistician replaced by a pharmacologist and toxicologist as an example of common role substitutions. The clinical operations or chemistry, manufacturing, and controls (CMC)/formulation roles could be replaced by Medicinal Chemistry as well. Typically, the timing of the investigational new drug application (IND) (if the candidate progresses that far) is a common point at which the team composition is reconsidered. For late-stage programs (near or post-approval), DMPK and Bioanalysis roles may depart in favor of Medical Affairs/Pharmacovigilance and Marketing team members. In any case, project team composition is intended to be fluid, with the right skill sets joining the team when they are most valuable.

Good team dynamics require that the team communicate well with transparency and trust and that their functional area leadership adequately support them

Table 11.1 Common project team representatives supporting an early development candidate (excludes project team leader and project manager).

Functional group	Member qualifications	Role on project team
Clinical	Usually MD or Pharm D; common to expect training in a specific clinical discipline though exact alignment with project usually not required.	Primary author of clinical development plan, clinical strategy, protocols, synopsis, CRF/ICF, etc. Therapeutic area expertise Due diligence on clinical aspects for in-license opportunities Ensure patient safety Engagement with PI's that enroll patients in clinical trials
Biostatistics	MS or PhD in Statistics or Biostatistics (could be Epidemiology also); experience with trial design support usually as well	Contribute to strategy development, development plans for using quantitative tools to inform strategy development Inform pros/cons of various study designs that meet PT objectives Write SAP of protocols, perform CTS to justify analyses to be performed and sample sizes of clinical studies Provide Biostats oversight of CROs managing studies
DMPK/M&S (Drug metabolism and pharmacokinetics)/ (Modeling and simulation)	MS/PhD in quantitative science and/or pharmacy	Incorporate DMPK input and expertise into project strategy, timelines, and deliverables Summarize key DMPK and M&S findings, provide updates to PT, propose model strategy and assumptions, data inputs Evaluate and identify novel approaches to priority questions/efforts Availability of data sources to inform a model for decision making

Table 11.1 (Continued)

Functional group	Member qualifications	Role on project team
Bioanalysis	MS/PhD in chemistry, pharmacy, or related discipline	Implement biomarker (BM) plan and assay plan
		Provide BM/PK sample collection, processing, shipment, and analysis
		Identify BM/PK labs and assays
		Main interface with lab issues and resolution
		Implement Bio-storage process
CMC/formulation development	MS/PhD in chemistry, pharmacy, or related discipline	Update on CMC activities, highlight issues, provide a path forward
		Deliver clinical material to sites
		Preformulation, stability, etc. studies
		Support clinical strategy with appropriate product design
Clinical operations	BS/MS/PhD in life science discipline	Escalate issues and risks relating to timelines, budget, and quality
		Drive study execution
Regulatory	MD/PhD/PharmD with some clinical experience and knowledge of regulatory environment	Develop regulatory strategy in support of program goals, considering product characteristics, disease burden, intended indication, and population
		Engage/serve as an advisor on regulatory aspects
		Ensure timely, high-quality submissions (stakeholders/ IND party)

with critical review and resources. Likewise, governing bodies expect the project teams to be well informed and behave with the appropriate sense of urgency exposing assumptions, uncertainty in recommendations, and timing sensitivities. High-performing teams are always desired, but competing projects and unclear prioritization often stand in the way.

Terminating Projects and Project Teams

At some point in the development process, there may be a need to dissolve the team and move the team members to other projects. Reasons for project team termination can include any of the following:

- Disappointments in a product (unwanted side effects or marginal efficacy)
- Change in product environment (internal and external competition)
- Change in financing

Likewise, there are preplanned evaluations for consideration of project team termination based on key milestones timing that coincides with financial investment increases. Evaluations are usually made by one of the aforementioned governance bodies. Typically, these evaluation periods occur after Phase 0 (<15 subjects given a very small dose of a compound to make sure it isn't harmful to humans before higher doses are administered), after Phases 1–2, and before Phase 3. Of course, at any time, information regarding a development compound becomes known that suggests a low probability of development success, a candidate may be abandoned, and the project team disbanded. Luckily, this simply means a new opportunity for the company and a new project team for the team members.

References

Jekunen, A. (2014). Decision-making in product portfolios of pharmaceutical research and development – managing streams of innovation in highly regulated markets. *Drug Design, Development and Therapy* 8: 2009–2016.
Zeller, C. (2002). Project teams as means of restructuring research and development in the pharmaceutical industry. *Regional Studies* 36 (3): 275–289.

Chapter Self-Assessments: Check Your Knowledge

Questions:
- What is a project team?
- Why do project teams change team members over time?
- How does a project team interact with other decision-making bodies within a company?
- What is the role of the project team leader on a project team?

Answers:

- A project team is a team whose members belong to different groups or functions that drives the development of new drugs, devices, or vaccines. Project teams are responsible for the development plans, the achievement of critical project milestones, and the project budget and timelines.
- Project team membership is dependent upon the stage of development, the unique features of the project being developed, and any specific legal or commercial challenges the project may present. The timing of the IND (if the candidate progresses that far) is a common point at which the team composition is reconsidered. Typically, early-stage teams will focus on more basic science, such as chemistry and pharmacology, while later-stage teams will be staffed by clinical and commercial influences.
- The emphasis of the team is to drive the process to make informed recommendations for all Go/No Go decision milestones. The project team makes recommendations to decision-making bodies that review recommendations and ultimately decide the fate of the project.
- The project team leader leads team members to perform their roles efficiently so that the project goals are achieved. They review and give individual feedback to improve the skills of the team members, so that performance of the project improves overall.

Quiz:

1 True or false. Project teams make all key decisions for a pharmaceutical company given that they are typically matrix-based organizations.

2 General characteristics for every project team meeting include all but which:
   A Reiteration of the current meeting's objectives
   B Round table project update
   C Discussion of roadblocks and risks
   D Poor team behavior and dynamics
   E Discuss deadlines and major milestones
   F Budget considerations

3 Reasons for project team termination can include which (choose the best answer):
   A Disappointments in a product (unwanted side effects or marginal efficacy)
   B Lack of confidence in the project team leadership
   C Change in product environment (internal and external competition)
   D Change in financing
   E a, b, and c
   F a, c, and d
   G a and c

4 A type of bar chart that illustrates a project schedule commonly used to track project team timelines
  A Gantt chart
  B Gantt progression
  C Project tracker
  D Pie chart

5 True or false. Subject matter experts (SMEs) both within and external to the company are often also brought in to help define the target product profile (TPP) and help identify patient populations of interest.

12

Compound Progression and Go/No Go Criteria

Jeffrey S. Barrett

Aridhia Digital Research Environment

Introduction – History of Empiricism

As previously discussed in earlier chapters, drug development is a heavily regulated and extensively documented process with well-articulated expectations for drug sponsors to communicate their plans and status with regulatory authorities in a transparent manner. Its origins were truly based on an empirical approach with well-appreciated, overly optimistic approaches. Good examples for this historical practice exist for malaria (Riscoe et al. 2005) and oncology (Barrett et al. 2007) and certainly apply to other therapeutic areas. Much of the early decision-making regarding candidate selection and advancement of molecules through development stages was based on a hierarchical reporting structure as opposed to today's heavily matrixed organizations. While we can pride ourselves on more contemporary, rational, and data-driven approaches informed by conceptual and quantitative models in some cases, empiricism still proves very useful when challenging false "concepts." Many theoretical models are demonstrably wrong, and many more will be shown to be likewise. While we can look to evidence-based medicine (EBM) as an ideal that goes unchallenged, it may be well that committing to unexamined empiricism, EBM should take an example from Galen, who sought to synthesize the best elements of the empirical and rationalist traditions (Webb 2018).

Developing a new drug from inception to the launch of an approved product is a complex process that can take 12–15 years and cost in excess of $2.6 billion USD (DiMasi et al. 2016). The idea for a target can come from a variety of sources including academic and clinical research and from the commercial sector. It may take many years to build up a body of supporting evidence before selecting a target for an expensive drug discovery program. Once a target has been chosen, the pharmaceutical industry and, more recently, some academic centers have streamlined a number of early processes to identify molecules that possess suitable characteristics to make acceptable drugs. The intention for this effort is to look critically at key preclinical stages of the drug discovery process, from initial target identification and validation through assay development, high throughput screening, hit identification, lead optimization, and finally, the selection of a candidate molecule for clinical development in addition to critical stage gates in development to describe the desired conditions upon which a drug candidate should progress through the various stage gates. Crafting objective and meaningful quantitative criteria which should halt progression or kill the compound is also part of the process. Compound progression refers to a workflow process and mechanism to track the stage, status, and actions of compounds within a project to streamline drug discovery and development operations. This is often accomplished by providing tools and a workflow that facilitate transparent decision making, team accountability/empowerment, action-oriented meetings, and mentoring. Some companies actually create formal documentation for these processes drafting SOPs and/or compound progression manuals in an effort to both ensure the task gets documented so that some visible and transparent record of the process can be tracked and also to standardize the approach.

In this chapter, we will explore the various approaches to define and articulate compound progression strategies along with some exploration of the tools and workflows that are commonly used. We will do this in the context of therapeutic area-specific examples and expose the personnel, timing and cost involved.

Definitions

An important tenet for compound progression is the recognition that development candidates under investigation as potential new medicines must meet certain criteria, at each stage of development, in order to progress further. These criteria need to be agreed upon by the pharmaceutical sponsors developing them and incorporated into the target product profile (TPP). Likewise, Go/No-go criteria refers to the defined criteria (qualitative and quantitative) that defines each critical decision point. A "go" decision means that the compound meets the criteria and will be advanced to the next development step. Failure to meet the criteria will lead to a "no-go" decision, and the medicine development will stop. Stage gates describe points in a project or plan at which development can be examined, and any important changes or decisions relating to costs, resources, profits, etc., can be evaluated. Typical, recognized stage gates in drug development would include: lead identification, entry to clinical development including First in humans (FIH) testing, FIH dose selection, Proof of mechanism (PoM), Proof of concept (PoC), Phase 2/3 transition, Phase 2/3 transition (Differentiation), Submission, Approval, Risk Management Program, Pricing

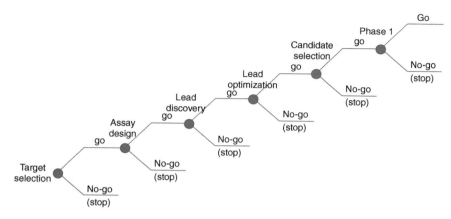

Figure 12.1 Representative implementation of Go/No-go criteria imposed on common early development stage gates.

and Approval, Pricing and Launch, and Post-market program. Figure 12.1 illustrates how Go/No-go decisions are evaluated at common early development stage gates to promote compound progression. Stage gates are obviously a focal point of discussion by project teams (see Chapter 11), but multiple levels of a company's decision-making hierarchy are involved. A stage-gate review goes significantly beyond a compound development team to include regulatory affairs, quality, manufacturing, supply chain management, reimbursement, medical affairs, finance, legal, marketing, etc.

Decision criteria refers to the principles, guidelines or requirements that are used to make a decision. This can include detailed specifications and scoring systems such as a decision matrix. Alternatively, a decision criterion can be a rule of thumb designed for flexibility.

By Phase Criteria

The aforementioned criteria are often captured in the TPP and evolve by the development phase coincident with stage gates. Figure 12.2 illustrates some of the commonly-accepted compound progression milestones across various development stages. Of course, the exact stage-gate criteria are linked to the specific development candidate and reflect the nuances of the therapeutic area, the target population, and the development candidate as well as the marketplace. There are good examples of such criteria put forth to support oncology (Roberts et al. 2003; Barrett et al. 2007) and HIV (Barrett et al. 2007).

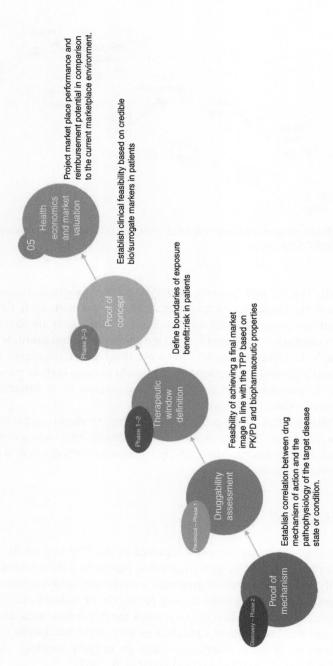

Figure 12.2 Commonly-accepted stage-gate criteria for compound progression by development phase.

Decision Theory

Decision theory is an interdisciplinary approach to arrive at the decisions that are the most advantageous given an uncertain environment; it brings together psychology, statistics, philosophy, and mathematics to analyze the decision-making process. Some have suggested procedures to optimize specific aspects of drug development programs, applying decision-theoretic approaches. In those approaches, pre-specified utility functions are typically optimized over a limited set of parameters. Another approach is to build a model flexible enough to work with a variety of situations and allow the calculation of metrics of interest-based on simulation results with an emphasis on team engagement and creation of viable alternatives is in line with the decision analysis process (Wiklund 2019).

At its core, decision theory provides a formal framework for making logical choices in the face of uncertainty. Given a set of alternatives, a set of consequences, and a correspondence between those sets, decision theory offers conceptually simple procedures for choice. Mathematically and statistically, this is accomplished by assigning probabilities to various outcomes and evaluating possible paths associated with the greatest probability of technical success. Applied to drug development, it would imply that we use such an approach to evaluate various stage-gate decisions using the various Go/No-go criteria with the data generated to assign a probability of proceeding or not. This can be visualized by looking at Figure 12.1 Go/No-go decision nodes and assigning a Go and No-go probability with presumably the decision to proceed (Go decision) having a higher probability for compounds that progress. Likewise, killing a compound at a certain stage gate would be associated with a highly probably No-go decision. While some have advocated such approaches and methods in drug development in practice, it happens seldomly and only with the proper leadership to guide the practice and interpretation (Lalonde et al. 2007; Frewer et al. 2016).

Early "Go/No go" decisions in drug development increasingly control the degree of subsequent investment in either a compound or a molecular mechanism for one or more indications. Ideally, one knows the specific mechanism(s) engaged by a compound, but this is often not the case. Early "No go" decisions are prevalent within the industry and are now being applied in academic research and even NIH-funded studies. The underlying assumptions and risks remain a matter of debate particularly when the motivation is not entirely driven by financial incentives. From the standpoint of science, it is especially challenging when the pathophysiology of a clinical syndrome is unknown, and animal models are poorly predictive, as is the case for many CNS diseases. How to optimize decision-making remains an open question and is largely dependent on which unproven assumptions are embraced.

The need for clear and evidence-based decision-making is essential, especially in early clinical development where decisions are not always made based on observing a significant p-value, and in fact, focus on statistical significance alone may be counter-productive. Many (Frewer et al. 2016; Wiklund 2019) have highlighted the need to improve productivity by improving the probability of technical success (POTS) in consecutive phases and how this can be enhanced by stopping the development of inferior compounds as early as possible and by accelerating the development of good compounds.

An essential element is, of course, committed projected team engagement, as we have discussed in the previous chapter. Representation by key project team stakeholders and their line management is critical for the generation of decision criteria used in the compound progression progress. Figure 12.3 provides a snapshot of the various functional groups involved and some of the more common activities that facilitate compound progression.

Of course, some groups are more involved than others with the direct assessment and analysis of such criteria, particularly with how it is summarized and presented to senior management decision-makers. Within R&D groups such as quantitative sciences, biostatistics, and analytics groups that may or may not reside with IS or IT would be involved with such analyses and presentations to senior managers with R&D. On the commercial side of the organization, groups such as marketing, epidemiology/surveillance, Real World Evidence, and portfolio management groups would have a similar responsibility. These groups typically have a role in de-risking decision-making via quantitative analyses and work directly with the data generating groups at the project team level to summarize data used for stage gate evaluations.

Algorithms, dashboards, and other quantitative tools including data visualization software, are critical to the evaluation. Common tools used in the tracking and evaluation of compound progression would include tools for modeling and simulation (NONMEM, SAS, R, Python, Monolix, etc.), project management and tracking (Smartsheets, Trello, Basecamp, etc.), and portfolio management (Cherwell software, Sciforma, Workzone, Keyedin Projects, Workfront, Wrike, Asana, etc.). It is also a recent phenomenon to attempt to co-locate the tools with critical data used for decision making into a platform that can efficiently manage such information, keep it secure and share with appropriate heuristics in place. Examples of such platforms include Palantir, Deloitte, Microsoft Azure, Cloudera, Sisense, Collibra, Tableau, MapR, Qualtrics, Oracle, etc. The main incentive for such platforms is to promote the collaboration of internal stakeholders, leveraging the internal expertise in quantitative science and data analytics. Some external sharing and connectivity, particularly by the commercial groups within the company, is often also desired. Of course, some sponsors prefer to outsource this effort entirely. This approach has the advantage of removing internal bias but at the expense of ignoring or minimizing internal

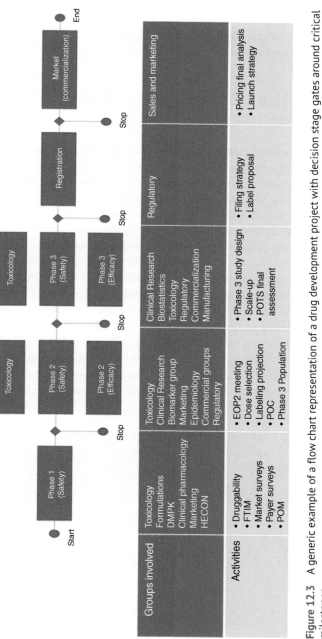

Figure 12.3 A generic example of a flow chart representation of a drug development project with decision stage gates around critical milestones.

expertise. Some of the CRO's engaged in this space supporting the pharmaceutical industry include the following: Certara (www.certara.com), Deloitte (www.deloitte. com), Boston Consulting Group (www.bcg.com), McKinsey (www.mckinsey.com), and Gartner (www.gartner.com/en/consulting).

Decision-making within a Company – Process and Interconnectedness

As previously discussed in other chapters, decision-making within an organization happens at many levels, with recommendations coming from some groups to inform decision-making by more senior members of the organization. Figure 12.4 projects a simplistic view of such interplay, illustrating the feedback mechanism that must exist and also the necessity of various groups to interact with data generators so that decisions can be informed by data (and models, hopefully).

One of the more esoteric tasks within a project team setting, as discussed in Chapter 11, is for the team to ask the right questions regarding the specific challenges that a development candidate presents. In a broader sense, we can look at decision-making performance based on similar questions asked about an entire therapeutic area and the company's entire portfolio. Likewise, all of these questions (and decisions) are interconnected and inform the company's valuation in addition to its culture for decision making. Ultimately, it's a reflection of how efficient its process for compound progression is performing. Figure 12.5 provides a simple example of how such broad questions regarding stage-gate decisions can connect over time within and across projects to inform a company's decision-making performance.

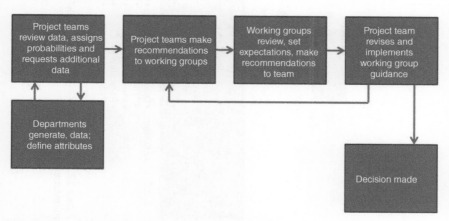

Figure 12.4 Decision-making process: stages and hierarchy.

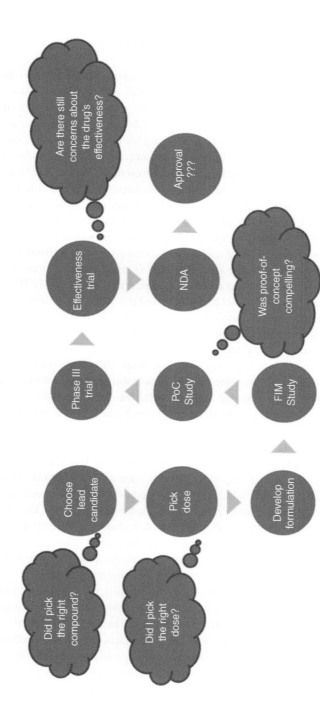

Figure 12.5 Inter-temporal stage gates linked to critical questions regarding individual candidates and ultimately therapeutic area and portfolio performance.

Impact on Portfolio Prioritization and Planning

Portfolio management ensures that projects and programs are reviewed to prioritize resource allocation and that the management of the portfolio is consistent with and aligned to organizational strategies. The main processes of portfolio management are the following: components identification and selection, assessment and prioritization of the components, portfolio monitoring, and control (Bode-Greuel and Nickisch 2008). There are different types of criteria (Frame 2003) that are used to evaluate and prioritize the portfolio components, such as financial criteria; technical criteria; risk-related criteria; resources-related criteria (human resources, equipment, etc.); contractual conditions criteria; experience and other qualitative criteria. Examples of financial criteria commonly evaluated include benefit-cost ratio, net present value, payback period, internal rate of return (IRR), the weighted average cost of capital, and terminal value. The limitations of these indicators have been discussed previously (Flanagan and Norman 1993; Yescombe 2002; Esty 2003; Fabozzi and Nevitt 2006; Phillips et al. 2007). The steps for developing a quantitative model to evaluate and prioritize the projects include the following: establish the evaluation criteria; establish a scoring scale for each criteria; establish the scoring method for each criterion; calculate the project score for each criterion and the total score; establish the project priority based on one single score (single-criteria approach) or total score (multicriteria approach). Project prioritization is usually done on the single profit-oriented criteria, rather than considering multiple criteria, both quantitative and qualitative. Another limitation of the existing practices in project prioritization is the deterministic approach. Most companies develop financial projections based on the deterministic estimation of project financial performance. For doing that, some basic assumptions are considered, such as the time frame (the financial projections cover the project implementation period plus 3–5 years after the project's completion), capital outlays, and financing costs (they include any up-front and ongoing capital needs during the reference period), revenues associated with the project, expenses, and capital structure. Figure 12.6 describes the interrelationships between the compound progression activities of the project team, various R&D and commercial decision-making groups, and those engaged in portfolio management. As is clearly visible, the business is never far away from science.

It should be clear many within an organization participate in the process of compound progression in one way or another. The nature of participation depends on the skill set of the functional group, their role as either data generators or data exploiters, and the phase of development. Likewise, the activities of the individual groups and project team members are aligned to the various stage gates, as shown in Figures 12.2 and 12.3. The costs of these efforts are typically absorbed within the various functional groups, so it is hard to get an exact estimate of the expenses

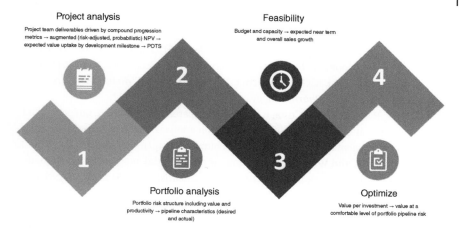

Figure 12.6 Interplay of compound progression (stage-gate evaluation of decision criteria and Go/No-go criteria) with portfolio management and company value proposition.

associated with compound progression. What should be clear is that the cost of drug development should be less when appropriate stage gate criteria are put forward and when companies adhere to the Go / No-go decision rules they propose. The efficiencies gained from a "quick kill" approach go a long way to reduced development costs, especially when well-defined goals are in place.

References

Barrett, J.S. (2007). Facilitating compound progression of antiretroviral agents via modeling and simulation. *Journal of Neuroimmune Pharmacology* 2: 58–71.

Barrett, J.S., Gupta, M., and Mondick, J.T. (2007). Model-based drug development for oncology agents. *Expert Opinion on Drug Discovery* 2 (2): 185–209.

Bode-Greuel, K.M. and Nickisch, K.J. (2008). Value-driven project and portfolio management in the pharmaceutical industry: drug discovery versus drug development – Commonalities and differences in portfolio management practice. *Journal of Commercial Biotechnology* 14: 307–325. https://doi.org/10.1057/jcb.2008.6.

DiMasi, J.A., Grabowski, H.G., and Hansen, R.W. (2016). Innovation in the pharmaceutical industry: new estimates of R&D costs. *Journal of Health Economics* 47: 20–33. https://doi.org/10.1016/j.jhealeco.2016.01.012.

Esty, B.C. (2003). *Modern Project Finance: A Casebook*. New York: Wiley.

Fabozzi, F. and Nevitt, P.K. (2006). *Project Financing*, 7e. London: Euromoney.

Flanagan, R. and Norman, G. (1993). *Risk Management and Construction*. Oxford, U.K.: Blackwell Scientific.

Frame, J.D. (2003). *Project Finance: Tools and Techniques*. University of Management & Technology.

Frewer, P., Mitchell, P., Watkins, C., and Matcham, J. (2016). Decision-making in early clinical drug development. *Pharmaceutical Statistics* 15: 255–263. doi: 10.1002/pst.1746.

Lalonde, R.L., Kowalski, K.G., Hutmacher, M.M. et al. (2007). Model-based drug development. *Clinical Pharmacology and Therapeutics* 82: 21–32.

Phillips, L.D., Bana, C.A., and Costa, E. (2007). Transparent prioritisation, budgeting and resource allocation with multi-criteria decision analysis and decision conferencing. *Annals of Operations Research* 154 (1): 51–68. https://doi.org/10.1007/s10479-007-0183-3.

Riscoe M, Kelly JX and Winter R. Xanthones as antimalarial agents: discovery, mode of action, and optimization. *Current Medicinal Chemistry* (2005) 12: 2539. https://doi.org/10.2174/092986705774370709.

Roberts, T.G. Jr., Lynch, T.J. Jr., and Chabner, B.A. (2003). The phase III trial in the era of targeted therapy: unraveling the "go or no go" decision. *Journal of Clinical Oncology* 21 (19): 3683–3695.

Webb, W.M. (2018). Rationalism, empiricism, and evidence-based medicine: a call for a new galenic synthesis. *Medicines (Basel)* 5 (2): 40. https://doi.org/10.3390/medicines5020040. PMID: 29693563; PMCID: PMC6023440.

Wiklund, S.J. (2019). A modelling framework for improved design and decision making in drug development. *PLoS ONE* 14 (8): e0220812. https://doi.org/10.1371/journal.pone.0220812.

Yescombe, E.R. (2002). *Principles of Project Finance*. San Diego, CA: Academic Press.

Chapter Self-Assessments: Check Your Knowledge

Questions:

- Describe and define compound progression and why it is useful for drug development?
- How does decision theory apply to drug development decision-making?
- What is the value of large and expensive data-sharing platforms for drug sponsors?
- What is portfolio management and how does it facilitate efficiencies in drug development?

Answers:

- Compound progression refers to a workflow process and mechanism to track stage, status and actions of compounds within a project to streamline drug discovery and development operations. This is often accomplished by providing tools and a workflow that facilitate transparent decision making, team accountability/empowerment, action-oriented meetings, and mentoring.

- Decision theory provides a formal framework for making logical choices in the face of uncertainty. Given a set of alternatives, a set of consequences, and a correspondence between those sets, decision theory offers conceptually simple procedures for choice. Mathematically and statistically, this is accomplished by assigning probabilities to various outcomes and evaluating possible paths associated with the greatest probability of technical success. Applied to drug development it would imply that we use such approach to evaluate various stage-gate decisions using the various Go/No-Go criteria with the data generated to assign a probability of proceeding or not.
- The main incentive for such platforms is to promote the collaboration of internal stakeholders, leveraging the internal expertise in quantitative science and data analytics. Some external sharing and connectivity, particularly by the commercial groups within the company, is often also desired.
- Portfolio management ensures that projects and programs are reviewed to prioritize resource allocation and that the management of the portfolio is consistent with and aligned to organizational strategies. The main processes of portfolio management are the following: components identification and selection, assessment and prioritization of the components, portfolio monitoring, and control.

Quiz:

1 True or false. Crafting objective and meaningful quantitative criteria which should halt progression or kill the compound is also part of the compound progression process. (true)

2 Go/No-go criteria refers to the defined criteria (_____ and _____) that defines each critical decision point. Choose the best answer: (a)
   A qualitative and quantitative
   B early and late phase
   C R&D and commercial
   D subjective and objective
   E None are correct

3 Typical, recognized stage gates in drug development would include all but which of the following. Choose the best answer: (d)
   A Lead identification
   B FIH dose selection
   C Phase 2/3 transition
   D Drug metabolism
   E Pricing and Approval

4 Examples of financial criteria commonly evaluated as part of a portfolio management exercise include *all but which* of the following. Choose the best answer: (a)
 A Coupons and rebates
 B Benefit-cost ratio
 C Net present value, payback period
 D Internal rate of return (IRR)
 E Weighted average cost of capital

13

Regulatory Milestones and the Submission Process
Eileen (Doyle)Castranova

Certara, Inc., Princeton, NJ, USA

Starting at the Finish: Applying for Marketing Authorization

Drug development is run with an end goal in mind: product approval. To get any product approved, it must undergo comprehensive testing in humans. To support testing in humans, testing in vitro, in silico, and in vivo (animal models) must be performed. Each regulatory agency has its own rules governing the use of experimental drug candidates in animals and humans. In the United States, the FDA oversees more than 20000 prescription drug products approved for marketing. In Europe, the pharmaceutical industry is a key asset of the economy. During the period 2013–2018, approximately 65% of sales of new medicines launched on the US market, followed by 18% in Europe, 6% in Japan, and 9% in the rest of the world (EPFA 2019). This chapter, therefore, focuses on the submission of marketing applications to the United States and leaves other markets and their processes for discussion elsewhere. Details regarding regulatory milestones, the timing and content of regulatory submissions, and the engagement required on behalf of the pharmaceutical sponsor are also described.

The Submission Process: United States

Regulatory Milestone: The IND

From a regulatory standpoint, the first major milestone is the submission of the investigational new drug application (IND). Receipt of the IND allows for testing in humans. The IND is a heterogeneous document with several types. Which type one applies for depends on the status of the sponsor. Commercial INDs are submitted by drug companies and are typically what people think of when one says "IND." Other common types are investigator INDs submitted by a physician who initiates

Fundamentals of Drug Development, First Edition. Edited by Jeffrey S. Barrett.
© 2022 John Wiley & Sons, Inc. Published 2022 by John Wiley & Sons, Inc.
Companion website: www.wiley.com/go/Barrett/FundamentalsDrugDevelopment

and investigates (typically clinical involving patients), and under whose immediate direction the investigational drug is administered or dispensed. A research IND might be submitted to propose the study of an unapproved drug or study of an approved product for a new indication or for use in a new patient population. Emergency use INDs allow the use of an experimental drug in an emergency situation that does not allow time for submission of a typical commercial IND. Treatment IND or single-patient IND is submitted for experimental drugs that show promise in clinical testing for serious or immediately life-threatening conditions while the final clinical work is conducted and during the FDA (or other agency) review. Expanded access INDs allow for access to investigational drugs (not yet approved) outside of clinical trials (for example, for cancer patients who have exhausted all other treatment options).

Each quarter for the past ten years, the FDA received between 100 and 200 original IND applications, leading to a yearly application of 400–800 IND applications. In 2019, 626 original investigational new drug applications were received by the FDA. In 2020 that number skyrocketed in response to the SARS-CoV-2 pandemic; the FDA received 7000 original IND applications (FDA n.d.). Common examples of emergency use INDs include many of the diagnostic tests that were authorized on an emergency basis (before conversion to full authorization): for example, Abbott Laboratories Inc. AdviseDx SARS-CoV-2 IgG II (Alinity), which uses semi-quantitative high throughput technology to detect the SARS-CoV-2 spike protein (FDA 2019a) during the recent COVID-19 pandemic. In addition to medical devices, drugs such as chloroquine and remdesivir were granted emergency use authorization based on emergency use INDs in 2020. Though chloroquine and remdesivir were not FDA-approved for COVID-19 indications, expanded access allowed patients with serious or life-threatening cases of the virus to have access to them as investigational medicinal products (FDA 2019b). Ultimately the FDA revoked emergency use authorization for chloroquine to treat certain hospitalized patients with COVID-19 when a clinical trial was unavailable, or participation in a clinical trial was not feasible. The agency determined that the legal criteria for issuing a EUA were no longer met, as they were unlikely to be effective in treating COVID-19 for the authorized uses in the EUA (FDA n.d.). The FDA ultimately granted approval of remdesivir for use in adults and pediatric patients (12 years of age and older and weighing at least 40 kg) for the treatment of COVID-19 requiring hospitalization (FDA n.d.).

In more typical times, expanded access programs (EAP) allow groups of patients with the same disease or condition (such as boys with Duchenne muscular dystrophy) to access a medical product before approval. To qualify for an EAP, patients must meet certain criteria. These criteria are typically less rigid than those for clinical trials, as the purpose of an EAP is to provide treatment (as opposed to research) (Code of Federal Regulations (CFR) n.d.).

For example, if an investigational drug is in late-stage clinical trials, and the trial data suggest that the drug is safe and effective, the sponsoring company may open an EAP to make the drug available to those patients who will be harmed by waiting until the drug receives FDA approval to use it. An EAP also covers the time lag between FDA approval and when it is stocked in pharmacies (or when it is covered by insurance). In such instances, patients may be able to access the approved but not easily available drug through an EAP during this lag. Expanded access programs end once a drug is available on the market. If no expanded access program exists or a patient does not qualify for an EAP, that patient, through a physician, may request the investigational product via single-patient expanded access. In 2019, the FDA accepted 1755 expanded access requests. Of these, 1108 were for non-emergency situations for individual patients; 444 were for emergency situations for individual patients; 51 were for treatment of multiple patients (FDA n.d.).

Original IND

An IND is opened by applying to the regulatory authority, including submission of an application, a briefing book describing what is known about the investigational new drug (in vitro, in silico, in animals), and includes a copy of the proposed first-in-human protocol. Many regulatory authorities have guidance documents describing the content of materials needed to receive regulatory approval (FDA n.d.; EMA n.d.). A pre-IND assessment can be organized with the FDA to discuss issues such as the design and appropriateness of animal research (sufficiency to support trials in humans), the first in the human protocol (design, assessments, safety), and the chemistry, manufacturing, and control of the investigational drug (CMC). The sponsor submits the investigational new drug application, a statement of the investigator, certification of compliance, introductory statement and general investigational plan (investigator's brochure), protocols, CMC information, pharmacology and toxicology information, previous human experience with the investigational drug (if applicable), other important information, relevant information, and submission information.

The sponsor sends the application to the appropriate division (CBER, CDER), and the FDA forwards the application to the appropriate review team. The first protocol submission to open an IND is subject to a 30-day review clock. Studies may not be initiated until 30 days after the date of receipt of the IND by the FDA. During that time, the protocol is reviewed for safety and to assure that research subjects will not be subjected to unreasonable risk. Protection of the safety and rights of human subjects are key responsibilities of regulatory agencies. The safety assessment looks for evidence that the drug is reasonably safe to administer to humans. The dose must be justified (via extrapolation from no observed adverse effect level (NOAEL) in animals if this is the first-in-human experience or via previous clinical experience if this is a new application for use in a different patient population or indication). Use in healthy volunteers, or in human subjects with the disease of interest in some cases must be considered. Intrinsic and extrinsic factors such as sex, race, organ impairment, and use of concomitant medications (drug interactions) are assessed. Any extra safety risks are documented and discussed. The agency will also assess the protocol to ensure the patient population being proposed is reasonable, that safety monitoring is sufficient, and that there is sufficient scientific quality of the clinical investigation so as to permit the evaluation of the drug's safety and activity. The review team then provides a positive response, negative response, or no response to the application. If positive or no response after 30 days, the sponsor may begin trials in humans. If the IND is denied, the sponsor must resubmit.

The IND is Continuously Reviewed During Clinical Development

Approval of the IND allows the submitted protocol to be opened and the clinical trial to begin in humans. Approval is a dynamic process, however – the IND is continuously reviewed for safety throughout the drug development process. If, at any time, the FDA feels there is a safety concern, they can issue a clinical hold.

Clinical holds may occur at any point in the IND life and may affect a single study or the entire IND. Holds are placed when there is an unreasonable and significant risk of illness or injury; when unqualified clinical investigators are found to be operating in the trial; when the investigator's brochure (IB) is found to be misleading, erroneous, or incomplete; or when there is insufficient information to assess risks to subjects. Once a program addresses these deficiencies, an application to remove the hold is made. If the deficiencies are addressed, the clinical hold is lifted (FDA n.d.).

News regarding clinical holds can be found by searching the FDA database or via press releases. In 2021, the FDA temporarily halted an early-stage trial of experimental gene therapy (Rocket Pharmaceuticals) for Danon disease, a rare and deadly heart condition. Gene therapies have, still in their adolescence, have suffered multiple regulatory setbacks due to safety and manufacturing concerns across companies (Fidler 2021).

Clinical holds can occur because of clinical findings, as in the example above, or as a result of new nonclinical findings. Recently the FDA placed a clinical hold on Larimar Therapeutics' investigational new drug indicated for the treatment of

Friedreich's ataxia following deaths of non-human primates in a toxicology study (Endpoints News n.d.).

In all cases, the finding(s) will be investigated, and inadequacies addressed. If the risk outweighs the potential benefit to patients, the compound may not proceed further in drug development.

Regulatory Milestones During Clinical Development

Assuming the IND remains open, and the program does not experience clinical holds, clinical research continues on the path to proving the safety and efficacy of the investigational new drug. During this time, sponsors have the opportunity to meet with the FDA to discuss the progress of their program and ask questions.

Typically, throughout drug development (preclinical – Phase 1 – Phase 2 – Phase 3 – application (NDA or biologics license application [BLA])), meetings are held with the FDA as regulatory milestones are approached. This often happens at phase changes, such as the transition from preclinical to Phase 1, at the end of Phase 2, and pre-application (NDA or BLA).

Types of Meetings

The FDA provides direction for drug development in the form of Guidance for Industry (FDA n.d.). There is a Guidance for Industry on nearly every subject one can imagine. Each guidance is structured to be the answer to "frequently asked questions" of sorts. In order to understand how to communicate with the FDA, a sponsor should consult the Guidance for Industry: Formal Meetings Between the FDA and Sponsors or Applicants (FDA n.d.).

The guidance describes three types of meetings: A, B, and C. The purpose of a type A meeting is to help a stalled product development program proceed. A common example of a reason for a type A meeting is to address questions surrounding clinical holds. Type B meetings are milestone meetings and include the pre-IND meeting, some end of Phase 1 meetings, end of Phase 2, end of Phase 3, and pre-NDA or pre-BLA meetings. Type C meetings are any meetings that do not fall into the categories of A or B. These meetings are requested when the company has a question about development that cannot wait to become a review issue. For example, discussion of endpoints, clarifications based on interim analyses, considerations of new biomarkers could all be reasons to request at type C meeting. A meeting is held whenever the sponsor feels FDA guidance is needed for development to proceed.

Communication with the FDA is structured. The meeting request, regardless of the method of submission, should include adequate information for the FDA to assess the potential utility of the meeting and to identify FDA staff necessary to discuss the proposed agenda items. The meeting request is actually a package of information that contains the product name, application number, chemical name, and structure, proposed indications or context of product development, the type of meeting being requested (and if it's a type A meeting, the rationale for the meeting), a brief statement of the purpose and objectives of the meeting, including a brief background of the issues underlying the agenda. It can include a brief summary of completed or planned studies and clinical trials or data that the sponsor or applicant intends to discuss at the meeting, the general nature of the critical questions to be asked, and where the meeting fits in the sponsor's overall development plans. Also included are a proposed agenda, a list of proposed questions grouped by discipline, and operational information (a list of all individuals with their titles and affiliation who will attend the meeting (sponsor side) and a list of FDA staff, if known (FDA side) asked to participate in the meeting, and the format of the meeting: face-to-face or videoconference/teleconference).

A sponsor requests the format it believes will be most beneficial, and the FDA grants the meeting in the format it believes is most efficient. In-person meetings occur when across-the-table discussions between the company team and the FDA review team are deemed critical to the program's progress. VC/TC often occurs with smaller groups of people with more directed questions (i.e. if only one subteam is needed for the discussion, such as if a CMC question or a pharmacology question were to be discussed, but not both). Written responses are often delivered when the questions to the agency are straightforward, and the agency feels the answers are straightforward and would not benefit from the discussion.

Meetings throughout drug development benefit everyone. The sponsor is assured that issues and questions that come up are solved in real-time instead of becoming a reason to reject marketing authorization at the end of a (7–15 year) process. The utility of regular meetings includes the FDA review team having an overview of the drug development plan, allowing for the sponsor to comply with regulations and work in accordance with current (and often evolving) standards. They allow for early FDA feedback that keeps the program moving in a direction toward marketing authorization.

Timing of Milestone Meetings

A pre-IND meeting is not required but sponsors often avail themselves of the opportunity to meet with the agency to make sure they are on the right track. Following an application for the IND, the first required milestone meeting is the end of Phase 2 (EOP2) meeting.

The EOP2 meeting is required one month before ending Phase 2 trials. At the EOP2 meeting, the pathway for proceeding to Phase 3 is decided. The Phase 3 plan is evaluated, including assessing the protocol(s) for adequacy; assessing pediatric plans for safety and effectiveness; and to identify any additional information that would be necessary to support a marketing application. At least one month prior to the meeting, the sponsor should submit background information on the sponsor's plan for Phase 3, including summaries of Phase 1 and 2 investigations, the specific protocols for Phase 3 studies, plans for additional nonclinical studies, plans for pediatric studies (including a timeline for pediatric protocol finalization, enrollment, completion, data analysis; or any information to support a request for waiver or deferral of pediatric studies), and tentative labeling for the drug.

Once the sponsor is prepared to discuss the final filing, the sponsor initiates the pre-NDA or pre-BLA meeting (9–12 months before NDA submission is planned). The FDA has found that delays associated with the initial review of a marketing application can be reduced by exchanges of information about a proposed marketing application. The primary purpose of the exchange is to uncover any major unresolved problems (anything that could lead to a "refuse to file"). The pivotal studies (the ones the sponsor is relying on as adequate and well-controlled to establish the drug's effectiveness), pediatric studies, and any unresolved issues (i.e. manufacturing) are discussed. By the end of this meeting, a sponsor should understand exactly what is expected of them in order to receive marketing authorization for their compound.

Types of Applications

The first application for marketing approval with either be an NDA or a BLA, depending on the format of the drug candidate. The type of product determines which division will review the application (i.e. small molecules and antibody therapeutics are reviewed by the Center for Drug Evaluation and Research [CDER], while gene therapies and blood products are reviewed by the Center for Biologics Evaluation and Research [CBER]).

Using the NDA as an example, there are ten types of applications. Type 1 is a new molecular entity (NME); type 2 is a chemistry change (new salt, new noncovalent derivative, new ester); type 3 is a new dosage form; type 4 is a new combination; type 5 is a new formulation or new manufacturer; type 7 is a drug already marketed without an approved NDA (i.e. drug on the market prior to 1938); type 8 is an over-the-counter (OTC) switch; type 9 is a new indication submitted as a distinct NDA, consolidated with the original NDA after approval; and type 10 is a new indication submitted as a distinct NDA but not consolidated with the original NDA after approval. Follow-on indications are typically handled as supplements to the original NDA and become sNDAs.

Focus here will be on new or type 1 applications.

The NDA is a vehicle through which drug sponsors (pharmaceutical or biotechnology companies) formally propose that the FDA approve a new pharmaceutical for sale and marketing in the United States. The data gathered during the in vitro, animal, and human studies become part of the application. Since 1938, every new drug has been the subject of an approved NDA before US commercialization.

The goals of the NDA are to provide enough information to permit the FDA reviewers to reach several key decisions:

- Whether the drug is safe and effective in its proposed use(s), and whether the benefits of the drug outweigh the risks
- Whether the drug's proposed labeling (package insert) is appropriate and what it should contain
- Whether the methods used in manufacturing the drug and the controls used to maintain the drug's quality are adequate to preserve the drug's identity, strength, quality, and purity

Unmet medical needs and safety drive all marketing authorizations. Everything is a risk-benefit calculation. Different conditions lend themselves to different burdens of proof. For these reasons, development programs are tailored to the indication a sponsor is seeking to achieve.

Orphan Drug Designation

Pharmaceutical companies are for-profit companies. Even though the development of medicinal products is noble, at the end of the day, if the company does not make a profit, the company will fold. And then there will be no more development. As risk-benefit is a balance, the cost of drug development is a balance.

This balance shifted away from the development of drugs used to treat rare conditions. Promising "orphan drugs" – drugs that would be used to treat syndromes which affect small numbers of individuals residing in the United States – were being dropped for corporate portfolios because they were costing companies too much money (they cost more to discover and develop than the revenue they would generate over the life of their use). It became clear that promising drugs were not being developed and would not be developed unless changes were made in Federal laws to reduce the costs of developing these drugs. By the 1980s, "rare diseases" affected 20–25 million patients, who, together, suffered from approximately 5000 rare diseases – some of which affected only a handful of individuals.

In order to shift this paradigm, legislation was passed in the United States to facilitate the development of orphan drugs. The Orphan Drug Act of 1983 provided for financial incentives to attract industry to orphan drug development. It included provisions such as a seven-year period of market exclusivity (with or without a patent), tax credits of up to 50% for research and development expenses, grants for clinical testing, and lower application costs. Prior to 2017, orphan drug status also allowed for exemptions from pediatric testing.

The passing of the Orphan Drug Act did not decrease the rigor with which drug development proceeds in populations of people who are afflicted with rare diseases. On the contrary, it allowed for practical, rational, and logistical considerations to be addressed based on the size of the target population.

For example, recombinant human alpha-galactosidase (Genzyme Corporation) received orphan drug designation for the treatment of Fabry's disease. Fabry's disease is one of a group of lysosomal storage diseases. It is an inherited condition caused by a genetic variation in which the body is unable to make enough alpha-galactosidase A. Without enough of this enzyme, a particular fat builds up in cells, causing damage. The disease has a wide range of symptoms, including life-threatening ones such as heart attack, stroke, and kidney disease. With proper care, the disease is manageable (Fabry Disease - NORD n.d.). Type 1 classic Fabry disease affects an estimated 1 in 40 000 males (prevalence in females unknown). Prevalence for late-onset disease is higher (up to 1 in 3000 males). In 2001 Sanofi started a Fabry Registry (ClinicalTrials.gov n.d.) that currently follows approximately 7000 patients in over 40 countries (Sanofi n.d.). Genzyme completed four clinical trials of recombinant human alpha-galactosidase in subjects with Fabry's disease. Study 1 was a randomized, double-blind, placebo-controlled, multinational, multi-center study of 58 Fabry patients (56 males and 2 females), ages 16–61 years, all naïve to enzyme replacement therapy. Patients received either 1 mg/kg of Fabrazyme or placebo every two weeks for five months (20 weeks) for a total of 11 infusions. Study 2 was a randomized (2:1 Fabrazyme to placebo), double-blind, placebo-controlled, multinational, multi-center study of 82 patients (72 males and 10 females), ages 20–72 years, all naïve to enzyme replacement therapy. Patients received either 1 mg/kg of Fabrazyme or placebo every two weeks for up to a maximum of 35 months (median 18.5 months). Sixty-seven patients who participated in Study 2 were subsequently entered into an open-label extension study in which all patients received 1 mg/kg of Fabrazyme every two weeks for up to a maximum of 18 months. Study 3 (Pediatric Study) was an open-label, uncontrolled, multinational, multi-center study to evaluate the safety, pharmacokinetics, and pharmacodynamics of Fabrazyme treatment in 16 pediatric patients with Fabry disease (14 males, 2 females), who were ages 8–16 years at first treatment. All patients received Fabrazyme 1 mg/kg every two weeks for up to 48 weeks. Study 4 was an open-label, re-challenge study to evaluate the safety of Fabrazyme treatment in patients who had a positive skin test to Fabrazyme or who had tested positive for Fabrazyme-specific IgE antibodies.

In this study, six adult male patients, who had experienced multiple or recurrent infusion reactions during previous clinical trials with Fabrazyme, were re-challenged with Fabrazyme administered as a graded infusion for up to 52 weeks of treatment (FDA 2010).

The study sizes were large enough to show a statistically significant decrease in the surrogate endpoint (fat buildup in lysosomes), and the FDA approved Fabrazyme for marketing authorization [FDA 2003].

Expediting Development and Review of Drugs

In addition to incentivizing the development of drugs that would not be profitable for a pharmaceutical company, the FDA has mechanisms for shepherding drugs through the development process. Two of these mechanisms are breakthrough therapy designation and fast track designation.

Fast Track Designation

Fast track (FT) is a process designed to facilitate the development and expedite the review of drugs to treat serious conditions and fill an unmet medical need. The purpose is to get important new drugs to the patient earlier. Determining whether or not a condition is "serious" is based on whether the drug will have an impact on survival, daily functioning, or the likelihood of the condition progressing from less severe to more severe. AIDS, Alzheimer's disease, heart failure, and cancer are examples of serious conditions, as are epilepsy, depression, and diabetes. The ability of a treatment to

A drug that receives FT designation is eligible for more frequent meetings and communications with the FDA, heavier guidance through the drug's development, and rolling review of the NDA or BLA (section by section instead of waiting for the entire application to be finished). Once a drug receives FT designation, early and frequent communication between the FDA and a drug company is encouraged throughout the entire drug development and review process. The frequency of communication assures that questions and issues are resolved quickly, often leading to earlier drug approval and access by patients.

Breakthrough Therapy Designation

Breakthrough therapy (BT) designation is a process designed to expedite the development and review of drugs that are intended to treat a serious condition, where preliminary clinical evidence indicates the drug may demonstrate substantial improvement over the available therapy (as measured by one or more clinically significant endpoints). The determination of "substantial improvement" is subjective and depends on the magnitude of the treatment effect and the gravity of the observed outcome. Preliminary clinical evidence must show a clear advantage over available therapy. The clinically significant endpoint is one that measures an effect on irreversible morbidity or mortality or on symptoms that are a serious consequence of the disease. In 2020, CDER reported the approval of 34 drugs that had BT designation: 25 were original applications (16 NDAs, 9 BLAs) and 9 were supplemental applications (4 sNDAs, 5 sBLAs). The indications spanned multiple therapeutic areas, including oncology (20 of the 34 approvals), metabolic disease, immunology and inflammation, infectious disease (HIV, malaria, Ebola), and rare diseases (FDA n.d.). Ideally, a BT designation request should be received by FDA no later than the EOP2 meetings to allow the features of the designation to be useful. The primary intent of BT designation is to develop the evidence needed to support approval as efficiently as possible. A drug that receives BT designation is eligible for all the fast-track designation features, plus intensive guidance on an efficient drug development program, beginning as early as Phase 1. It also carries an organizational commitment from the sponsors, including senior managers. FDA responds to BT designation requests within sixty days of receipt of the request.

Drug candidates that meet the eligibility for orphan drug status, fast track, or breakthrough therapy status all have the commonality of meeting a serious unmet medical need. In addition to designations that allow for a partnership of sorts with the FDA, and intensive guidance from an early stage, these drug candidates may qualify for expedited review, to make sure the treatment gets to patients as quickly as possible.

Standard Review

Prior to the Prescription Drug User Fee Act (PDUFA) of 1992, it could take years to review an NDA or BLA. To address this issue, Congress passed a law that allowed the FDA to collect an application fee from drug companies to fund the application review. To continue collecting the fees, the FDA is required to meet performance benchmarks. PDUFA was renewed in 1997 (PDUFA II), 2002 (PDUFA III), 2007 (PDUFA IV), 2012 (PDUFA V), and 2017 (PDUFA VI). For the fiscal year 2021, the drug application fee is approximately $2.9 million for a full application requiring clinical data and $1.4 million per application not requiring clinical data or per supplemental application requiring clinical data.

A major PDUFA goal is for the FDA to review and provide a ruling on applications within one year unless significant changes are made to the application during the last three months of the review cycle. Therefore, "Standard Review" timelines are set at 10 months from the date that an NDA or BLA is accepted by the FDA as complete.

Priority Review Designation

For conditions that have been deemed serious, or for which there is no therapy, a year-long process is often too long. For this reason, a mechanism for expedited application review was included. Drug candidates with breakthrough therapy or fast track designation often qualify for priority review. In order to qualify, an application must treat a serious condition that, if approved, would provide a significant improvement in safety or effectiveness; or be a supplementary application that proposes a labeling change pursuant to a pediatric study under 505A; or is an application for a drug that has been designated as a qualified infectious disease product; or is submitted with a priority review voucher. The designation is assigned at the time of NDA or BLA filing.

Designations are not static. They may be rescinded if the drug candidate no longer meets the qualifying criteria (for fast track or breakthrough therapy designation).

Approval Pathways: Full Approval and Accelerated Approval

To get key drugs to patients more quickly, that is, to maximize the benefit-risk ratio, there are two types of marketing authorization: Full approval and accelerated approval. Accelerated approval is an initial approval based on some effect on a surrogate endpoint (or intermediate clinical endpoint) that is reasonably likely to predict a drug's clinical benefit. A surrogate endpoint is a marker, such as a laboratory measurement, radiographic image, physical sign, or other measures that are thought to predict clinical benefit but are not itself a measure of clinical benefit. The use of a surrogate endpoint can considerably shorten the time required prior to receiving FDA approval.

Accelerated approval allows for the introduction of the drug to the market to fulfill an unmet medical need. The drug company is still required to conduct studies to confirm clinical benefits. These studies are performed post-marketing or as "Phase 4 confirmatory trials" and are typically underway at the time of accelerated approval. If the confirmatory trial shows the drug provides a clinical benefit, the accelerated approval is converted to full (traditional) approval. If the confirmatory trial does not show a clinical benefit, the drug could be removed from the market, or the indication can be removed from the drug label.

The sponsor should discuss their intention to seek accelerated approval as early as possible in development. Agreement on the suitability of the planned endpoint(s) as a basis for accelerated approval and agreement on the makeup of the confirmatory trial(s) to convert to full approval will need to be reached.

Once the NDA or BLA is received, it is reviewed (on the Standard or Priority review timeline) and, if it meets all the requirements set forth by the FDA, is granted (Accelerated or Full) marketing approval. Within about one year, the Summary Basis of Approval is published (by the FDA), detailing information from each division about the review process and any (non-proprietary) information used to reach that decision. The approval letter details requirements that need to be met by the company (i.e. submission of final labeling) and details any post-marketing commitments and requirements. Periodic updates are required to keep receiving marketing authorization (i.e. continued safety).

For a corporation, the drug goes into "life cycle management"; new indications are considered, safety is tracked, and the requirements for maintaining active approval are met. Investigator brochures are updated yearly, and other administrative tasks are completed and reported to the FDA. Pediatric trials are run, and clinical programs are overseen. In many ways, achieving marketing authorization is the beginning of a drug's journey, not the end.

References

ClinicalTrials.gov (n.d.). National Library of Medicine (U.S.). https://clinicaltrials.gov/ct2/show/NCT00196742.

Code of Federal Regulations (CFR) (n.d.). Title 21, Part 312.315. https://www.accessdata.fda.gov/scripts/cdrh/cfdocs/cfcfr/CFRSearch.cfm?fr=312.315.

EMA (n.d.). https://www.ema.europa.eu/en/human-regulatory/post-authorisation/data-medicines-iso-idmp-standards/reporting-requirements-authorised-medicines/guidance-documents.

Endpoints News (n.d.). https://endpts.com/deaths-of-monkeys-spur-clinical-hold-for-rare-disease-biotechs-lead-and-only-drug-and-a-95m-cash-infusion-goes-out-the-window.

EPFA (2019). https://www.efpia.eu/media/412931/the-pharmaceutical-industry-in-figures-2019.pdf.

Fabry Disease - NORD (n.d.). https://rarediseases.org/rare-diseases/fabry-disease.

FDA (2003). Clinical review: agalsidase beta (Fabrazyme), Genzyme Corp, Department of Health & Human Services, FDA/CBER. https://www.accessdata.fda.gov/drugsatfda_docs/nda/2003/agalgen042403r5.pdf.

FDA (2010). https://www.accessdata.fda.gov/drugsatfda_docs/label/2010/103979s5135lbl.pdf.

FDA (2019a). https://www.fda.gov/medical-devices/coronavirus-disease-2019-covid-19-emergency-use-authorizations-medical-devices/eua-authorized-serology-test-performance.

FDA (2019b). https://www.fda.gov/news-events/public-health-focus/expanded-access Updated 6 May 2019. (accessed 19 March 2020).

FDA (n.d.a). https://www.accessdata.fda.gov/scripts/fdatrack/view/track.cfm?program=cber&status=public&id=CBER-All-IND-and-IDEs-recieved-and-actions&fy=All.

FDA (n.d.b). https://www.fda.gov/news-events/press-announcements/coronavirus-covid-19-update-fda-revokes-emergency-use-authorization-chloroquine-and#:~:text=Today%2C%20the%20U.S.%20Food%20and,clinical%20trial%20was%20unavailable%2C%20or.

FDA (n.d.c). https://www.fda.gov/media/137574/download#:~:text=On%20October%2022%2C%202020%2C%20FDA,of%20COVID%2D19%20requiring%20hospitalization.

FDA (n.d.d). https://www.fda.gov/news-events/expanded-access/expanded-access-compassionate-use-submission-data.

FDA (n.d.e). https://www.fda.gov/regulatory-information/search-fda-guidance-documents.

FDA (n.d.f). https://www.accessdata.fda.gov/scripts/cdrh/cfdocs/cfcfr/cfrsearch.cfm?fr=312.42.

FDA (n.d.g). https://www.fda.gov/media/97001/download.

Fidler B. (2021). FDA unexpectedly grounds a gene therapy for a rare heart disease. Biopharma Dive. https://www.biopharmadive.com/news/rocket-fda-clinical-hold-danon-gene-therapy/599941.

Sanofi Pharmaceuticals (n.d.). How registries accelerate rare disease research. https://www.sanofi.com/en/science-and-innovation/stories/how-registries-accelerate-rare-disease-research.

Chapter Self-Assessments: Check Your Knowledge

Questions:
- Describe the process for a sponsor opening an IND with FDA?
- What does a "clinical hold" refer to?
- Describe the various types of meetings that sponsors can have with FDA and discuss the differences?
- Define the Orphan Drug Act and discuss its purpose?

Answers:
- An IND is opened by applying to the regulatory authority, including submission of an application, a briefing book describing what is known about the

investigational new drug (in vitro, in silico, in animals), and includes a copy of the proposed first-in-human protocol. A pre-IND assessment can be organized with the FDA to discuss issues such as the design and appropriateness of animal research (sufficiency to support trials in humans), the first in human protocol (design, assessments, safety), and the chemistry, manufacturing, and control of the investigational drug (CMC). The sponsor submits the investigational new drug application, a statement of the investigator, certification of compliance, introductory statement, and general investigational plan (investigator's brochure), protocols, CMC information, pharmacology and toxicology information, previous human experience with the investigational drug (if applicable), other important information, relevant information, and submission information.

- Clinical holds may occur at any point in the IND life and may affect a single study or the entire IND. Holds are placed when there is an unreasonable and significant risk of illness or injury; when unqualified clinical investigators are found to be operating in the trial; when the investigator's brochure (IB) is found to be misleading, erroneous, or incomplete; or when there is insufficient information to assess risks to subjects. Once a program addresses these deficiencies, an application to remove the hold is made. If the deficiencies are addressed, the clinical hold is lifted.

- FDA Guidance describes three types of meetings: A, B, and C. The purpose of a Type A meeting is to help a stalled product development program proceed. A common example of a reason for a Type A meeting is to address questions surrounding clinical holds. Type B meetings are milestone meetings and include the pre-IND meeting, some end of phase 1 meetings, end of phase 2, end of phase 3, and pre-NDA or pre-BLA meetings. Type C meetings are any meetings that do not fall into the categories of A or B. These meetings are requested when the company has a question about development that cannot wait to become a review issue. For example, discussion of endpoints, clarifications based on interim analyses, considerations of new biomarkers could all be reasons to request at Type C meeting. A meeting is held whenever the sponsor feels FDA guidance is needed for development to proceed.

- The Orphan Drug Act of 1983 provided for financial incentives to attract industry to orphan drug development. It included provisions such as a seven-year period of market exclusivity (with or without a patent), tax credits of up to 50% for research and development expenses, grants for clinical testing, and lower application costs. Prior to 2017, orphan drug status also allowed for exemptions from pediatric testing. Its purpose was to incentivize pharmaceutical sponsors to develop drugs for rare diseases. Promising "orphan drugs" – drugs that would be used to treat syndromes which affect small numbers of individuals residing in the United

States – were being dropped for corporate portfolios because they were costing companies too much money (they cost more to discover and develop than the revenue they would generate over the life of their use). It became clear that promising drugs were not being developed and would not be developed unless changes were made in Federal laws to reduce the costs of developing these drugs.

Quiz:

1   From a regulatory standpoint, the first major milestone is the submission of the _____ (choose the best answer to fill in the blank)
   A New Drug Application (NDA)
   B Investigational new drug application (IND)
   C Investigator's Brochure (IB)
   D Informed Consent Form (ICF)

2   How long does FDA have to review the sponsor's first protocol submission to open an IND?
   A 30 days
   B 15 days
   C 45 days
   D 90 days

3   True or False. A pre-IND meeting is not required but sponsors often avail themselves of the opportunity to meet with the agency to make sure they are on the right track.

4   True or False. Fast track (FT) is a process designed to facilitate the approval of drugs to treat serious conditions and fill an unmet medical need. The purpose is to get important new drugs to the patient earlier by ramping up external FDA reviewers to speed up the review times.

14

Life Cycle Management

Jeffrey S. Barrett

Aridhia Digital Research Environment

Introduction

Product life-cycle management (LCM) is the succession of strategies by business management as a product goes through its life cycle. The conditions in which a product is sold change over time and must be managed as it moves through its succession of the stage. The main task in pharmaceutical life cycle management is identifying external opportunities and threats, such as foreign competition and regulatory demands. Pharmaceutical companies must manage the entire life cycle of a drug in order to make the smartest investment decisions and maximize ROI. Doing so not only ensures healthy revenues, but it also increases competitive advantage and helps maintain strong R&D capabilities for future drug development efforts. There are five distinct product life cycle stages: Product Development, Introduction (sales slowly grow as the product is introduced in the market), growth, maturity, and decline. As indicated in Figure 14.1, until a development candidate is approved and becomes a marketed product, it is not revenue-generating. Based on the 20-year patent life, the innovator company has a limited window within which it can recoup its R&D investment and return profits which also have to support R&D efforts that do not yield marketed products. LCM strategies are focused on extending this window, and planning begins at the time the initial candidate is developed (i.e. during drug discovery). As more about the drug candidate and populations likely to benefit from the drug becomes known, these strategies evolve, and those with the highest probability of technical success are advanced as part of both R&D and commercial strategies.

The motivation for LCM should be clear – the so-called "patent-cliff" refers to the phenomenon of the approach of patent expiration dates and the abrupt drop in sales that follows for a group of products capturing a high percentage of a market. The reality for some sponsors is that their blockbuster pharmaceuticals are "at

Fundamentals of Drug Development, First Edition. Edited by Jeffrey S. Barrett.
© 2022 John Wiley & Sons, Inc. Published 2022 by John Wiley & Sons, Inc.
Companion website: www.wiley.com/go/Barrett/FundamentalsDrugDevelopment

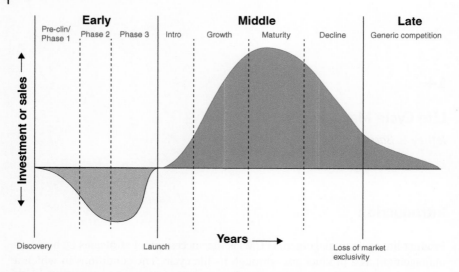

Figure 14.1 Drug life-cycle in relation to product maturity and marketplace performance.

risk" of losing as much as 90% of their sales revenues to generic competition as the steady flow of "patent cliff" expiries continues. Life cycle management in the context of drug development comprises activities to maximize the effective life of a product. Life cycle approaches can involve new formulations, new routes of delivery, new indications or expansion of the population for whom the product is indicated, or the development of combination products. Life cycle management may provide an opportunity to improve upon the current product through enhanced efficacy or reduced side effects and could expand the therapeutic market for the product. Successful life cycle management may include the potential for superior efficacy, improved tolerability, or a better prescriber or patient acceptance. Unlike generic products where bioequivalence to an innovator product may be sufficient for drug approval, life cycle management typically requires a series of studies to characterize the value of the product.

Drug manufacturers may employ various LCM patent strategies, which may impact managed care decision-making regarding formulary planning and management strategies when single-source, branded oral pharmaceutical products move to generic status. Passage of the Hatch-Waxman Act enabled more rapid access to generic medications through the abbreviated new drug application process. Patent expirations of small-molecule medications and approvals of generic versions have led to substantial cost savings for health plans, government programs, insurers, pharmacy benefits managers, and their customers (Berger et al. 2016). However, considering that the cost of developing a single medication is currently estimated at $2.6B (2013 USD estimate), pharmaceutical patent protection enables companies to recoup investments, creating an incentive for innovation. Under current law, patent protection holds for 20 years from the time of patent filing, although much of this time is spent in product development and regulatory review, leaving an effective remaining patent life of 7–10 years at the time of approval. To extend the product life cycle, drug manufacturers may develop variations of originator products and file for patents on isomers, metabolites, prodrugs, new drug formulations (e.g. extended-release versions), and fixed-dose combinations. These additional patents and the complexities surrounding the timing of generic availability create challenges for managed care stakeholders attempting to gauge when generics may enter the market. Of course, a drug sponsor may choose to manufacture their own generic (so-called "authorized generics") to extend control or at least some influence over the pricing of the generic version as a competitor in that market as well. An understanding of pharmaceutical patents and how intellectual property protection may be extended would benefit managed care stakeholders and help inform decisions regarding benefit management. The goal of every sponsor is to reshape the life-cycle curve, as shown in Figure 14.1 so that profitability starts earlier and maturity ends later. This is typically viewed as a matter of survival across all industries.

The goal of this chapter is to define the various approaches that are typically employed in the LCM of different pharmaceutical products and to illustrate how these approaches are tailed to the product modality, the disease therapeutic area, and the marketplace. The timelines for engagement on LCM will be exposed, as will the emphasis on establishing return on investment (ROI). Several examples will be examined in detail.

Common Strategies and Strategies Tailored to Certain Product Types

Common strategies used in LCM include the following: new indications, new formulations, pediatric market exclusivity, disease management programs, strategic pricing changes, authorized generics, combination products, next-generation products, new dosing regimens, patent litigation, Rx-to-OTC switch. The choice of strategy to pursue is dictated to a large extent by the attributes of the drug candidate along with the marketplace assessment. Much of the opportunity evaluation happens very early on in the drug development process and is often articulated to some extent in the initial target product profile.

With respect to compound attributes, basic characteristics such as the physiochemical properties of the drug (e.g. molecular weight, permeability, and solubility) will provide guidance for routes of administration that are possible or not (e.g. transdermal or ocular delivery). Physiologic and ADME parameters such as pre-systemic metabolism, site-specific absorption, etc. will determine if novel

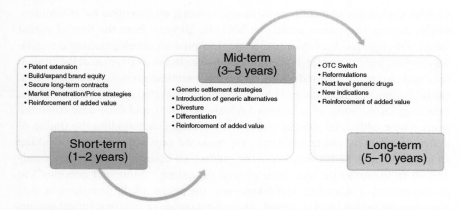

Figure 14.2 Common LCM strategies linked to the duration of their potential impact.

oral delivery systems (e.g. fast-dissolving tablets or extended-release formulations) are possible. Pharmacologic data such as sensitivity and specificity of drug actions and the presence or absence of so-called "off-target effects" may shed light on the potential for drug synergy and both combination product potential as well as the potential for additional target populations and indications. Finally, knowledge of the marketplace can provide data on the acceptability of certain routes of administration, formulations, and benefit: risk characteristics for potential populations of interest that an LCM strategy may target. All of these factors influence the overall strategy, which will be both time and marketplace-sensitive.

Figure 14.2 provides a projection of the various duration of the strategies in relation to the duration of their impact. As the figure suggests, many of these are more closely linked to short- versus long-term strategies, and in only a few cases are they mutually exclusive. In most instances, the entire LCM strategy will incorporate short-, mid-, and long-term opportunities. The specific strategies are described in greater detail in later sections in the chapter along with recent examples.

New Indications – Requirements and Regulatory Considerations

One common strategy depending on the mechanism of action of the drug candidate and the potential medical conditions and clinical benefit derived from the drug's actions is to increase the number of claims regarding these actions. Of course, the regulatory standards are the same – the drug must be shown to be safe and efficacious in the various patient populations, and market access is typically granted only after the successful completion of two adequately powered, well-controlled, Phase 3 trials supporting the intended claims (Holbein 2009). Much of the early-stage development activity (pharmacology, toxicology, formulation development, etc.) can be conserved across indications, but there may be instances when it cannot (e.g. different route of administration or dose required or different biomarkers required to show activity/efficacy or safety) or when it is advantageous to repeat certain studies or design new studies (e.g. new formulation specific to indication) especially if they have patent implications which may offer further protection against competition. Based on recent analysis (Tiene 2017), the development of new indications is currently the most common strategy adopted by pharmaceutical sponsors to extend a product's life cycle. It should be clear that clinical value through life-cycle management can be also be demonstrated by improving efficacy, reducing side effects, simplifying dosing, and increasing patient compliance in addition to proposing a new indication.

There are many options available to a sponsor regarding the choice of indication to pursue a compound that may have clinical benefit to more than one potential population. As all indications pursued carry with them the regulatory requirement for two adequately powered, well-controlled clinical trials that demonstrate safety and efficacy in the intended population of interest, the sponsor must manage the timing and cost associated with conducting such trials. The order indeed may promote certain efficiencies and take advantage of marketplace features. Likewise, factors such as the size of the various target populations, the likely duration of the therapy for certain indications, the price that can be charged based on the marketplace (standard of care cost considerations and other factors), and reimbursement likelihood all contribute to the choice of which indicated to pursue first. Additional operational factors such as the duration and cost of the clinical trials are also relevant and as the sponsor is still reliant on regulatory authorities to review and approve the application. There are additional factors regarding the likelihood of additional requirements including Phase 4 studies that may also be more likely based on the order of filing. As the sponsors motivation is to both recoup their initial investment quickly and extend the duration and extent of market penetration, the fastest route to product launch may not always be the best.

One of the more intriguing developments in drug discovery, genomic profiling, may also have a role in life-cycle management in the future, particularly in the search to examine new indications for current drugs in development and even approved drugs (as in the case of repurposing). Genomic profiling is beginning to extend current applications toward direct medical applications that hold the promise of more precise and individualized healthcare delivery (Mousses et al. 2008). It is likewise appreciated that the wide spectrum of scientific strategies, bioinformatics approaches, IT tools, and knowledge resources developed to support discovery research may have broader applications. Based on a translational engineering approach and intelligence for interpreting genomic information from an individual case, the idea of biological intelligence-based knowledge recovery can be broadly applied for personal genomics across many indications. Such a paradigm can also support the engagement of complex diseases and is particularly suited for supporting therapeutic decisions. Likewise, a complementary tool beyond mechanistic plausibility can be brought to bear to support the extension of a drug candidate's use beyond the primary indication. In the near future, this approach can likewise be a component of the early LCM strategy as well.

New Formulations, Combinations, and Over the Counter (OTC) Switch

A new formulation strategy is intended to increase sales and extend the product life of an existing drug by expanding patient preferences and perhaps targeting certain populations favoring the new formulation (e.g. easy to swallow oral formulations for the elderly). A detailed analysis of such strategies of the innovator pharmaceutical companies is important to the efficient development and marketing of generic drugs by generic pharmaceutical companies. One of the primary drivers for the choice of new formulation as a life-cycle management strategy is the gains to be made via market exclusivity. There are six types of regulatory exclusivity in the United States. The exclusivity periods for a new chemical entity (NCE), a new formulation, a new indication, an orphan drug, a pediatric clinical trial, and the first ANDA filer (submission) with a Paragraph IV certification are five years, three years, three years, seven years, six months, and 180 days, respectively. Market entry by generic pharmaceutical companies is generally not permitted during these exclusivity periods.

A compelling look at new formulations as a life cycle management strategy (Daidoji et al. 2013) assessed 301 approvals and 180 new formulation approvals for which ANDA approvals were granted for the NCEs, to analyze the pharmaceutical life cycle extension period (LEP) as an indicator of the effectiveness of the strategy. For NCEs, approximately 70% of approvals were granted for the dosage forms "Oral-General Formulation" (comprising formulations such as tablets and capsules that do not require any special formulation technology) and "Injection." In contrast, for new formulation approvals, the top three dosage forms accounted for only about 29% of all approvals, but the proportion of each formulation other than those in the top three is less than 5%, indicating that more than 70% of the total comprises formulations other than those in the top three. The trend that more diverse dosage forms were selected for new formulations rather than NCE approvals was confirmed. Looking at the LEP for all 180 new formulation approvals showed a significant difference between "Oral" and "Non-Oral" formulations in the median survival. Survival in this context refers to the length of time the innovator product "survives" or is unchallenged until ANDA approval (the endpoint). Drilling down to look at the trend of oral dosage forms preventing generic drug competition among the oral formulations, dosage forms such as "Extended/Delayed Release," which require advanced technology, significantly extend the LEP. Chronological changes in survival probability show a significant difference between "Oral-General" and "Extended/Delayed Release" formulations in the median survival at 229 days and 1498 days, respectively. While these results are in many ways not surprising, they also illustrate the criticality of acquiring high-quality patents for formulation techniques that pose considerable technical barriers to perfect so as to protect products in new formulation strategies.

Combinations and OTC Switch

Combination drugs are an increasingly popular life-cycle extension strategy (e.g. Advair, Caduet, Vytorin), with impressive worldwide sales since the practice became popularized in the 1990s. Combination drugs are those in which two or more active ingredients are physically or chemically combined to produce a single oral dosage form, inhaler, injection, or transdermal system. In some circumstances, companies are allowed to co-package drugs; however, this is more common for OTC drugs. Combination drugs are often based on two or more ingredients already on the market that qualify (based on their coadministration and new formulation) for a new patent. It is required for approval that the combination drugs provide an improved treatment for at least some type of patient, compared to the single components. Improvement can be based on improved efficacy, greater safety (e.g. fewer side effects) or patient convenience (e.g. once daily versus twice daily dosing). The approval process of a combination drug depends on the experience with the single components. If the single components are already approved, drug agencies move more swiftly. For example, the Food and Drug Administration (FDA) may allow the combination drug to start testing in Phase 3. While empirical research on the sales success of combination drugs is lacking, they can be considered as a form of product bundling (Stremersch and Tellis 2002). Bundling is often pursued to leverage market power from one to another market, but also to provide a quality signal, which lowers the informational costs for customers.

Regulatory authorities sometimes allow drugs with a proven safety under self-medication circumstances to be available over the counter. Similar to prescription drugs, in the United States, the FDA regulates the approval and marketing claims for OTC drugs, and they require an approved label with drug facts for patient education (Ling et al. 2002). There are two different forms of OTC products: (1) those that may only be dispensed after a pharmacy employee has assessed the needs of the patients and has given some patient education, and (2) those that are just like any other consumer product and are freely available in store. In case the OTC drug has a new indication, dosage, or form, it is eligible for three years of market exclusivity. OTC products made up 28% of unit prescriptions in the United States in 2002, comparable to other countries. The advantage of OTC drugs is that they can have high sales for a prolonged period of time. For example, Listerine (oral mouthwash) has been available for over 100 years and is still successful. The number of prescription drugs approved for OTC usage has risen since the nineties. Older classes of H2-antagonists (e.g. Tagamet, Zantac) and antacids are well-known examples of prescription drugs that are converted to OTC. In Europe, switching a prescription drug to OTC can be a good strategy as then it can then be advertised to consumers.

Partnering, Joint Ventures, and Selling Assets

Device and pharmaceutical companies face many non-economy-related pressures as well. Large, fully integrated pharmaceutical manufacturers are undergoing tremendous restructuring and cost-cutting efforts driven primarily by upcoming patent expirations and a lack of strong products in development pipelines. For example, in 2012, the following major drugs were scheduled to go off patent: Astra Zeneca's cholesterol drug Crestor; Forest Laboratories' antidepressant Lexapro; GlaxoSmithKline's diabetes drug Avandia; and Merck's asthma drug Singulair representing a significant future deficit to their companies' sales and profits forecast. Device and pharmaceutical companies also are confronted with a more conservative FDA that is taking longer to approve new products and typically taking a more conservative approach to effectiveness review. Throughout the industry, companies are looking for new ways to extend patent life as they race to bring new products to market before their cash reserves are depleted (Bhat 2005).

The result of these challenges has been a renewed focus on partnering efforts. For example, in 2008, 63% of the year's top forty biotechnology transactions involved licensing or other types of partnering arrangements rather than an outright acquisition of a target company. Partnering can help life science firms deal with these twin challenges of a difficult financial market and an increasingly conservative FDA. "Partnering efforts" is a general term used to describe the broad range of collaborations between life science companies. These strategies can provide needed capital or access to the skills necessary to bring products to market or keep research programs viable when outside capital is otherwise unavailable. Such efforts can include sharing facilities, sponsored research, co-marketing, licensing arrangements, co-development, joint ventures, and a variety of other structures. The level of integration required to implement each varies from low to high, as does the technology development stage at which a given strategy is best implemented.

Some recent prominent examples of each of these categories illustrate the variety of the approaches. With respect to shared facilities, Pfizer launched several Centers for Therapeutic Innovation, a type of open-innovation partnering model putting their scientists side-by-side with academic teams, and now operating in Boston, New York City, San Francisco, and San Diego. To date, more than 20 partnerships have been forged with major academic medical centers, which have yielded over 300 research proposals. With respect to sponsored research examples, Merck allocated \$92M in 2012 to the California Institute of Biomedical Research (Calibr) to bring a drug concept to proof of principle. Harvard University estimates that at least 5% of its funding is derived from industry partnerships. In the category of co-marketing, Pfizer, Yamanouchi, and several other companies teamed up to drive nearly \$13B in global sales of Lipitor.

For the purpose of life cycle management specifically, partnering can assist in all of the aforementioned areas. Table 14.1 below illustrates how the various common

Table 14.1 Benefit of partnering, joint ventures, or the buying/selling of assets to enhance common LCM strategies.

LCM strategy	Partnering, joint ventures, or the asset changeover benefit
New indication	• Partnership may rely on therapeutic area expertise that doesn't reside at innovator company; new indication may hinge upon JV or sublicense agreement.
New formulation	• Separate companies with novel delivery systems (e.g. rapidly dissolvable tablet or extended-release technology) may assist with the development in either a fee-for-service or co-development agreement.
Combinations	• Many situations where combination therapy is warranted particularly polypharmacy settings like HIV and oncology. Many options for co-development or other arrangements.
OTC switch	• Partnering can be based on product development, marketing, or distribution considerations (many examples).

(Continued)

Table 14.1 (Continued)

LCM strategy	Partnering, joint ventures, or the asset changeover benefit
New delivery route	• Especially if the new route involves a new formulation (likely) where expertise comes from external partner.
Disease management	• Costly chronic conditions, including asthma, diabetes, congestive heart failure, coronary heart disease, end-stage renal disease, depression, high-risk pregnancy, hypertension, and arthritis, have been the focus of these programs. Disease management generally entails using a multidisciplinary team of providers, including physicians, nurses, pharmacists, dieticians, respiratory therapists, and psychologists, to educate and help individuals manage their conditions. These services are likely to come from those other than the innovator company, likely via partnership agreement.
Strategic pricing	• Partners in this capacity both advise on strategy or facilitate the marketing, sales, or distribution aspects.
Pediatric market exclusivity	• By running a pediatric clinical trial with FDA's approval, a sponsor can earn an extra six months of market exclusivity whether or not the drug turns out to be effective in children. Partners could assist in many capacities including pediatric formulation development or conducting the development of the pediatric indications.
Authorized (branded) generics	• Can be an agreement with an established generic developer or a formulation developer depending on what and how the generic is to be marketed and/or distributed.

life cycle management strategies can be supported/augmented by partnering, joint ventures, or the buying/selling of assets.

The economic incentives that a sponsor can provide with respect to LCM are often overlooked when we discuss drug development as we tend to focus on science. Strategic pricing as a category may seem vague as a strategy and approach, but it reflects the reality of the product life-cycle in the context of the various stages of post-approval efforts to recoup R&D investment and maximize profits. Figure 14.1 illustrates the trajectories of sales, marketing, and competition in relation to pricing strategies as a product matures. The final stage in a product's life cycle is decline. There is less demand for the product, and businesses must decide if they want to discontinue the product or keep producing and selling it. Companies are unlikely to pull the product entirely and instead add features (formulations, routes, indications, etc.) to make it stand out more and give it fresh life. Discount pricing is another strategy to increase customer traffic. This also has the intention to free up space for new products. Another pricing strategy option is bundling. With bundling, a company includes a declining product in a deal with other products. This can help get rid of the declining product and increase sales.

Another approach is for the innovator to implement authorized generics strategies as part of their overall life cycle management. The "authorized generic" is a recent development in post-expiry strategy. An authorized generic is created when the manufacturer of a drug soon to lose exclusivity contracts with a generic company to sell an "authorized" version of the molecule, in some cases supplying the product to the authorized generic company. Particularly in those instances where a legal challenge (under paragraph IV of the Hatch-Waxman Act) creates the potential for the first generic to enjoy a 180-day exclusivity period, brand owners have begun to utilize this "join rather than fight" approach. Pharmaceutical companies have increasingly pursued authorized generics strategies to retain revenue flow and protect market share. When a generic manufacturer obtains an ANDA approval, they are given 180 days to exclusively market the product. This means that the first to enter the market has the potential for large financial returns. The challenge, however, is that during the 180-day period, the existing branded products are also allowed to market their own generic medications without additional need for further authorization. This results in a gap that can be filled by subsidiaries producing and subsequently marketing generics, or they provide a license to new generic firms, taking over market share that would otherwise go to new entrants to the market (Kvesic 2008; Suri and Banerji 2016). This is a common strategy of sponsor companies who want to maintain their market share and profitability and combat new entrants to the sector (Anusha et al. 2017). While these generic products from branded pharmaceutical companies may impact the profit achieved from their branded products, the net impact is that as a company, the revenue stream is maintained. Although not a long-term strategy, it does prevent some generic firms from entering the market in the short term and, as a result, can be a viable strategy for extending, for a short period, the market share of branded pharmaceuticals.

Drug Repurposing

Drug repurposing refers to the evaluation of existing drugs for their potential use for new therapeutic purposes. The approach capitalizes on the fact that approved drugs and many abandoned compounds have already been tested in humans, and detailed information is available on their pharmacology, formulation, dose, and potential toxicity. Drug repositioning is supported by the fact that common molecular pathways contribute to many different diseases. Drug repositioning has many advantages over traditional de novo drug discovery approaches in that it can significantly reduce the cost and development time, and as many compounds have demonstrated safety in humans, it often negates the need for Phase 1 clinical trials.

There are usually three kinds of repurposing approaches: computational approaches, biological experimental approaches, and mixed approaches, all of which are widely used in drug repositioning (Xue et al. 2018; Pushpakom et al. 2019).

A number of prominent examples exist including sildenafil (Viagra) for erectile dysfunction and pulmonary hypertension and thalidomide for leprosy and multiple myeloma. In each case, the initial indication for which the drug was developed was different from the clinical use in a distinctly different patient population. Likewise, new data was generated to support a new or extended use patent that extended the product's exclusivity for the new indication. In all cases, this patent life extension came with the expectation that the sponsor would conduct the required clinical trials to demonstrate safety and efficacy in the new patient population.

Pricing Considerations

The current market scenario makes it imperative for pharmaceutical companies to focus on comprehensive life-cycle management planning to make the most of each branded product at every stage of its life. This should ideally begin early in the life-cycle of product, possibly in the pre-launch period. Such early planning and monitoring of progress can facilitate the evaluation of a product's economic potential and aid in planning and successful implementation of other LCM strategies. As we have discussed, a multitude of strategies, often used in combination, are available to mitigate the impending revenue loss when the innovator patent expires. Launch of new formulations and identifying newer indications, and drug repositioning are among the most effective and preferred strategies, although they are expensive and take several years for implementation.

When a patent expires for a branded drug, there is an end to the exclusive rights to the market, allowing generic products to use the formulations and offer lower-priced products. Another approach that can be used to combat this challenge is to decrease the price in direct competition with the generics. However, an alternative, competitive pricing approach can lead firms to focus on segments of the market which are not sensitive to price, and thus there is maintenance or increase of prices. This is not, however, an option in regions where medication prices are regulated, such as Europe but can occur in countries such as America. In other words, the market is segmented into those areas where either there are long-term arrangements with hospitals or fee for service general practice. This approach may have the effect of decreasing the elasticity of price, which means increasing a branded price drug is an optimal solution.

Use Case Examples of Life Cycle Management

The LCM strategies include various techniques such as new indications, new formulations, new combinations, authorized generic, OTC drug switch, and pricing strategies among others. The choice of strategy depends on the compound and product attributes as well as the marketplace, as previously stated. Not all of these strategies are equally successful, obviously and many strategies include multiple approaches. It is useful to look more in depth at a few of these to understand the approach. Two past approaches for LCM strategies are discussed in the context of timing and rationale.

Allegra (fexofenadine) is an antihistamine used in the treatment of allergy symptoms, such as hay fever and urticaria. Therapeutically, fexofenadine is a selective peripheral H1-blocker and is classified as a second-generation antihistamine because it is less able to pass the blood–brain barrier and cause sedation, compared to first-generation antihistamines. It was patented in 1979 and was first prescribed for medical use in 1996. Fexofenadine has been manufactured in generic form since 2011. It achieved blockbuster drug status with global sales of $1.87B USD in 2004 (with $1.49B USD coming from the United States). On 25 January 2011, the FDA approved over-the-counter sales of fexofenadine in the United States, and Sanofi Aventis' version became available on 4 March 2011. As of January 2017, it was marketed as a combination drug with pseudoephedrine (under brand names including Alerfedine D, Allegra-D, Allergyna-D, Altiva-D, Dellegra, Fexo Plus, Fexofed, Fixal Plus, Ridrinal D, Rinolast D, and Telfast D) and as a combination drug with montelukast (under brand names including Fexokast, Histakind-M, Monten-FX, Montolife-FX, and Novamont-FX). It is commercially available as an oral immediate-release tablet, capsule (gelcaps), and pediatric solution and suspension.

In summary, fexofenedine's physiochemical properties (good solubility and permeability) facilitated both the ability to create multiple oral formulations and combination products, which enhanced penetration on the marketplace and expanded the product line, indications, and drug label. The product itself can be viewed as an extension of the original product Seldane, which was eventually removed from the market by FDA (fexofenadine is the active metabolite of Seldane but does not exhibit the QT prolongation and drug interaction concerns). The OTC approval obviously substantiated the safety of the drug and further expanded the market and extended the product life-cycle.

Perhaps the most visible LCM strategy belongs to the most successful pharmaceutical product of all time, Lipitor (atorvastatin). At its peak, Lipitor provided approximately $13B USD in annual sales for Pfizer - making it the best-selling product ever, turning Pfizer into the world's largest pharmaceutical company. In the current climate, it is hard to imagine a recurrence of this pharmaceutical blockbuster dream. Long before the development of the first statin, the publication of the lipid hypothesis in the 1850s postulated the science. This was followed by the first landmark study named the Framingham Heart Study – ongoing since 1948 – which highlighted to the world the major heart attack risk factors, such as smoking, hypertension, obesity, and high cholesterol. It marked the beginning of research in deciphering the role of cholesterol and overall lipids in cardio-metabolic disorders. It would take another 40 years from the publication of the Framingham Heart Study for the first statin to make it to the market, and the credit goes primarily to the work of biochemist Akira Endo, who isolated the first statin, mevastatin, from fermentation broths while at Sankyo, a Japanese pharmaceutical company. Ironically, Sankyo never developed statins commercially. Merck, in 1987 was the first company to develop and commercialize the first two statins (Mevacor and Zocor). With statins already on the market, Pfizer was able to enter the market and turn what could have been a poorly differentiated "me-too" drug into a commercial phenomenon (for a drug, it did not develop).

In 1970, Warner-Lambert acquired Park Davis and felt that a marketing partner would help them compete more effectively. In 1997, Warner-Lambert signed a co-marketing deal with Pfizer to market Lipitor. The two companies finally merged in 2000, forming the present-day Pfizer. However, it would take a coordinated strategic approach on Pfizer's management to push Lipitor to the market. Atorvastatin was patented in 1986 and approved for medical use in the United States in 1996. Lipitor was first launched in the United States on 17 December 1996 and, by the end of 2010, had generated cumulative sales of $118B USD for Pfizer. Much of its success is attributed to a unified strategy consisting of hyper-aggressive marketing, deft timing, financial power, and luck. While Pfizer was renowned for its world-class sales force, simple marketing was not enough to push the product. To create momentum, Pfizer invested $800M USD in various studies to highlight the importance of driving LDL cholesterol as low as possible. Pfizer pushed the boundaries of advertising budgets in promoting the potency of Lipitor over already available statins. Pfizer's patent on atorvastatin expired in November 2011. Initially, generic atorvastatin was manufactured only by Watson Pharmaceuticals (authorized generic) and India's Ranbaxy Laboratories. Prices for the generic version did not drop to the level of other generics – $10 or less for a month's supply – until other manufacturers began to supply the medication in May 2012. A big part of Pfizer's LCM strategy was focused on offering patients, and insurance plans big discounts and rebates if they stay on Lipitor until more generics were approved (after Watson's exclusivity expired). The other LCM emphasis was to expand its promotion in geographic areas where its patent was still in force (many other countries where Pfizer heavily promoted Lipitor, especially in emerging markets such as China).

As we have seen, it is clear that branded pharmaceuticals are under high pressure to find ways of improving and extending the profitability stage of the product life cycle. The lengthy R&D process, which can take up to 10 years given the length of some clinical trials and drug approval are taken into consideration, means that the time that the drug is on the market needs to be extended with strategies that encourage continued purchase. These strategies can come from a legal standpoint, either licensing to generic manufacturers so that some level of revenue is retained during the post-patent expiry stage or through the development of lower cost generic medications from the branded firms themselves. Evidence suggests that the majority of branded pharmaceuticals firms will follow at least one of these routes to combat the challenge from generic manufacturers. What is less clear is what level of return can be achieved, and this is an area that would need further investigation in future works.

Similarly, legal approaches to enhancing protection through extensions of the patent are also widely recognized and utilized in the industry. Patents can be applied to cover ingredients, composition, and branding, but also processes. As such, if a new process is developed, that does not require further approval, this can be patent protected, creating further barriers to generic entry to the market. Specific data on the level of improvement in the longevity of the life-cycle is difficult to come by, and legal strategies can take time to implement. What this means for branded firms is that they may still face strong competition as generics find innovative means to bypass the patents once the core patent for selling the medication has expired.

Potentially more effective is the adoption of a marketing focus to life-cycle management. Due to the diversity of approaches that were identified as coming under the marketing strand, such as pricing strategies, promotional approaches, differentiation, and divestiture, it appears that the marketing focus for LCM has a wider impact on improving the longevity of branded pharmaceuticals. In essence, building on company reputation and customer loyalty, i.e. using the brand equity for leverage, appears to deliver the highest potential. The lower research time and costs for reformulations are increasing the usage of this approach. When this approach is aligned with a research and development focus, the knowledge within the firm and the introduction of, for example, combination drugs and next-generation drugs, then the use of a cohesive marketing focus for promoting these products can potentially have a significant impact on the profit-making stage of a product, and thus a greater return on investment. Future LCM strategies will likely have to incorporate the viewpoints of the payer, provider and insurance communities more specifically as the industry transforms due to the growing focus on a value-based healthcare system. This will be heavily influenced by greater scrutiny on real-world data that supports the effectiveness of some treatment options over others.

References

Anusha, K., Priya, P.K., and Kumar, V.P. (2017). Pharmaceutical product management. *The Pharma Innovation* 6 (11, Part B): 112.

Berger, J., Dunn, J.D., Johnson, M.M. et al. (2016). How drug life-cycle management patent strategies may impact formulary management. *The American Journal of Managed Care* 22 (16 Suppl): S487–S495.

Bhat, V.N. (2005). Patent term extension strategies in the pharmaceutical industry. *Pharmaceuticals Policy and Law* 6: 109–122.

Daidoji, K., Yasukawa, S., and Kano, S. (2013). Effects of new formulation strategy on life cycle management in the US pharmaceutical industry. *Journal of Generic Medicines* 10 (3–4): 172–179.

Holbein, M.E. (2009). Understanding FDA regulatory requirements for investigational new drug applications for sponsor-investigators. *Journal of Investigative Medicine* 57 (6): 688–694. https://doi.org/10.2310/ JIM.0b013e3181afdb26. PMID: 19602987; PMCID: PMC4435682.

Kvesic, D.Z. (2008). Product life-cycle management: marketing strategies for the pharmaceutical industry. *Journal of Medical Marketing* 8 (4): 293–301.

Ling, D.C., Berndt, E.R., and Kyle, M.K. (2002). Deregulating direct-to consumer marketing of prescription drugs: effects on prescription and over-the counter product sales. *Journal of Law and Economics* 45 (2): 691–723.

Mousses, S., Kiefer, J., Von Hoff, D. et al. (2008). Using biointelligence to search the cancer genome: an epistemological perspective on knowledge recovery strategies to enable precision medical genomics. *Oncogene* 27: S58–S66. https://doi. org/10.1038/onc.2009.354.

Pushpakom, S., Iorio, F., Eyers, P. et al. (2019). Drug repurposing: progress, challenges and recommendations. *Nature Reviews Drug Discovery* 18: 41–58. https://doi.org/10.1038/nrd.2018.168.

Stremersch, S. and Tellis, G.J. (2002). Strategic bundling of products and prices: a new synthesis for marketing. *Journal of Marketing* 66 (1): 55–72.

Suri, F.K. and Banerji, A. (2016). Super generics – First step of Indian pharmaceutical industry in the innovative space in US market. *Journal of Health Management* 18 (1): 161–171.

Tiene, G. (2017). Lifecycle management strategies can uncover hidden value. *Pharma Manufacturing*. https://www.pharmamanufacturing.com/articles/2017/lifecycle-management-strategies-can-uncover-hidden-value.

Xue, H., Li, J., Xie, H., and Wang, Y. (2018). Review of Drug Repositioning Approaches and Resources. *International Journal of Biological Sciences* 14 (10): 1232–1244. https://doi.org/10.7150/ijbs.24612. PMID: 30123072; PMCID: PMC6097480.

Chapter Self-Assessments: Check Your Knowledge

Questions:

- Define product life cycle management (LCM) as discuss its primary purpose?
- Explain how LCM attempts to combat the so-called patent cliff?
- Explain combination products and OTC switches offer a reasonable LCM strategy?
- Define and explain the authorized generic approach including the final benefit to the company?

Answers:

- Product life-cycle management (LCM) is the succession of strategies by business management as a product goes through its lifecycle. The conditions in which a product is sold changes over time and must be managed as it moves through its succession of stage. The main task in pharmaceutical life cycle management is identifying external opportunities and threats, such as foreign competition and regulatory demands.
- The motivation for life-cycle management should be clear – the so-called "patent-cliff" refers to the phenomenon of the approach of patent expiration dates and the abrupt drop in sales that follows for a group of products capturing a high percentage of a market. The reality for some sponsors is that their blockbuster pharmaceuticals are "at risk" of losing as much as 90% of their sales revenues to generic competition as the steady flow of "patent cliff" expiries continues. Life cycle management in the context of drug development comprises activities to maximize the effective life of a product. Life cycle approaches can involve new formulations, new routes of delivery, new indications, or expansion of the population for whom the product is indicated, or development of combination products.

- Combination drugs are an increasingly popular lifecycle extension strategy (e.g. Advair, Caduet, Vytorin), with impressive worldwide sales since the practice became popularized in the 1990s. Combination drugs are those in which two or more active ingredients are physically or chemically combined to produce a single oral dosage form, inhaler, injection, or transdermal system. In some circumstances, companies can co-package drugs; however, this is more common for OTC drugs. Combination drugs are often based on two or more ingredients already on the market that qualify (based on their coadministration and new formulation) for a new patent. It is required for approval that the combination drugs provide an improved treatment for at least some type of patient, compared to the single components. Improvement can be based on improved efficacy, greater safety (e.g. fewer side effects), or patient convenience (e.g. once daily vs. twice daily dosing). The approval process of a combination drug depends on the experience with the single components.

- An authorized generic is created when the manufacturer of a drug soon to lose exclusivity contracts with a generic company to sell an "authorized" version of the molecule, in some cases supplying the product to the authorized generic company. Particularly in those instances where a legal challenge (under paragraph IV of the Hatch-Waxman Act) creates the potential for the first generic to enjoy a 180-day exclusivity period, brand owners have begun to utilize this "join rather than fight" approach. Pharmaceutical companies have increasingly pursued authorized generics strategies to retain revenue flow and protect market share. When a generic manufacturer obtains an ANDA approval they are given 180 days to exclusively marketing the product. This means that the first to enter the market has the potential for large financial returns.

Quiz:

1   The five distinct product life cycle stages include the following: (select the correct list):
    A   Product development, introduction, growth, maturity, and decline.
    B   Introduction, growth, valuation, maturity, and decline.
    C   Introduction, growth, valuation, maturity, and decline.
    D   Product development, growth, valuation, maturity, and decline.

2   Under current law, patent protection holds for 20 years from the time of patent filing, although much of this time is spent in product development and regulatory review. What is the typical effective remaining patent life at the time of approval?
    A   1–5 years
    B   7–10 years

C 10–15 years

D None of the above

3 Common strategies used in LCM include the following except which: Choose the best answer.

A New indications, new formulations, pediatric market exclusivity,

B Disease management programs, strategic pricing changes,

C Authorized generics, combination products, next-generation products,

D New dosing regimens, patent litigation, Rx-to-OTC switch.

E All are correct

4 True or False. Drug repositioning has many advantages over traditional de novo drug discovery approaches in that it can significantly reduce the cost and development time, and as many compounds have demonstrated safety in humans, it often negates the need for Phase 3 clinical trials.

15

Pre-formulation and Formulation Development
Robert G. Bell, PhD

Drug and Biotechnology Development, LLC, Clearwater, FL, USA

Introduction

In common terms, a pharmaceutical formulation can be viewed as a recipe of your favorite food or drink, like's Mom's apple pie or a bartender's special concoction. However, in the highly regulated pharmaceutical industry, a pharmaceutical formulation is a drug preparation, process, and manufacturing recipe that have a continuing lifecycle that demonstrates the drug product can be made consistently with known quality attributes throughout development to commercialization and beyond. Formulation development encompasses a very wide range of activities that include pre-formulation, analytical assay development, testing and chemical characterization, excipient screening, dosage form development, delivery, design and manufacture, and the stability of the chosen dosage form such as a solid, topical, aerosol, liquid, or lyophilized dosage form.

The point of pharmaceutical formulation development is to design a quality drug dosage product and its manufacturing process to consistently deliver the intended performance and specifications of the product. This is usually evaluated by the design of experiments (DoE) with the drug substance (known also as the active pharmaceutical ingredient (API)), excipients (excipients are pharmaceutical inactive ingredients that solubilize, suspend, thicken, dilute, bind, emulsify, stabilize, preserve, color, and flavor drug products into safe, efficacious and appealing dosage forms), container closure systems, and the subsequent manufacturing processes. The information and knowledge gained from pharmaceutical formulation development studies and associated manufacturing experience provide a scientific understanding and information to support the establishment of the design space, specifications, process, manufacturing controls and is the basis for quality risk management. It is important for the reader to recognize that

Fundamentals of Drug Development, First Edition. Edited by Jeffrey S. Barrett.
© 2022 John Wiley & Sons, Inc. Published 2022 by John Wiley & Sons, Inc.
Companion website: www.wiley.com/go/Barrett/FundamentalsDrugDevelopment

quality cannot be tested into drug products, its quality needs to be built in by design. Changes will occur in formulation and manufacturing processes during the development and lifecycle management of the drug product as the pharmaceutical manufacturer gains additional knowledge about the product's behavior to further establish the design space and control strategies that are critical to product quality.

The appropriate design and formulation of a dosage form require consideration of the physical, chemical, and biological characteristics of all the drug substances and pharmaceutical ingredients to be used in manufacturing the final drug product. The drug, pharmaceutical excipients, and container closures must be compatible with one another to produce a drug product that is safe, efficacious, stable, organoleptically appealing, easy to administer and manufacture with quality, packaged, and labeled in the appropriate containers that keep the product stable. These critical quality attributes should be defined in a Product Development Plan and a Quality Target Product Profile as described in International Conference on Harmonization (ICH) guidance Q8, Pharmaceutical Development (FDA 2009), ICH Q9, and Quality Risk Management (FDA 2006).

Prior to the formulation of any drug product, it is essential that the fundamental physical and chemical properties of the API, potential excipients, and container closures are thoroughly understood. These properties will determine the subsequent formulation development approaches. This phase of formulation development is known as preformulation. Once these properties are elucidated, the formulations can be designed for the phases of drug development – pre-clinical animal studies, Phase 1 first in human studies, pharmacokinetic and dose-ranging studies, Phase 2 efficacy and proof of concept studies, Phase 3 pivotal clinical studies, and Phase 4 post-approval studies. During these phases of product development, the formulation will undergo changes during the product's lifecycle. These changes could include different API salt forms, changes in the dosage form (from an IV to an oral liquid to an oral solid), changes in the manufacturing process, changes in analytical and quality control testing, and changes in clinical indications. Formulation development, as with product development, is risk-based which has a lifecycle that evolves over time as more product knowledge and experience is gained. This knowledge is used to improve the product over its' lifecycle – the only constant in life and pharmaceutical development is change and thus, being proactively prepared for change.

According to the FDA's website, there are over 150 dosage form types (FDA n.d.). The most common dosage forms are oral solids (tablets and capsules), oral liquids, injections, suppositories, and pessaries, topical (ointments, creams, lotions, etc.), transdermals, ophthalmic preparations, and aerosols (inhalation, nasal). It is not possible for one chapter to define and discuss all of the aspects of formulation development of the many dosages forms; however, this chapter will provide an overview of the basic principles and quality aspects associated with small molecule (not large molecule biologics) formulation development, focusing mainly on preformulation and formulation activities related to oral solids such as tablets and capsules (since they are the most common and preferred route of drug administration, mainly due to patient compliance, cost-effectiveness, design of dosage form and ease of production), oral liquids and intravenous (IV) injections and the quality aspects of associated with these drug formulations. The interested reader should seek additional sources such as Remington's The Science and Practice of Pharmacy (Adejare 2020), Martindale: The Complete Drug Reference (Brayfield 2017), Encyclopedia of Pharmaceutical Technology (Swarbrick 2013), Ansel's Pharmaceutical Dosage Forms and Drug Delivery Systems (Allen and Ansel 2014), Lachman's The Theory and Practice of Industrial Pharmacy (Lachman et al. 2016), and other books (Ofoefule 2002; Ghosh and Jasti 2005; Jones 2008; Shayne 2008; Felton 2012; Dash et al. 2014; Hoag 2017; Aulton and Taylor 2018; Niazi 2019), pharmaceutical journal and articles regarding the formulations of other pharmaceutical dosage forms.

Preformulation

Preformulation is the associated research and development process performed to determine the physical, chemical, and mechanical properties of an API, the potential excipients and container closure to develop a stable, safe, and effective dosage form. It is usually desired to eventually produce an oral solid dosage form such as a tablet or capsule for patient compliance, cost-effectiveness, and ease of production. However, an intravenous (IV) formulation is usually required during early toxicology, metabolism (absorption, distribution, metabolism, and excretion (ADME)), bioavailability/bioequivalence (BA/BE), and first in human clinical studies to assure precise drug dosing and deposition. The understanding of the physical, chemical, and biological considerations is essential for formulation development and is the initial priority in formulation development. Usually, at the early stages of development, there is a limited amount of API available, so it is imperative that critical chemical information on the API be prioritized, such as identity, purity, and solubility. Table 15.1 lists some of the critical attributes of the drug substance that should be examined.

As mentioned previously, it is not unusual that at the beginning of preformulation studies that there may be a limited quantity of the drug substance available. The initial critical preformulation activities are to accurately identify and assay the molecule and to understand its' solubility. This requires analytical methods, usually simple spectroscopic analysis initially followed by chromatographic methods capable of assessing purity, impurities, and stability of the drug substance and

Table 15.1 Preformulation drug substance characterization.

Activity	Rationale/Methods/Characterization
Literature and patent searches	Prior relayed relevant information
Qualification of analytical instrumentation	United States Pharmacopeia <1058>
Spectroscopy (Identity and characterization)	Ultraviolet-visible-infrared, fluorescence, phosphorescence, atomic absorption, inductively coupled plasma, nuclear magnetic resonance, Carbon 13 nuclear magnetic resonance, mass spectroscopy, X-ray, optical rotation, hyphenated chromatographic-spectroscopic methods
Melting point (Polymorphs, hydrates, solvates)	Capillary melt, differential scanning calorimetry (DSC), thermogravimetric analysis, hot stage microscopy
Solubility Aqueous Nonaqueous	Purity, phase solubility, dissolution, hygroscopicity intrinsic solubility, pH effects, pKa, salts, stability vehicles, extraction, partition coefficient, lipophilicity
Appearance, odor, solution color	Visual, smell, spectroscopy
Assay development	Chromatography, spectroscopy, titration
Impurities (Organic, inorganic, heavy metals)	Chromatography, spectroscopy
Microscopy (Morphology, particle size)	Optical/Electron microscopy, laser diffraction
Stability (Solid state and in solution)	Thermal, acid, base, oxidation, photolysis, metal ions, pH 1–14
Excipient compatibility	DSC, assay (Chromatography)
Powder flow	Bulk density, angle of repose
Compression properties	Oral solid excipient choices
Solution properties	Oral liquid excipient choices

product. It is important that all instruments used in the characterization and analysis of drug substances and products are fully qualified and validated as per the United States Pharmacopeia (USP) <1058> Analytical Instrument Qualification (United States Pharmacopeia 2020a).

Identification of the drug molecule is usually conducted by spectroscopic methods (e.g. nuclear magnetic resonance (NMR), carbon[13] nuclear magnetic resonance (C^{13}NMR), mass spectroscopy (MS), ultraviolet (UV), and infrared spectroscopy (IR)), and establish purity with a stability-indicating analytical method, usually with a chromatographic method (e.g. high-performance liquid chromatography (HPLC), with spectroscopic detection (UV, MS), gas chromatography (GC), etc.) which enables an understanding of the quality attributes of the suitability of both the drug substance and subsequent drug product for its intended use. The solubility of the drug substance is perhaps the most critical quality attribute that needs to be determined since it will dictate the initial dosage form. Usually, the initial studies require an IV injection, so an understanding of the aqueous solubility (especially in saline), its dissociation parameters, and behavior at various pH's is critical. The pKa (the negative base-10 logarithm of the acid dissociation constant) of the drug molecule is important since it provides an understanding of the pH needed to maintain solubility and points to salt forms that would be required to achieve bioavailability from the solid state for oral solid dosage forms. It is desired that the drug substance have a solubility of greater than 1% (10 mg/mL) over the pH range of 1–7 at 37 °C and an intrinsic dissolution rate greater than 1 mg/cm·min (Kaplan 1972). However, keep in mind that the drug substance may be insoluble or unstable in water. If this is the case, miscible cosolvents (e.g. propylene glycol, ethanol, glycerol, polyethylene glycol (PEG), etc.) may be required to improve the stability and solubility and facilitate extraction and assist in analytical analysis through an understanding of the octanol-water partition coefficient (K_{ow}). These initial solubility evaluations form the basis for the structure-activity relationship, which depends on a thorough understanding of the physiochemical properties of the drug substance molecule.

One of the analytical techniques applied to the characterization of pharmaceutical drug substances is the melting point (MP) determination. In addition, the identifying aspects of chemical and crystal purity, it helps assist with the identification of polymorphs, which is a solid material with at least two different molecular arrangements resulting in distinct crystal species. The main concern of polymorphs is their differences in formulation stability and solubility. This compliments microscopy, x-ray, and laser diffraction to understand the crystallography, crystal structure and habit (morphology), polymorphs, solvates, and particle size of the drug substance.

Understanding the stability of your drug substance and formulation is critical to the future commercialization of the intended drug product. Commercial APIs usually should have a shelf life of five years, the final drug product should have a shelf life of at least two years, and the potency of both should not fall below 90% at the recommended storage conditions. In addition, the drug substance and drug product should look, perform, and maintain the same quality attributes that it had when first manufactured.

Drug substance and drug product decomposition occurs mainly through thermal (with and without humidity), acid-base hydrolysis, oxidation, photolysis, and trace metal catalysis. Thermal effects typically lead to drug substance and drug product degradation. The effects of temperature on thermal degradation of a substance, especially in solution, can be appreciated through Arrhenius equation:

$$K = Ae - Ea / RT \left(\text{or } \log K = \log A - Ea / 2.303 \, RT \right) \tag{15.1}$$

where K = specific reaction rate;
A = frequency factor;
Ea = energy of activation;
R is the universal gas constant; and
T is the absolute temperature (in Kelvins).

With the Arrhenius equation, it can be roughly estimated that the rate of reaction increases by a factor of about 2 to 5 times for every 10 °C rise in temperature [Conners 1990; Helmenstine 2020; Aulton 1988; Wikipedia n.d.]. This assumption forms the basis for accelerated testing (40 °C/75% Relative Humidity) in the ICH Stability Testing of New Drug Substances and Products Guidance Q1A(R2) [ICH n.d.]. Accelerated testing for six (6) months at 40 °C with 75% relative humidity typically demonstrates that the drug substance and drug product should be stable for up to two (2) years. Testing at room temperature conditions (25 °C/60% relative humidity) for two to five years will be needed to support this claim; however, if the drug substance and drug product demonstrate they are stable at accelerated temperatures and humidity, it is likely they will be stable for at least two or more years at room temperature. In addition, the time course of drug degradation will elucidate the drug reaction kinetics, half-life of the drug substance and product, further supporting the stability and quality attributes of the pharmaceutical product.

Excipient selection and compatibility with the drug substance is essential to any formulation – if the inactive ingredients interact with the drug substance, there will be no viable stable or effective formulation. Thermal analysis such as differential scanning calorimetry (DSC) with 50% mixtures of the drug substance and excipients can determine if drug-excipient reactions will occur through thermal changes in their melting points. In addition, simple organoleptic testing such as the visual appearance, odor, color, and taste of the formulation at room and accelerated temperatures and humidity will assist in the determination of drug-excipient compatibility.

Hydrolysis and solvolysis are likely culprits associated with drug instability for both solid and liquid products, whether it is associated with promoting pH-based acid-base reactions in the presence (or absence) of metal ions or changing the ionic strength or polarity of the drug substance's local environment. Hydrolytic reactions usually involve nucleophilic attack of labile bonds by water or changes in pH or by a solvent (solvolysis). Oral solid formulations such as tablets and capsules can be susceptible to moisture (e.g. they are hygroscopic) to which they become unstable and could become deliquescent (e.g. dissolve itself), so the formulation needs a balance to be "wettable" to dissolve in the gastrointestinal tract but has limit hygroscopicity to ensure physical-chemical stability. Solid oral formulation manufacture also requires the drug product powder to flow, and moisture affects the bulk density, angle of repose, and tablet compression properties. The appropriate preformulation studies with the drug and excipients will assist in determining these factors as well as the appropriate container closures (e.g. with or without a desiccant or a heat induction seal) to prevent air, moisture, and light from affecting the product during shelf life.

Oxidation (red-ox reactions), light (photolysis), and trace metals (complexation) are usually reactions introduced by the environment to the drug substances or products when manufactured or stored. The use of antioxidants (e.g. ascorbic acid, metabisulfite, etc.), yellow manufacturing lights, light-resistant containers, and chelating agents (e.g. ethylene diamine tetraacetic acid [EDTA]) can be used to minimize formulation oxidation, photolysis and trace metal degradation, respectively.

Appropriately performed preformulation studies will provide a roadmap to a successful formulation for identifying both liquid and solid dosage forms. The physical-chemical properties such as solubility and crystallography will allow for formulation design and drug substance salt formation. Careful excipient selection and drug-excipient interactions identified through thermal analysis are critical for a viable formulation. Stability studies in drug solutions will guide parenteral and liquid dosage forms as well as identify instability and methods of stabilization. The analytical techniques measuring the drug formulations must be accurate, precise, and robust. This is where quality and risk analysis start - it must be built into the product and the design space of the formulation at the beginning of product development which will assist and assure the quality of manufacture throughout the product's lifecycle.

Formulation of Parenteral Products

Parenteral dosage forms are injected directly into body tissue through the skin and mucous membranes and must be sterile, pure, and free from physical (e.g. particulate matter, etc.), chemical (e.g. impurities), and biological contaminants (e.g. pyrogens, endotoxins), which requires the pharmaceutical industry to practice strict aseptic and sterile controls and current good manufacturing practices (cGMPs) to assure sterility and quality of the product. All parenteral products (e.g. solutions, suspensions, emulsions, lyophilized sterile powders, liposomes, implantable devices (including microparticles and products that consist of both a drug and a device such as drug-eluting stents) must be stable and free from microbial contamination throughout the shelf-life of the product and be compatible with intravenous (IV) diluents, delivery systems, and other drug products that may be co-administered with the drug product. Parenteral products can be immediate release (e.g. solutions, reconstituted lyophilized powders) or sustained release drug products (e.g. suspensions, depots, implants).

The volume of the parenteral injection depends on the solubility determined during the preformulation studies and the preferred route of administration. Parenteral products should be isotonic (approximately 308 mOsm/l) for intravenous (IV) infusions, cerebrospinal administration, etc., but this depends on the intended site and route of administration of the injectable product (e.g. intramuscular (IM), intravenous (IV), subcutaneous (SC), intra-arterial (IA), epidural, intraventricular, intra-articular, subcutaneous, intrathecal, intracisternal, intraocular). Volumes of injections can range from large volume parenteral (LVP), which applies to single-dose injections intended for intravenous use and packaged in containers in volumes greater than 100 ml, and small volume parenterals (SVPs) which are injections packaged in containers in volumes of 100 ml or less. Volumes greater than 15 ml for IV use should not contain bactericides.

Aqueous solutions are the most common parenteral formulation and can be used by any route of administration. Aqueous parenteral drugs are formulated mainly as solutions (immediate-release formulations) and suspensions (sustained release formulations), but there are many types of parenteral formulations such as emulsions, liposomes, microspheres, nanosystems, sustained-release depots and oils, and reconstitutional powders (for chemically unstable drugs). As determined in the preformulation activities, the drug substance and final product should have good water solubility to be administered in small volumes for IM and SC administration. The IV route permits drug doses in a larger range of volumes. Cosolvent solutions are formulated with water for injection and combined with cosolvents such as ethanol, propylene glycol, polyethylene glycols (PEG) to make solution dosage forms from drugs with poor water solubility. The use of cosolvents typically makes the injection hypertonic (>308 mOsm/l), which can cause irritation at the injection site or hemolytic reaction. These dosage forms can generally be used IV or IM, but the drugs in cosolvent solutions, when used for IM administration, may precipitate, and the undissolved drug precipitate may behave like a depot dosage form and be absorbed very slowly (like a sustained release product). Drugs formulated as emulsions for water-insoluble drugs for IV administration are less irritating to veins than cosolvent solutions and offer a less toxic alternative to surfactant solubilized solutions, which can cause adverse events such as bronchospasm, hypotension, nephrotoxicity, and anaphylactic reactions.

Injectable solutions are clear, homogeneous liquid dosage forms that contain one or more drug substances and excipients dissolved in an aqueous or nonaqueous solvent or a mixture of mutually miscible solvents. Suspensions for injections are sterile solid particles of uniform dispersion once reconstituted (resuspended) in the appropriate solvent. Sustained-release parenteral products can be formulated as suspensions and oily solutions as well as a number of novel delivery systems such as polymeric gels, liposomes, PEGylation (polyethylene glycol polymers), and microspheres. Suspensions are dispersing systems with the solid undissolved drug in a sterile aqueous or oily vehicle for drugs not stable in water. Suspensions have greater chemical stability than solutions and the ability to release drug as it dissolves over time from an injected depot and can be administered by IM, SC, IA, epidural and intrathecal routes. Oil parenteral formulations are irritating to use SC and are mainly used by the IM route. Microparticle sterile suspensions generally range from 20 to 100 µm in diameter, consisting of drug substances embedded within a biocompatible, bioresorbable polymeric excipient and are usually suspensions used for extended-release injections. Sterile powders for injection consist of drug substances and other excipients to ensure the stability of the ingredients. Usually, a sterile diluent is provided to facilitate reconstitution to the desired final volume. The sterile powders for injection can be formulated as lyophilized (freeze dried) powders, powdered solids or dry solids that form viscous liquids upon constitution. Liposomes for injection are drug products with unique properties that can be either solutions or suspensions that are aqueous dispersions of amphiphilic lipids with low water solubility that have an aqueous core surrounded by phospholipid layers. Emulsions for injections are liquid preparations of drug substances dissolved or dispersed in a suitable emulsion medium such as oil-in-water or water-in-oil emulsions that typically entrap the drug substance. Implants consist of a sterile matrix of drug substance and polymeric excipient that may or may not have an outer rate-controlling membrane used for extended-release injectable formulations. Drug-eluting stents are small metal or polymer scaffolds used to keep arteries open following a medical intervention where the drug substance is incorporated into or onto the stent that elutes from the device over time.

All solvents and vehicles used in the delivery of injectable products should be soluble, inert, nontoxic, physically, and chemically stable, especially to changes in pH. If hydrolysis is shown not to be an issue through preformulation studies, water is the ideal vehicle for most injections. The water should be pure, sterile, free of contaminants, chemically, microbially, pyrogens, and comply with such compendial standards such as the USP's monographs for water for injection (United States Pharmacopeia 2020b, 2020c). If the drug substance is not water soluble, cosolvents (e.g. ethanol, propylene glycol, etc.), cyclodextrins (α and β cyclodextrins) oils (e.g. sesame, soybean, castor, etc.), suspension or lyophilized dosage forms can be considered. These sterile dosage forms can generally be used IV or IM, but the drugs in cosolvent solutions may also when used IM and the undissolved drug precipitate may behave like a depot dosage form and absorbed very slowly. Drugs formulated as emulsions for water-insoluble drugs for IV administration are less irritating to veins than cosolvent solutions and offer a less toxic alternative to surfactant solubilized solutions, which can cause adverse events such as bronchospasm, hypotension, nephrotoxicity, and anaphylactic reactions.

Parenteral product formulation is essential to the safety and efficacy of the injectable drug product. In addition to identity, purity, impurities, strength, uniformity, appearance, foreign and particulate matter (USP Visual Particulates in Injections <790>), elemental impurities (USP Elemental Impurities-Limits <232>), vehicle specifications (e.g. USP Water for Injection, USP Sterile Water for Injection, Fats and Fixed Oils <401>, etc.), the container closure, labeling, sterility, and freedom from microbial contamination must be assured. The sterility of all drug products intended for parenteral administration should be confirmed using methods described in the USP Sterility Tests <71> and prepared in a manner designed to limit bacterial endotoxins as described in USP Bacterial Endotoxin Tests <85> and/ or Pyrogen Tests <151>. If microbial preservatives are used (i.e. multi-dose containers), products must meet the requirements of USP Antimicrobial Testing <51> and Antimicrobial Agents-Content <341>. Parenteral quality attributes can be assured by examining their stability regulatory specifications over their shelf life. The USP monograph for Phenytoin Sodium Injection USP requires that the product demonstrate its' identity, purity, and strength over its expiry period, which is 24 months. This includes identity testing by chromatographic retention and infrared (IR) spectroscopy; the assay needs to be 95–105% by HPLC, analysis of alcohol (methanol and ethanol) and glycol content (propylene and ethylene glycol), bacterial endotoxins by USP <85>, pH of 10.0–12.3, particulate matter by USP <788>, meets the requirements of Injections and Implanted Drug Products as per USP <1> and have the appropriate package, labeling, and storage (United States Pharmacopeia 2020d). In addition, there are USP-specific product tests for other parenteral dosage forms such as emulsions, liposomes, suspensions, and implants. Interested readers should consult the USP (or other international pharmacopeias), FDA and ICH guidances, books such as Sterile Drug Products – Formulation, Packaging, Manufacturing, and Quality (Akers 2016) journals (i.e. Parenteral Drug Associations Journal), and professional associations such as the Parenteral Drug Association.

Oral Liquids and Suspensions

Pharmaceutical oral liquids are liquid mixture preparations consisting of one or more non-sterile medicaments dissolved, suspended or diffused an aqueous vehicle that include solutions, suspensions, draughts, elixirs, linctuses, mouthwashes, nasal sprays, ear drops, syrups, enemas, etc. This section will focus only on aqueous oral solutions and suspensions. The interested reader should seek additional sources such as Remington's The Science and Practice of Pharmacy (Adejare 2020), Martindale: The Complete Drug Reference (Brayfield 2017), Encyclopedia of Pharmaceutical Technology (Swarbrick 2013), Ansel's Pharmaceutical Dosage Forms and Drug Delivery Systems (Allen and Ansel 2014), Lachman's The Theory and Practice of Industrial Pharmacy (Lachman et al. 2016), and other books (Ofoefule 2002; Ghosh and Jasti 2005; Jones 2008; Shayne 2008; Felton 2012; Dash et al. 2014; Hoag 2017; Aulton and Taylor 2018; Niazi 2019), other books, journals and articles regarding the various liquid dosage forms.

Although tablets and capsules are the most common oral dosage forms, liquids, since they are easier to swallow, are generally preferred for pediatric and geriatric populations. A drug must be in solution before it can be absorbed, so drugs administered in a liquid solution are immediately available for absorption, usually with a faster therapeutic response. However, liquid solutions can be unstable (due to hydrolysis), potentially reducing their shelf life, require volumetric dosing and are bulkier to ship and store than tablets or capsules.

The simplest form of a solution is a mixture of two or more components that form a single homogenous phase. The solute is the component that gets dissolved, and the solvent is the medium in which solute dissolves. The solvent usually constitutes the largest portion of the solution, and the solubility of a substance is the amount of compound that dissolves in the solvent, which determines the concentration, usually expressed as the weight or volume of solute that is contained in the weight or volume of a solution. USP Purified Water is the most widely used solvent or vehicle because it is nontoxic, physiologically compatible, and dissolves a wide variety of drug substances. However, not all substances are water-soluble, and the formulation may require additional components such as cosolvents (e.g. ethanol, sorbital, glycerol, etc.), pH control for drug substances that are weak acids or bases (e.g. citrate to maintain pH's between 3 and 5, phosphate to maintain pH's between 6 and 8.), micelles (e.g. polysorbates, etc.), complexation agents (e.g. polyvinylpyrrolidine, etc.), particle size reduction and even a change in the drug substance's salt form. If it is not possible to use these technologies, nonaqueous solutions using oils or polyethylene glycols are possible. Oral solutions usually require formulation additives such as buffers to maintain the appropriate pH, flavors and sweeteners to mask unpleasant tasting drugs, coloring (usually to compliment the flavor (yellow [FD&C #5 or #10] coloring if the flavor is lemon)), antioxidants such as ascorbic acid, tocopherol, and sodium metabisulfite (to prevent oxidative decomposition), viscosity modifying agents such as methyl cellulose (to assure accurate dosing), chelating agents such as EDTA to protect the drug substance from catalysts that could accelerate oxidative reactions, and preservatives such as antimicrobial agents such as benzoic acid or parabens. The stability of oral solutions in their intended container closure system is critical. The oral solution must retain its' potency as well as its' other quality attributes such as clarity, color, odor, taste, and consistency (viscosity) over its intended shelf life (United States Pharmacopeia, 2020e).

Oral solution quality attributes can be assured by examining their stability regulatory specifications over their shelf life. The USP monograph for Acetaminophen Oral Solution USP requires that the product demonstrate its' identity, purity, and strength over its expiry period, which is 24 months. This includes identity testing by chromatographic retention and thin-layer chromatography; the assay needs to be 90–110% by HPLC, deliverable volume, uniformity of dosage, impurity testing of 4-aminophenol, pH (3.8–6.1), alcohol determination, and appropriate package, labeling and storage (United States Pharmacopeia 2020f).

Oral Suspensions

A suspension is a dispersion of insoluble solid particles (dispersion phase) in usually an aqueous liquid medium (the dispersion medium), although in some formulations, the dispersion media can be nonaqueous or an oil. The suspension dosage form provides a liquid dosage form for insoluble or poorly soluble drugs and drugs that are unstable in an oral solution. The physical characteristics, especially the particle size of the drug substance (usually very fine (colloidal suspensions) to micronized (less than 25 microns)) and content uniformity (due to the potential for particle segregation), are critical quality attributes for suspensions. Other critical quality attributes of suspensions include viscosity, pH, and dissolution. Viscosity can be an important aspect to minimize segregation and has been shown to be associated with bioequivalence. The pH of the suspension affects the preservative systems and influences the amount of drug in the solution. Many suspensions such as phenytoin, carbamazepine, and sulfamethoxazole, and trimethoprim suspensions have dissolution specifications to assure the suspension is absorbed (Oral Solutions and Suspensions (8/94) n.d.).

In an ideal suspension, insoluble drug particles are uniformly suspended the dispersion medium for a prolonged period to assure the suspension will contain the same amount of drug (content uniformity) and will give the same therapeutic effect. A flocculated suspension is a suspension in which the dispersion medium quickly becomes clear due to the formation of large flocs that settle rapidly. Flocculated suspensions form loose sediments which are easily resuspendable, but the sedimentation rate is fast so there maybe issues with inaccurate dosing as well as the suspension's unappealing appearance. A deflocculated suspension is a suspension in which the suspended particles remain as separated units, leaving the dispersion medium cloudy for an appreciable time after shaking, due to the very slow settling rate of the particles in the product, preventing entrapment of liquid within the sediment, which can become compacted and can be very difficult to redisperse.

As with oral solutions, suspensions should be physically, chemically, and microbiologically stable and have the appropriate organoleptic characteristics of taste, color, odor. Suspensions should also not be too viscous (not too thick) and have good resuspendability properties and sedimentation rates. The most commonly used vehicle to disperse suspensions is purified water, but some suspension formulations can contain nonaqueous solvents that impart stability, such as propylene glycol and polyethylene glycols. The choice of the dispersion medium depends on the nature and physical and chemical properties of the drug substance, which is determined by prior preformulation assessments. As with oral solutions, suspension formulations usually include buffering agents, preservatives, antioxidants, flavoring, colorants, sweeteners, and chelating agents. In addition, suspensions usually contain wetting agents such as polysorbates or sorbitan esters to improve the homogeneity and vehicle flow with the insoluble particles, excipients such as simethicone to prevent foaming, suspending agents such as cellulose derivatives and acacia to prevent agglomeration, decrease sedimentation and increase viscosity, flocculation modifiers such as sodium chloride to prevent caking, and humectants such as glycerol to prevent the evaporation of the aqueous vehicle during storage.

Oral suspension quality attributes can be assured by examining their stability regulatory specifications over their shelf life. The USP monograph for Phenytoin Oral Suspension USP requires that the product demonstrate its' identity, purity, and strength over its expiry period, which is 24 months. This includes identity testing by infrared spectroscopy and chromatographic retention; the assay needs to be 95–105% by HPLC, dissolution (80% in one hour), deliverable volume, uniformity of dosage, organic impurity testing of phenytoin related substances A, B, and others ($\leq 0.9\%$), and the appropriate package, labeling, and storage (United States Pharmacopeia, 2020g; Mahato & Narang, 2018).

Oral Solid Tablets and Capsules

Solid oral dosage forms include tablets (i.e. immediate release, sustained release, chewable, enteric coated, buccal, sublingual, effervescent, pellets, vaginal, etc.) and capsules (i.e. hard shell and soft gelatin). This chapter will only focus on immediate-release tablets and hard and soft gelatin capsules. The interested reader should seek additional sources as previously referenced (Ofoefule 2002; Ghosh and Jasti 2005; Jones 2008; Shayne 2008; Felton 2012; Swarbrick 2013; Allen and Ansel 2014; Dash et al. 2014; Lachman et al. 2016; Brayfield 2017;

Hoag 2017; Aulton and Taylor 2018; Niazi 2019; Adejare 2020) such as Remington's The Science and Practice of Pharmacy, Martindale: The Complete Drug Reference, Encyclopedia of Pharmaceutical Technology, other books, journals and articles regarding these other oral solid pharmaceutical dosage forms and formulations.

Oral solid tablets are the most popular dosage form, and 70% of the total medicines are dispensed in the form of a tablet (Patil et al. 2016). A tablet is a compressed solid dosage form containing drug substances (APIs) with or without excipients that vary in size, shape, color, coatings, weight, hardness, thickness, disintegration, and dissolution depending on their intended use and method of manufacture. Most tablets are used for the oral administration of drugs but also may be administered sublingually, buccally, or vaginally.

Tablets are prepared primarily by direct compression, molding, and more recently by three-dimension (3D) printing (one FDA-approved drug – Spritam® – is manufactured using 3D printing technology (CDER Researchers Explore the Promise and Potential of 3D Printed Pharmaceuticals n.d.)). Compressed tablets are manufactured with tablet machines with various shaped punches and dies capable of compacting the powdered or granulated material into a tablet. Molded tablets are prepared by forcing dampened powder material into a mold from which is then ejected and dried.

Tablets are solid dosage forms usually prepared with the aid of suitable pharmaceutical excipients. They may vary in size, shape, weight, hardness, thickness, disintegration, and dissolution characteristics and in other aspects, depending on their intended use and method of manufacture. Most tablets are used in the oral administration of drugs, and many of these are prepared with colorants and coatings of various types. Other tablets, such as those administered sublingually, buccally, or vaginally, are prepared to have features most applicable to their particular route of administration. Tablets are prepared primarily by compression, with a limited number prepared by molding. Compressed tablets are manufactured with tablet machines capable of exerting great pressure in compaction the powdered or granulated material. Their shape and dimensions are determined using various shaped punches and dies. Molded tablets are prepared on a large scale by tablet machinery or on a small scale by manually forcing dampened powder material into a mold from which the formed tablet is then ejected and allowed to dry.

Tablet and capsule properties must provide accurate dosing, bioavailability, chemically and physically stable, uniform in weight, drug substance content and particle size, dissolution rate and appearance, free from physical defects and should have the appropriate hardness to withstand mechanical shock during manufacturing, packing, shipping, dispensing and use. Tablet formulations generally have very accurate dosing, very good chemical, physical and microbiological stability when compared to other dosage forms, easy to transport, convenient to use and generally inexpensive to manufacture. However, the manufacture of tablets involves weighing, milling, drying, mixing, compression, filling, and packaging, and there can be an increased level of product loss at each stage in the manufacturing process.

Tablets are compressed dosage forms comprised of fluid and compressible powders. Fluidity is required so the powder can flow through the tablet manufacturing equipment. Compressibility is the ability of forming a stable, intact tablet when pressure (compression) is applied to the granulated drug ingredients. Tablet compression methods include wet and dry powder methods such as direct compression, roller compaction, and slugging for dry powders and wet granulations. Dry powder excipients should be free-flowing, inert, tasteless, colorless, and be ascetically pleasing. Such dry compression vehicles include diluents such as lactose and microcrystalline cellulose, disintegrates such as sodium starch glycolate and crospovidone, pH adjusters such as citric acid, surfactants such as sodium lauryl sulfate, binders such as starch, lubricants such as magnesium stearate, glidants such as colloidal silicon dioxide, anti-adherents such as talc, coating agents such as sugar, shellac, hydroxypropyl methylcellulose, and plasticizers such as glycerin. These excipients are also used for capsule formulation and manufacture.

Capsules are solid preparations in which the drug substance and excipients are enclosed in soft gelatin or hard gelatin or polymer-based water-soluble shell. Gelatin is the major component in capsules since it is a nontoxic film-forming material and readily soluble in biological fluids. The hard capsule shells are two-piece units – a longer cylindrical piece (capsule body) and capped with a shorter piece (cap). Soft gelatin capsules are comprised of a single piece. Capsules for oral administration can be filled with multiple formulation types including dry powders, granules, semisolids, nonaqueous liquids such as oils, beads, and small tablets. Soft gelatin capsules are made of gelatin and plasticizers such as glycerol, polyols, or other suitable biocompatible polymers. The distinctive feature of soft gelatin capsules is its one-piece construction. Soft gelatin capsules usually contain excipients such as oils, cosolvents, surfactants, and suspending agents. Capsules are easy to swallow, tasteless, odorless, and stable. However, capsules are usually unsuitable for children, can become brittle and crack under dry conditions, soften under high humidity conditions, and some drug substances (e.g. primary and secondary amines, liquid filled capsules, etc.) can react with or dissolve gelatin causing instability, and are more costly to manufacture, especially soft gelatin capsules, than oral solid tablets.

Oral tablet and capsule quality attributes can be assured by examining their stability regulatory specifications over their shelf life. The USP monograph for Acetaminophen Tablets USP requires that the product demonstrate its' identity, purity, and strength over its expiry period, which is 24 months. This includes identity testing by ultraviolet spectroscopy and chromatographic retention; the assay needs to be 90–110% by HPLC, dissolution (80% in thirty (30) minutes), uniformity of dosage, organic impurity testing of 4-aminophenol (≤0.15%), and total impurities (0.60%) and the appropriate package, labeling and storage (United States Pharmacopeia 2020h). The USP monograph for Acetaminophen Capsules USP requires that the product demonstrate its' identity, purity, and strength over its expiry period, which is 24 months. This includes identity testing by thin layer and high-performance liquid chromatographic retention; the assay needs to be 90–110% by HPLC, dissolution (75% in 45 minutes), uniformity of dosage, organic impurity testing of 4-aminophenol (≤0.15%), and the appropriate package, labeling, and storage (United States Pharmacopeia 2020i).

Conclusions

Preformulation and the subsequent formulation of pharmaceutical dosage forms is a critical and major aspect of the lifecycle of any drug product since it defines the type of dosage form (solid, liquid, injectable, etc.), specifications, and the associated quality it must maintain throughout its' shelf life to consistently produce the desired therapeutic effect. A thorough understanding of the physical, chemical, and mechanical properties of the drug substance and excipients during the preformulation phase is required to develop stable, safe, and effective practical delivery system and dosage form. Risk-based preformulation activities investigate and confirm that there are no significant barriers to the compound's development through commercialization and use this information to develop final dosage forms. The pharmaceutical formulation is the process in which the active drug substance is combined with functional excipients to produce a final drug product for medicinal use, which requires utilizing the information gained from preformulation to optimize the formulation, process, and commercial manufacture of the drug product. Preformulation and formulation studies consider such drug substance and drug product quality attributes such as crystallography, particle size, polymorphism, pH, solubility, impurities, stability, excipient, and container closure compatibility since all these factors can influence the bioavailability and therapeutic effect of the drug product. The drug dosage form must have the appropriate identity, purity, strength, uniformity, organoleptic characteristics such as appearance, odor, consistency and taste (if orally delivered), sterility (in delivered parenterally), and quality throughout the product's shelf life. There is no drug product without the critical preformulation activities to guide formulation development to produce the appropriate dosage form and delivery of such dosage form to the final customer, the patient in need of a reliable drug product that will deliver the required amount of medication in a consistent and convenient way to produce the desired therapeutic effect.

References

Adejare, A. (ed.) (2020). *Remington's The Science and Practice of Pharmacy*, 23e. Philadelphia: Academic Press.

Akers, M. (2016). *Sterile Drug Products - Formulation, Packaging, Manufacturing and Quality*. New York: CRC Press/Informa Healthcare.

Allen, L. and Ansel, H. (2014). *Ansel's Pharmaceutical Dosage Forms and Drug Delivery Systems*. Philadelphia: Lipincott Williams and Wilkins.

Aulton, M. (1988). Pharmaceutical applications of microbiological techniques. In: *Pharmaceutics: The Science of Dosage Form Design*, 243. New York: Churchill Livingston.

Aulton, M. and Taylor, K. (2018). *Aulton's Pharmaceutics: The Design and Manufacture of Medicines*, 5e. London: Elsiver.

Alison Brayfield (2017). Martindale: The Complete Drug Reference, 39, Alison Brayfield, Editor, Pharmaceutical Press, London.

CDER Researchers Explore the Promise and Potential of 3D Printed Pharmaceuticals (n.d.). FDA. http://www.fda.gov/drugs/news-events-human-drugs/ cder-researchers-explore-promise-and-potential-3d-printed-pharmaceuticals.

Conners, K. (1990). *Chemical Kinetics: The Study of Reaction Rates in Solution*, 245–250. NY: VCH Publishers.

Dash, A., Singh, S., and Tolman, J. (2014). *Pharmaceutics – Basic Principles and Application to Pharmacy Practice*. USA: Academic Press.

FDA (2006). Guidance for Industry, ICH Q9 Quality Risk Management. U.S. Department of Health and Human Services Food and Drug Administration Center for Drug Evaluation and Research (CDER) Center for Biologics Evaluation and Research (CBER). http://www.fda.gov/media/71543/download.

FDA (2009). Guidance for Industry Q8(R2) Pharmaceutical Development. U.S. Department of Health and Human Services Food and Drug Administration Center for Drug Evaluation and Research (CDER) Center for Biologics Evaluation and Research (CBER). http://www.fda.gov/media/71535/download.

FDA (n.d.). http://www.fda.gov/industry/structured-product-labeling-resources/ dosage-forms.

Felton, L. (2012). *Remington Essentials of Pharmaceutics*. UK: Pharmaceutical press.

Ghosh, T. and Jasti, B. (2005). *Theory and Practice of Contemporary Pharmaceutics*. USA: CRC Press LLC.

Helmenstine, Anne Marie (2020). The Arrhenius Equation Formula and Example. ThoughtCo. (http://www.thoughtco.com/arrhenius-equation-4138629).

Hoag, S. (2017). Capsules dosage form: formulation and manufacturing considerations. In: *Developing Solid Oral Dosage Forms – Pharmaceutical Theory and Practice*, 2e (ed. Y. Qui, Y. Chen, G. Zhang, et al.), 723–747. UK: Elsevier Inc.

ICH (n.d.). International Conference on Harmisation (ICH) Stability Testing of New Drug Substances and Products Guidance Q1A(R2) (http://database.ich.org/sites/default/files/Q1A%28R2%29%20Guideline.pdf).

Jones, D. (2008). *Fasttrack Pharmaceutics – Dosage Form and Design*. London: Pharmaceutical Press.

Kaplan, S. (1972). Biopharmaceutical considerations in drug formulation design and evaluation. *Drug Metabolism Reviews* 1: 15–32.

Lachman, L., Lieberman, H., and Kangi, J. (2016). *The Theory and Practice of Industrial Pharmacy*, 4e. USA: Lea & Febiger.

Mahato, R. and Narang, A. (2018). *Pharmaceutical Dosage Forms and Drug Delivery*, 3e. New York: Taylor & Francis Group, LLC.

Niazi, S. (2019). *Handbook of Preformulation: Chemical, Biological, and Botanical Drugs*. Boca Raton, Fl: CRC Press.

Ofoefule, S. (2002). *Textbook of Pharmaceutical Technology and Industrial Pharmacy*. Nigeria: Samakin (Nig) Enterprise.

Oral Solutions and Suspensions (8/94) (n.d.). FDA. http://www.fda.gov/inspections-compliance-enforcement-and-criminal-investigations/inspection-guides/oral-solutions-and-suspensions-894.

Patil, P.R., Bobade, V.D., Sawant, P.L., and Marathe, R.P. (2016). Emerging Trends in Compression Coated Tablet Dosage Forms: A Review. *International Journal of Pharmaceutical Sciences and Research* 7 (3): 930–938.

Shayne, C. (2008). *Pharmaceutical Manufacturing Handbook: Production and Processes*. New Jersey: John Wiley & Sons, Inc.

James Swarbrick (2013). Encyclopedia of Pharmaceutical Technology, 4, James Swarbrick, Editor, Informa Healthcare USA, Inc. New York.

United States Pharmacopeia (2020a). United States Pharmacopeia 43-National Formulary 38, <1058> Analytical Instrument Qualification.

United States Pharmacopeia (2020b). United States Pharmacopeia 43-National Formulary 38, <1> Injections.

United States Pharmacopeia (2020c). United States Pharmacopeia 43-National Formulary 38, <1231> Water for Pharmaceutical Purposes.

United States Pharmacopeia (2020d). United States Pharmacopeia 43-National Formulary 38, Phenytoin Sodium Injection Monograph.

United States Pharmacopeia (2020e). United States Pharmacopeia 43-National Formulary 38, <1151> Pharmaceutical Dosage Forms.

United States Pharmacopeia (2020f). United States Pharmacopeia 43-National Formulary 38, Acetaminophen Oral Solution Monograph.

United States Pharmacopeia (2020g). United States Pharmacopeia 43-National Formulary 38, Phenytoin Oral Suspension Monograph.

United States Pharmacopeia (2020h). United States Pharmacopeia 43-National Formulary 38, Acetaminophen Tablet Monograph.

United States Pharmacopeia (2020i). United States Pharmacopeia 43-National
Formulary 38, Acetaminophen Capsule Monograph.
Wikipedia (n.d.). Arrhenius equation. http://en.wikipedia.org/wiki/
Arrhenius_equation.

Chapter Self-Assessments: Check Your Knowledge

Questions:
- Define CMC and CMC compliance?
- Describe the scope of CMC compliance regulatory activities?
- Explain the role of the pharmaceutical science unit and how it must work with other groups to complete necessary CMC activities?
- Describe the time course of product testing that occurs for a drug product as its being developed?

Answers:
- Chemistry, Manufacturing and Controls (CMC) of a medicinal product is the body of information that defines not only the manufacturing process itself but also the quality control release testing, specifications, and stability of the product together with the manufacturing facility and all of its support utilities, including their design, qualification, operation, and maintenance. CMC regulatory compliance is seen as a process of governance that ensures CMC practices are carried out in agreement with regulatory agencies requirements and expectations.
- CMC regulatory compliance provides the collated information which defines both the manufacturing process and the quality control release testing. It also defines the stability and specifications of the product, the manufacturing facility, and its support utilities – including the facilities' design, qualification, operation, and maintenance. CMC regulatory compliance is best understood as a process of governance, ensuring that CMC practices are undertaken in line with the expectations and requirements of regulatory agencies.
- The pharmaceutical science unit works collaboratively with various departments in the organization, such as drug metabolism and pharmacokinetics (DMPK), safety–toxicology, pharmacology–biochemistry, quality assurance, regulatory affairs (RA), project management, and clinical supplies. Its responsibilities are the technical and process development of all CTM and the documentation of the CMC sections in regulatory filings before new drug application (NDA) registration.
- All stages of the drug development life cycle after drug discovery involve CMC. During preclinical drug development, the proper analytical methods are validated to monitor the product. Stability testing may be initiated, the physico-chemical properties of the product are determined, raw materials are chosen

and tested. When the drug development process moves into the clinical stage, further analytical method validation is required, and additional characterization of the drug product is needed. After clinical trials, the scale-up process must ensure that the larger batches of products are the same and meet the same specifications as the drug tested in the clinical trials. After the manufacturing process is qualified, lot release and in-process testing will continue to take place.

Quiz:

1   True or false. CMC regulatory compliance activities last through the lifetime of the product, with much of the post-marketing activity focused on lifecycle management including all necessary process changes required along the way.

2   The change management system should include the following elements as appropriate for the stage of the lifecycle except which. Choose the best answer:
   A Quality risk management (QRM – As per ICH Q9) system both early and late phase
   B Proposed changes log evaluated relative to the market authorization
   C Proposed changes log evaluated by teams with expertise and knowledge from relevant areas (e.g. Pharmaceutical Development, Manufacturing, Quality, Regulatory Affairs, and Medical) to ensure technical justification
   D All are correct

3   The main types of regulatory affairs CMC activities are involved with regulatory documentation, including _____, _____ and _____, and _____ activities. Choose the best answer:
   A authoring, curating, and editing, and publishing
   B curating, editing, and publishing, and submitting
   C compiling, editing, and publishing
   D authoring, compiling, and coordinating, and submitting
   E None of the above are correct

4   The pharmaceutical science unit is generally responsible for the CMC functions and has three typical subgroups. These include which groups? Choose the best answer:
   A process chemistry, formulation or pharmaceutics, analytical development groups
   B process chemistry, DMPK, analytical development groups
   C formulation or pharmaceutics, DMPK, analytical development groups
   D process chemistry, formulation, or pharmaceutics, DMPK groups

16

Chemistry, Manufacturing, and Controls
Jeffrey S. Barrett

Aridhia Digital Research Environment

Introduction

Chemistry, manufacturing, and controls (CMC) of a medicinal product is the body of information that defines not only the manufacturing process itself but also the quality control release testing, specifications, and stability of the product together with the manufacturing facility and all its support utilities, including their design, qualification, operation, and maintenance. CMC regulatory compliance is seen as a process of governance that ensures CMC practices are carried out in agreement with regulatory agencies requirements and expectations. Since such requirements and expectations change with time, a function of CMC regulatory compliance is to ensure that all CMC practices are updated accordingly. Within a pharmaceutical company, CMC regulatory compliance personnel ensure that, if the pharmaceutical organization has made any CMC-specific commitment to regulatory agencies, either verbally or in writing, such practices are carried out.

CMC regulatory compliance activities last through the lifetime of the product, with much of the post-marketing activity, focused on lifecycle management including all necessary process changes required along the way. Changes are inevitable and necessary to ensure business continuity. Changes are important for many reasons, such as to improve the process, reduce costs, improve efficiency, reduce potential failures, and meet state of the art process or to improve compliance against regulations/ good manufacturing practice (GMP). Likewise, making changes is highly complex and requires systematic planning, documentation, and coordination to functional owners including quality and regulatory fronts, the details of which are typically captured in a change management system that is maintained by the CMC Regulatory Compliance group. The change

Fundamentals of Drug Development, First Edition. Edited by Jeffrey S. Barrett.
© 2022 John Wiley & Sons, Inc. Published 2022 by John Wiley & Sons, Inc.
Companion website: www.wiley.com/go/Barrett/FundamentalsDrugDevelopment

management system should include the following elements as appropriate for the stage of the lifecycle:

- Quality risk management (QRM – as per ICH Q9) should be utilized to evaluate proposed changes. The level of effort and formality of the evaluation is typically commensurate with the level of risk.
- Proposed changes should be evaluated relatively to the market authorization adhering to local geographical considerations, including design space, where established, and/or current product and process understanding. There should be an assessment to determine whether a change to the regulatory filing is required under regional requirements. As stated in ICH Q8, working within the design space is not considered as a change (from a regulatory filing perspective). However, from a pharmaceutical quality system standpoint, all changes should be evaluated by a company's change management system.
- Proposed changes should be evaluated by teams having appropriate expertise and knowledge from relevant areas (e.g. Pharmaceutical Development, Manufacturing, Quality, Regulatory Affairs, and Medical) to ensure the change is technically justified. Prospective evaluation criteria for a proposed change should be set.
- After implementation, an evaluation of the change should be undertaken to confirm the change objectives were achieved and that there was no deleterious impact on product quality.

The goal of this chapter is to define the roles and responsibilities of the CMC Regulatory Compliance organization within a pharmaceutical company, describe the major activities of this group at various stages of development and emphasize the regulatory components of their deliverables, focusing at the investigational new drug application (IND) and new drug application (NDA) stage in particular.

CMC Definitions and Role in a Pharmaceutical Company

CMC regulatory compliance provides the collated information which defines both the manufacturing process and the quality control release testing. It also defines the stability and specifications of the product, the manufacturing facility, and its support utilities – including the facilities' design, qualification, operation, and maintenance. CMC regulatory compliance is best understood as a process of governance, ensuring that CMC practices are undertaken in line with the expectations and requirements of regulatory agencies. Because these expectations and requirements are likely to change over time, a key function of CMC regulatory compliance is ensuring that all CMC practices are kept up to date. CMC personnel typically populate both early and especially late-stage development project teams prior to IND preparation (see Chapter 11) and are heavily involved in the post-marketing activities associated with Lifecycle management.

CMC Personnel and Infrastructure

An individual in a regulatory affairs CMC role provides the strategy and knowledge needed to ensure that CMC practices are carried out in accordance with the requirements of regulatory bodies, such as FDA (US Food and Drug Administration) and EMA (European Medicines Agency). CMC regulatory is pivotal in ensuring that drugs and treatments being manufactured are safe, effective, and of high quality for patients. CMC roles are present in all stages of the drug lifecycle, from development and manufacturing to licensing and marketing. At each stage, CMC staff provide knowledge, understanding, and interpretation of regulations so that the drug has the best chance of being approved. Regulatory affairs personnel specializing in CMC principally involve authoring and compiling regulatory submissions and interacting directly with complimentary staff members at regulatory authorities.

The main types of regulatory affairs CMC activities are involved with regulatory documentation, including authoring, compiling, coordinating, and submitting activities. Authoring refers to the creation of CMC documentation that requires scientific understanding and the ability to interpret raw data concerning the API (active pharmaceutical ingredient) of drugs. The CMC professional must be able to write clear arguments that will ensure that the drug substance (active chemicals) and dry product (leaflet and packaging) are manufactured and stored according to regulatory requirements. These individuals are responsible for ensuring that other information, such as the composition of the dosage form, raw materials used to manufacture the product, description of the manufacturing process and stability data, is clearly communicated in the MAA (Marketing Authorization Application). Each country will have different specifications, so the individual must have a thorough understanding of the relevant requirements.

Compiling and coordinating CMC coordination/collating involves collecting and transferring documentation from the laboratory, pharmaceutical/CRO partners, any other companies (if they are manufacturing generic drugs) to help to produce the dossier license application for submission. Usually, entry-level, and junior professionals undertake CMC coordination roles. Submitting involves the actual process of submitting dossiers to the regulatory bodies for approval. These individuals respond to any questions that regulatory bodies may have about the drug. Questions could involve where and how the drug is manufactured and can involve a particularly lengthy correspondence when dealing with a novel drug compared with a generic drug (FDA 2003; FDA 2011; ICH Q1A(R2) 2003).

The pharmaceutical science unit is generally responsible for the CMC functions and has three typical subgroups (1) process chemistry group to scale up the

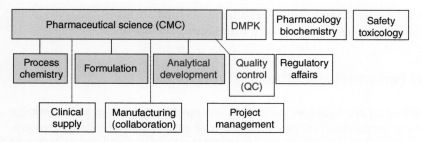

Figure 16.1 A schematic diagram of a typical organization structure in the pharmaceutical company for technical development, generally categorized as CMC activities described in a regulatory dossier.

synthetic route of the drug substances, (2) a formulation or pharmaceutics group for formulation development and production of drug products for the clinic, and (3) an analytical development group to develop methodologies to characterize and assess the quality of drug substance and drug product lots and to support process development (see Figure 16.1).

The pharmaceutical science unit works collaboratively with various departments in the organization, such as drug metabolism and pharmacokinetics (DMPK), safety–toxicology, pharmacology–biochemistry, quality assurance, regulatory affairs (RA), project management, and clinical supplies. Its responsibilities are the technical and process development of all CTM and the documentation of the CMC sections in regulatory filings before NDA registration.

IND Activities

CMC personnel must generate the required documentation for all regulatory submissions. In most situations, the IND represents the first introduction of CMC data and materials to regulatory authorities. The CMC section of the IND application should include (i.e. as appropriate for the particular clinical study (studies) covered by the IND) information describing the composition, manufacture, and control of the investigational drug substance and the investigational drug product. Specific IND sections include the following:

Section	Topic
F.	CMC information
1.	General principles
2.	CMC content and format
3.	cGMP compliance

The FDA provides many resources for those involved with CMC activities including a web-based training course, the "Chemistry, Manufacturing, and Controls (CMC) Perspective of the Investigational New Drug Application (IND)." This course focuses primarily on the CMC information for IND submissions and is not intended to include all the requirements applicable to INDs (https://www.accessdata.fda.gov/cder/cmc/index.htm). Regulations emphasize the graded nature of CMC information needed as drug development progresses under an IND. The amount of information needed depends on the following: Phase of investigation, Dosage form, Duration of study, Patient population, and the amount of information otherwise available.

The emphasis in an initial Phase 1 CMC submission should generally be placed on providing information that will allow evaluation of the safety of subjects in the proposed study. In actuality, all stages of the drug development life cycle after drug discovery involve CMC. During preclinical drug development, the proper analytical methods are validated to monitor the product. Stability testing may be initiated, the physicochemical properties of the product are determined, raw materials are chosen and tested. When the drug development process moves into the clinical stage, further analytical method validation is required, and additional characterization of the drug product is needed. After clinical trials, the scale-up process must ensure that the larger batches of products are the same and meet the same specifications as the drug tested in the clinical trials. After the manufacturing process is qualified, lot release and in process testing will continue to take place. Figure 16.2 illustrates a typical and generalized CMC Lifecycle approach. Differences exist for biologics as opposed to small molecules (see Figures 16.3 and 16.4), and the timing considerations are somewhat different and shifted for product candidates guaranteed expedited review.

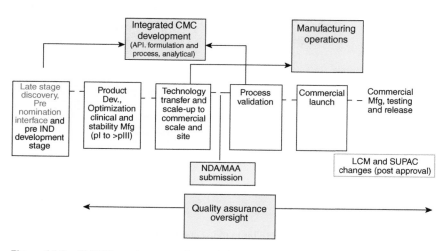

Figure 16.2 CMC life cycle approach: candidate nomination to commercialization.

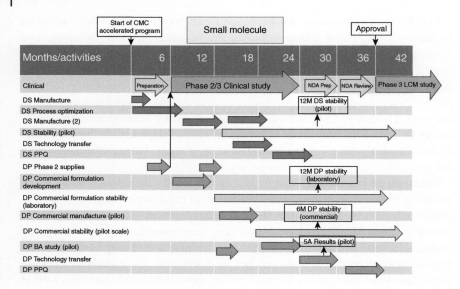

Figure 16.3 CMC activities related to small molecule development and Phase 1 entry (Dye et al. 2015).

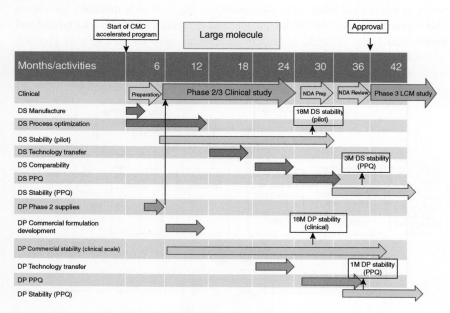

Figure 16.4 CMC activities related to large molecule development and Phase 1 entry (Dye et al. 2015).

CMC Role During Clinical Phase Development (Phase 1–3)

A typical CMC development plan for a small molecule drug candidate is shown in Figure 15.3. Drug substance development activities are given in orange, drug product activities in green and stability studies in yellow. CMC issues, which could arise from this plan include:

- Non-robust formulation used to supply Phase 2/3 study due to lack of time and drug substance.
- Formulation change required for commercial supply. Bioavailability study conducted using commercial formulation manufactured at pilot scale.
- Reduced data set on commercial formulation, e.g. stability data.
- Process Performance Qualification (PPQ) of drug products conducted in a phased manner and completed post-approval.

These challenges would be identified for discussion with the FDA early in the case of an accelerated CMC development program, approximately in the range of three to six months when submitting the IND for the Phase 2/3 study (Dye et al. 2015). A similar plan for a large molecule candidate (e.g. biologic or therapeutic protein) is shown in Figure 16.4.

CMC issues that may arise from this plan include:

- Stability data for drug substance and drug products do not comply with ICH Q5C, Stability Testing of biotechnological/biological products at the time of proposed filing of a marketing application
- Process performance qualification (PPQ) of drug substance and drug product is not complete at the time of proposed filing of the marketing application. It is conducted in a phased manner and completed post-approval.
- Patients are proposed to be supplied from PPQ batches.

Typically, these challenges would be identified for discussion with FDA early in the accelerated CMC development program, approximately in the range three to six months, when submitting the IND for the Phase 2/3 study.

NDA Submission and Support

All stages of the drug development life cycle after drug discovery involve CMC. During preclinical drug development, the proper analytical methods are validated to monitor the product. Stability testing may be initiated, the physico-chemical properties of the product are determined, raw materials are chosen and tested. When the drug development process moves into the clinical stage, further analytical method validation is required, and additional characterization of the drug product is needed. After clinical trials, the scale-up process must ensure that the larger batches of products are the same and meet the same specifications as the drug tested in the clinical trials. After the manufacturing process is qualified, lot release and in process testing will continue to take place. The ICH and FDA have provided several guidance documents designed to help pharmaceutical companies with the CMC expectations of their product (Van Buskirk et al. 2014). These include:

- FDA's Guidance for Industry, "INDs for Phase 2 and Phase 3 Studies: Chemistry, Manufacturing, and Controls Information"
- ICH Q2(R1) Validation of Analytical Procedures: Text and Methodology
- ICH Q3C(R6) Impurities: Residual Solvents
- ICH Q6A Specification: Test Procedures and Acceptance Criteria for New Drug Substances and
- New Drug Products: Chemical Substances

The FDA promotes the graded nature of CMC requirements. The further along in clinical trials, the more information that is needed. In this section, you will find a series of CMC requirements to support an IND or NDA. The amount of CMC information needed varies according to other clinical trial factors such as size, duration, dosage form, and prior usage. ICH Q6A breaks down the required CMC tests into two categories: Universal tests, which are applicable to all new drug substances or drug products, and specific tests, which are tests that can be considered on a case-by-case basis. To be clear on the definitions, a drug substance is the unformulated active substance while the drug product is the final product once formulated with excipients (non-active ingredients).

Many of the methods and assays currently used during pharmaceutical manufacturing to evaluate product quality generate an abundance of data, as in-process testing, as well as specification and stability testing parameters, must be closely monitored and data recorded throughout the process. Once collected and validated, the data generated by all manufacturing processes as well as quality assessments must subsequently be condensed and compiled into regulatory submissions to health authorities around the globe to obtain or maintain marketing approval for commercialization of human therapeutics or to initiate clinical trials with experimental therapeutics. From a regulatory perspective, much of these data are destined for the common technical document (CTD), an internationally recognized format created by the International Council for Harmonization of Technical Requirements for Pharmaceuticals for Human Use (ICH) for submitting documents for regulatory review. Use of the CTD is increasingly becoming a requirement in major markets, and in some regions, the CTD must be submitted electronically through an online portal.

This format is referred to as the electronic CTD (eCTD), and while the first version of eCTD was developed in 2002, the eCTD did not become mandatory in select regions until 2016. As of 2019, many agencies are still accepting paper submissions, with some regions predominantly utilizing paper submissions. Both the CTD and eCTD separate product regulatory information into five modules with predetermined, numbered, and itemized sections and subsections. Module 3 displays all quality and CMC data pertaining to product manufacturing, analytical methods, process development, specification testing, and stability of drug substances and drug products. This information is then summarized in the Quality Overall Summary (QOS), which appears in module 2. The current workflow for managing CMC regulatory submissions starts with output from experimental studies that are assembled into internal documents and reports (see Figure 16.5) (Algorri et al. 2020). From these internal documents, the data are summarized and repurposed for product quality system (PQS) documentation, technical reports, and regulatory filings. Regulatory filings vary based upon region and are sent to global health authorities (black lines). The initial submission process begins a series of back-and-forth submissions between health agencies and industry (red lines), until the submission is approved. Once finalized, the documents can then be organized as a series of regionally variable core dossiers.

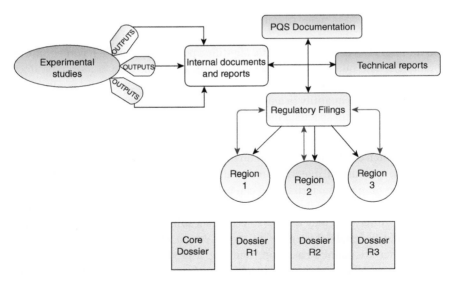

Figure 16.5 Generalized workflow for CMC component of regulatory submissions (Algorri et al. 2020).

Scale-up, Batch Size, and Quality Considerations

The challenge for the quality review and inspection for NDA is to assure that the characteristics and performance of the clinical batches will be replicated consistently in the commercial batches. To this end, the structure of the drug substance (API) must be proven, and controls must be used in the manufacturing process to ensure that the same structure is obtained in every batch. To accomplish this, a series of assays must be developed and routinely utilized to confirm identity and strength in various batches developed throughout the drug development process. These include assay of the drug substance (a validated stability-indicating assay or bioassay), an assay of the drug product (a validated stability-indicating assay or bioassay in which the assay is selective for the drug substance without interference from excipients, impurities, or degradants), a well-controlled manufacturing processes, in-process blend uniformity, uniformity of dosage units, container/closure (adsorption/absorption) assay, stability (expiration dating period) experiments.

The definition of a batch has regulatory implications, particularly with respect to current good manufacturing practice (CGMP), product recalls, and regulatory decisions. The terms batch and lot are defined in the regulations (21 CFR 210.3) as follows:

- a **batch** refers to a specific quantity of a drug or other material that is intended to have uniform character and quality, within specified limits, and is produced according to a single manufacturing order during the same cycle of manufacture.
- a **lot** refers to a batch, or a specific identified portion of a batch, having uniform character and quality within specified limits; or, in the case of a drug product produced by a continuous process, it is a specific identified amount produced in a unit of time or quantity in a manner that assures its having uniform character and quality within specified limits.

These definitions for both batch and lot are applicable to continuous manufacturing. A batch can be defined based on the production period, quantity of material processed, quantity of material produced, or production variation (e.g. different lots of incoming raw material), and can be flexible in size to meet variable market demands by leveraging the advantage of operating continuously over different periods of time. A lot may also be considered a sub-batch. The actual batch or lot size should be established prior to the initiation of each production run. For batches that are defined based on time (e.g. a production period), a connection between material traceability and batch must be established to identify the specific quantity of the drug (21 CFR 210.3).

Scale-up refers to the expected increase in batch size required as a drug candidate progresses through development. The coincidental expectation is the increasing batch size also results in unchanged product quality along with increased throughput, decreased cost of goods (COGs) hopefully, and a final product and process that is commercially viable. As with all stages of pharmaceutical production, scale-up requires careful planning and meticulous documentation of data. As the final step, a process development report is prepared as part of the submission package. Comparing data from each step in the process helps determine scalability requirements and identify critical process parameters at full scale. When successful physical and analytical testing at each stage is complete, the first full-scale feasibility batch can be produced and fine-tuned to maximize process efficiencies.

Commercial Obligations (Post-NDA) and Supply Chain

The number of CMC manufacturing supplements for NDAs and Abbreviated New Drug Applications (ANDAs) has continued to increase over the last several years. In connection with FDA's Pharmaceutical Product Quality Initiative and their risk-based approach to CMC review, FDA has evaluated the types of changes that have been submitted in CMC post-approval manufacturing supplements and determined that many of the changes being reported present low risk to the quality of the product and do not need to be submitted in supplements (FDA Guidance 2014). Based on FDA's risk-based evaluation, they have developed a list (see Table 16.1) to provide additional current recommendations to companies

Table 16.1 Examples of CMC post-approval manufacturing changes to be documented in the annual report if they have a minimal potential to have an adverse effect on product quality.

Category	Example
Components and composition	Elimination or reduction of an overage from the drug product manufacturing batch formula previously used to compensate for manufacturing losses.
	Change in coating formulation for immediate-release solid dosage forms if the coating material and quantity have been approved for another similar product and the change does not alter release of the drug, specification, or stability.
	In instances where the supplier of an inactive ingredient was specified in an approved application, change to a new supplier of that inactive ingredient (e.g. change from one drug master file (DMF) holder to other DMF holder or change to a new qualified supplier) – applicable only if the inactive ingredient's specification remains unchanged.

(Continued)

Table 16.1 (Continued)

Category	Example
Manufacturing sites	Minor structural modifications made in the sterile product manufacturing facility approved in an application that do not affect a product manufacturing area or sterility assurance and do not change product quality or specification.
	In the manufacturing of sterile products, the addition of barriers within a conventional fill area to prevent routine in-process human intervention in an existing filling or compounding area that is qualified and validated by established procedures.
Manufacturing process, batch size, and equipment	Addition of a sieving step(s) for aggregates removal if it occurs under non-aseptic conditions.
	Changes in mixing times (for blending powders, granules) for immediate-release solid oral dosage forms and solution products.
	Changes in drying times for immediate-release solid oral dosage forms.
	Changes in mixing times (for blending powders, granules) for immediate-release solid oral dosage forms and solution products
	Changes in drying times for immediate-release solid oral dosage forms.
	Manufacturing batch size or scale change that results from combining previously separated batches (or lots) of in-process material to perform the next step in the manufacturing process if all combined batches meet approved in-process control limits, the next step remains unaffected, and appropriate traceability is maintained.
	For equipment used in aseptic manufacturing processes, replacement of equipment with that of the same design and operating principle, when there is no change in the approved process methodology or in-process control limits.
	Addition of identical processing lines that operate parallel to each other in the drug substance and drug product manufacturing process with no change in in-process control limits or product specification
	For sterile drug products, addition of, deletion of, or change in a reprocessing protocol for refiltrations to control bioburden because of filter integrity test failures.
	Decrease in the number of open handling steps or manual operation procedures, when it reduces risk to product, and there is no other change to the process (e.g. implementation of aseptic connection devices to replace flame protection procedures).

Table 16.1 (Continued)

Category	Example
	For sterile drug products, changes to the ranges of filtration process parameters (such as flow rate, pressure, time, or volume, but not pore size) that are within currently validated parameters ranges and therefore would not warrant new validation studies for the new ranges.
	In the manufacture of sterile drug products, change from a qualified sterilization chamber (ethylene oxide [EtO], autoclave) to another of the same design and operating principle for the preparation of container/closure systems, sterilization of "change parts" for processing equipment, and terminal sterilization of product, when the new chamber and load configurations are validated to operate within the previously validated parameters. This does not include situations that change the validation parameters.
Specifications	Addition of a new test to the specification for an excipient.
	Change to the specification for a drug substance, drug product, or pharmacopeial excipient that is made to comply with the official compendia if it is a change that does not relax an acceptance criterion or delete a test.
	Change in the approved analytical procedure if the revised method maintains the original test methodology and provides equivalent or increased assurance that the drug substance or drug product will have the characteristics of identity, strength, quality, purity, or potency claims or is represented to possess and acceptance criteria remain unchanged.
	Replacement of a nonspecific identity test with a discriminating identity test that includes a change in acceptance criteria.
	Addition of an in-process test.
	Replacement of blend uniformity and in-process homogeneity tests with other appropriate testing that ensures adequacy of mix.
	Revision of tablet hardness if there is no change in the approved dissolution analytical procedure, criteria, or associated dissolution profile.
	Addition of a test for packaging material to provide increased assurance of quality.
	Tightening of an approved acceptance criterion for a drug substance, a drug product, drug product formulation components, and in-process material.

(Continued)

Table 16.1 (Continued)

Category	Example
Container/Closure system	A change in the container/closure system for the storage of a nonsterile drug substance (solid, semisolid, or liquid) when the proposed container/closure system has no increased risk of leachable substances in the extractable profile (for semisolids and liquids) and equivalent protection properties for the packaged material.
	Use of or transfer to a contract manufacturing organization (CMO) for the washing, drying, or/and siliconization of a drug product stopper or any part of a container closure system, provided the applicant certifies that the CMO's processes have been validated and the CMO's site has been audited and found CGMP compliant by the applicant (or by another party sponsored by the applicant).
	For solid oral dosage forms, when the change is to use another suitable primary packaging component used in any other CDER-approved drug product.
	For parenteral drug products, a change in glass supplier without a change in glass type or coating and without a change in container/closure dimensions.
	Changes to a crimp cap (ferrule and flip cap/overseal) if there are no changes to the color and that the container and closure integrity have been demonstrated using a validated test method. Note, however, that a change in the flip cap/overseal color to make it consistent with an established color-coding system for that class of drug products is to be documented in an annual report.
	Change to delete the company trademark or other markings on the crimp cap (ferrule and flip cap/overseal) to comply with the official compendium.
Labeling changes	Revision in drug product labeling to reflect the qualitative change in inactive ingredient(s) of coating formulation, as recommended above. The final structured product labeling (SPL) reflecting the qualitative change should be submitted to the Agency when implementing this change to allow for maintenance of the current product information in eLIST.
	A change in the drug product labeling to revise information related to CMC changes is discussed in this guidance. If the change involves associated revision of drug product labeling, the above would apply.

Table 16.1   (Continued)

Category	Example
Miscellaneous changes	Extension of the drug substance retest dating period or drug product expiration dating period based on real-time stability data from pilot-scale or larger/commercial-scale batches following an approved stability protocol.
	For immediate release solid oral dosage forms, if a dissolution test is performed, elimination of a test for identity or hardness from an approved stability protocol.
	For changes in an application that are fully consistent in scope and requirements with changes previously approved in a grouped supplement (also defined as a Bundled Supplement), the same applicant can make the same change to similar drug products.

regarding some post-approval manufacturing changes for NDAs and ANDAs that may be considered to have a minimal potential to have an adverse effect on product quality, and, therefore, may be classified as a change to be documented in the next annual report (i.e. notification of a change after implementation) rather than in a supplement.

Inspections and Follow-up

The FDA conducts inspections and assessments of regulated facilities to determine a firm's compliance with applicable laws and regulations, such as the Food, Drug, and Cosmetic Act. This typically involves an investigator visiting a company's manufacturing location. A pre-approval inspection is performed by FDA to ensure that a manufacturing establishment named in a drug application is capable of manufacturing a drug, and that submitted data are accurate and complete. The inspection has three objectives: (1) Readiness for Commercial Manufacturing (i.e. "determine whether the establishment has a quality system that is designed to achieve sufficient control over the facility and commercial manufacturing operation"); (2) conformance to application (i.e. "verify that the formulation, manufacturing or processing methods; analytical (or examination) methods; and batch records are consistent with descriptions contained in the CMC section of the application"); and (3) data integrity audit (i.e. "audit and verify raw data at the facility that are associated with the product.").

In the post-approval setting, the number of CMC manufacturing supplements for NDAs and ANDAs has continued to increase over the last several years. In connection with FDA's pharmaceutical product quality initiative and their risk-based approach to CMC review, FDA has evaluated the types of changes that have been submitted in CMC post-approval manufacturing supplements and determined that many of the changes being reported present low risk to the quality of the product and do not need to be submitted in supplements.

Based on FDA risk-based evaluation, they have developed a list (see Table 16.1 earlier summarized from Appendix A of FDA Guidance on CMC Post-approval Manufacturing Changes To Be Documented in Annual Reports) to provide additional current recommendations to companies regarding some post-approval manufacturing changes for NDAs and ANDAs that may be considered to have a minimal potential to have an adverse effect on product quality, and, therefore, may be classified as a change to be documented in the next annual report (i.e. notification of a change after implementation) rather than in a supplement.

References

Algorri, M., Cauchon, N.S., and Abernathy, M.J. (2020). Transitioning chemistry, manufacturing, and controls content with a structured data management solution: streamlining regulatory submissions. *Journal of Pharmaceutical Sciences* 109 (4): 1427–1438.

Dye, E.S., Groskoph, J.G., Kelley, B. et al. (2015). *CMC Considerations when a Drug Development Project is Assigned Breakthrough Therapy Status*. Pharmaceutical Engineering.

FDA (2003). *Guidance for Industry: INDs for Phase 2 and Phase 3 Studies, Chemistry, Manufacturing, and Controls Information*. U.S. Department of Health and Human Services, Food and Drug Administration, Center for Drug Evaluation and Research (CDER) https://www.fda.gov/media/70822/download (

FDA (2011). *FDA Guidance for Industry: Process Validation – General Principles and Practices*. U.S. Food and Drug Administration (FDA) www.fda.gov.

FDA (2014). *Guidance for Industry: CMC Post-approval Manufacturing Changes To Be Documented in Annual Reports*. U.S. Department of Health and Human Services, Food and Drug Administration Center for Drug Evaluation and Research (CDER) https://www.fda.gov/media/79182/download.

ICH Q1A(R2) (2003). Stability Testing of New Drug Substances and Products, Step 4. www.ich.org.

Van Buskirk, G.A., Asotra, S., Balducci, C. et al. (2014). Best practices for the development, scale-up, and post-approval change control of IR and MR dosage forms in the current quality-by-design paradigm. *AAPS Pharmaceutical Sciences and Technology* 15 (3): 665–693. https://doi.org/10.1208/s12249-014-0087-x. PMID: 24578237; PMCID: PMC4037495.

Chapter Self-Assessments: Check Your Knowledge

Questions:

- What is the reason(s) for formulation development?
- What are the appropriate design and formulation requirements of a dosage form?
- What is the function pre-formulation development?
- What are the requirements for an oral suspension?
- What are the essential tablet and capsule properties?

Answers

- The point of pharmaceutical formulation development is to design a quality drug dosage product and its manufacturing process to consistently deliver the intended performance and specifications of the product.
- The appropriate design and formulation of a dosage form require consideration of the physical, chemical, and biologic characteristics of all drug substances and pharmaceutical ingredients to be used in manufacturing the final drug product. The drug, pharmaceutical excipients, and container closures must be compatible with one another to produce a drug product that is safe, efficacious, stable, organoleptically appealing, easy to administer, and manufacture with quality, packaged, and labeled in the appropriate containers that keep the product stable. These critical quality attributes should be defined in a Product Development Plan and a Quality Target Product Profile.
- Preformulation is the associated research and development process performed to determine the physical, chemical, and mechanical properties of an API, the potential excipients, and container closure to develop stable, safe, and effective dosage form.
- A suspension is a dispersion of insoluble solid particles (dispersion phase) in usually an aqueous liquid medium (the dispersion medium), although in some formulations, the dispersion media can be non-aqueous or an oil. The suspension dosage form provides a liquid dosage form for insoluble or poorly soluble drugs and drugs that are unstable in an oral solution. The physical characteristics, especially the particle size of the drug substance (usually very fine (colloidal suspensions) to micronized (less than 25 microns)) and content uniformity (due to the potential for particle segregation) are critical quality

attributes for suspensions. Other critical quality attributes of suspensions include viscosity, pH, and dissolution. Viscosity can be important aspect to minimize segregation and has been shown to be associated with bioequivalence. The pH of the suspension affects the preservative systems and influences the amount of drug in the solution.

- Tablet and capsule properties must provide accurate dosing, bioavailability, chemically and physically stable, uniform in weight, drug substance content and particle size, dissolution rate and appearance, free from physical defects, and should have the appropriate hardness to withstand mechanical shock during manufacturing, packing, shipping, dispensing and use.

Quiz

1 According to the FDA's website, there are over ___ dosage form types.

  A 10
  B 100
  C 150
  D 1000

2 True or False: Hydrolysis and solvolysis are likely culprits associated with drug instability for both solid and liquid products, whether it is associated with promoting pH-based acid-base reactions in the presence (or absence) of metal ions or changing the ionic strength or polarity of the drug substance's local environment.

3 What is the most popular type of dosage form?

  A Oral liquids
  B Oral capsules
  C Sterile injections
  D Oral solid tablets
  E None of the above

4 Parenteral dosage forms should contain:

  A Impurities
  B Endotoxins
  C Pyrogens
  D Particulate matter
  E None of the above
  F All the above

17

Health Economics and the Healthcare Industry
Jeffrey S. Barrett

Aridhia Digital Research Environment

Introduction

While healthcare varies around the world, in general, healthcare is not free. Healthcare has become one of the largest industries in the world, and life expectancies in every corner of the globe continue to escalate. Adding more years to everyone's lifetime implies that people will have more years during which to become ill. This means more hospital stays and more prescription drugs being consumed. It also likely means more healthcare facilities will be built, resulting in opportunities for the engineering, architectural, and construction industries. Decisions about how to allocate doctors' and nurses' labor, whether to build a new hospital, whether to pass a new law, how much research to put into a new drug (and, once the research has been completed, decisions about how to price the drug) are made almost solely based on questions related to economic principles. Where will labor be most valuable? How can we find the most cost-effective materials for the hospital? How much did it cost to develop this drug, and what kind of long-term returns can we expect from it? People asking these and other healthcare-related questions turn to those with knowledge of economics for the answers. With exponential growth expected in the global healthcare industry, being conversant in the issues facing the industry and familiar with the players involved will prove invaluable. Healthcare decision-makers across the globe are often faced with the need to select therapeutic "interventions" from multiple treatment options, including biopharmaceuticals, medical devices, and healthcare services. However, the benefits and costs of these interventions can vary dramatically, and the benefits can be economic, clinical, both, or may include hard-to-measure costs or benefits the patient experiences directly. Health economics and outcomes research (HEOR) can help healthcare decision-makers – including clinicians, governments, payers, health ministries, patients, and more – to adequately compare and choose among the available options.

The objective of this chapter is to explain the components of health economic research and illustrate how it is used to evaluate the impact and cost-effectiveness of pharmaceutical products and the industry itself. We will also explore how the industry itself supports and utilizes health economics to create a value proposition from which it can illustrate the ROI to investors as well as policymakers.

Historical Perspective and Definitions

Health Economics is an applied field of study that pursues the systematic and rigorous examination of the problems faced in promoting health for all. By applying economic theories of consumer, producer, and social choice, health economics aims to understand the behavior of individuals, healthcare providers, public and private organizations, and governments in decision-making. Health economics is the leading interdisciplinary science that bridges the gap between the theory of economics and the practice of healthcare. It has experienced great development with earliest roots dating back for almost entire century. Its extensive diversification into various subdisciplines and areas of research endeavor is clearly visible today. Following its conception at the US National Bureau of Economic Research and Ivy League US Universities, this science has spread around the globe (Huynen et al. 2005; Jakovljevic and Ogura 2016). It has adapted to a myriad of local conditions and needs of the national health systems with diverse historical legacies, medical services provision, and financing patterns. Challenge of financial sustainability facing modern-day health systems remains primarily attributable to population aging, prosperity diseases, large-scale migrations, rapid urbanization, and technological innovation in medicine. Complex circumstances create strong drivers for the inevitable further development of health economics.

Health economics is used to promote health through the study of healthcare providers, hospitals and clinics, managed care, and public health promotion activities. Health economists apply the theories of production, efficiency, disparities, competition, and regulation to better inform the public and private sector on the most efficient, or cost-effective, and equitable course of action. Such research can include the economic evaluation of new technologies, as well as the study of appropriate prices, anti-trust policy, optimal public and private investment, and strategic behavior. Health economics can also be used to evaluate how certain social problems, such as market failure and inequitable allocation of resources, can impact on the health of a community or population. Health economics can then be used to directly inform the government on the best course of action with regards to regulation, national health packages, defining health insurance packages, and other national health programs.

In the Western world's healthcare systems, we often complain that preventative approaches are underfunded compared to treatment (Rheinberger et al. 2016). A standard practice in health economics that disadvantages prevention, "discounting" the value of future lives, may rest on weak empirical and moral considerations. Cost-effectiveness studies virtually never evaluate the overall efficiency of the prevailing mix of covered interventions (e.g. the mix currently covered by Medicare); they instead answer the narrower question of how much more expensive and/or more beneficial a new intervention is when compared to the current standard of care for a specific health problem. Thus, standard cost-effectiveness analyses (CEAs) evaluate new interventions in terms of their incremental cost-effectiveness, i.e. the ratio of the new intervention's additional costs to its additional benefits. In essence, the cost-effectiveness ratio compares two alternatives that could be applied to the same health problem. Often, the comparison is between a new approach, preventive or therapeutic, and the standard approach to the same problem, which again may be preventive or therapeutic.

In most of economic and business decision-making processes, private or public, the term 'cost' should always be considered while its counterpart varies from benefit (in cost-benefit analysis, CBA) to effectiveness (in CEA), and especially in healthcare fields, to quality-adjusted life years (QALYs) or latent utility (in cost-utility analysis, CUA). Though sometimes measured in different forms, costs are most commonly measured in monetary terms for direct comparison among alternative options. Cost of illness (COI), known as burden of disease (BOD), is a definition that encompasses various aspects of the disease impact on the health outcomes in a country, specific regions, communities, and even individuals. The category of COI can range from the incidence or prevalence of disease to its effect on longevity, morbidity along with the decrease in health status and quality of life (QoL), and financial aspects including direct and indirect expenditures that result from premature death, disability, or injury due to corresponding disease and/or its comorbidities.

Accurate knowledge about COI is essential and helps formulate and prioritize healthcare policies and interventions and eventually allocate healthcare resources in accordance with budget constraints in order to achieve policy efficiency. So, it is crucially important to understand how costs are defined, classified, and measured in the COI study. Figure 17.1 provides a schematic of healthcare factors associated with prevention and relationships to various HEOR approaches.

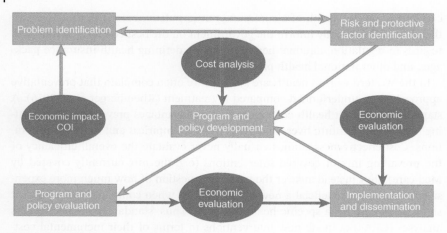

Figure 17.1 Public health model for prevention.

Health Economics Research: Who does it? When? Why?

Health economics outcomes research involves conducting economic analyses for healthcare interventions from several different perspectives, including that of the patient, the healthcare provider, and society as a whole. Methods ranging from cost-effectiveness, cost-utility, cost-benefit, cost of illness, and budget impact form the basis of much of the health economics research. Table 17.1 provides a list of

Table 17.1 Common types of health economic research methods and analyses.

Method	Definition and objectives	Example(s)
Cost-effectiveness	Outcomes measured in "natural units." Outcomes are usually clinically relevant e.g. life-years, mm Hg for BP, HbA1c for diabetes, etc. • Examines the costs of alternative approaches to achieving a specific (health) objective. Can be used to compare interventions to achieve the same outcomes e.g. the same clinical indication • Identifies the least cost way of achieving the objective to see how both cost and choice of technique vary as the magnitude of the objective varies.	(Babigumira et al. 2009b) Cost-effectiveness of facility-based care, home-based care, and mobile clinics for provision of antiretroviral therapy in Uganda.

Table 17.1 (Continued)

Method	Definition and objectives	Example(s)
Cost-utility	Uses non-financial common metrics that allow comparisons across health sector i.e. can compare different drugs or technologies • Metrics are combination of length of life and quality of life (e.g. QALY and/or Disability-adjusted life-year [DALY]) • CUA[a]s may not capture inter-health sector comparisons completely. Some health interventions have other outcomes which must be explicitly listed as inputs to the decision-making process • CUAs require studies to estimate utility (for QALY measurement) or disability weights (for DALY measurement)	(Sempa et al. 2012) Cost-effectiveness of early initiation of first-line combination antiretroviral therapy in Uganda.
Cost-minimization	Used when outcomes are equal or assumed to be equal (owing to outcomes being roughly identical) • Historically recommended for economic evaluations of trials showing no statistical significance in effectiveness • Conduct separate and sequential hypothesis tests on costs and effects to determine whether incremental cost-effectiveness is necessary	(Babigumira et al. 2009a) Impact of task-shifting on costs of antiretroviral therapy and physician supply in Uganda.
Cost-benefit	Places monetary values on inputs (costs) and outcomes thereby allowing comparison of projects (or interventions or investments) across the economy. • Allows the assessment of intrinsic value i.e. if benefits exceed costs the intervention is worth doing (ignoring deadweight loss from taxation and fiscal constraints).	(Brent 2009). A cost-benefit analysis of female primary education as a means of reducing HIV/AIDS in Tanzania

(Continued)

Table 17.1 (Continued)

Method	Definition and objectives	Example(s)
Cost-threshold	The cost-effectiveness threshold is the maximum amount a decision-maker is willing to pay for a unit of health outcome.	(Brouwer et al. 2019) When is it too expensive? Cost-effectiveness thresholds and healthcare decision-making.
	If the cost-effectiveness (ICER) of a new therapy (compared with a relevant alternative) is estimated to be below the threshold, then it is likely that the decision-maker will recommend the new therapy. For values near the threshold, the level of uncertainty may become important. Thresholds are often established by analysis of previous (reimbursement) decisions: they are not themselves outputs of cost-effectiveness analyses, but guides (or rules) to the interpretation of these outputs for decision-making, and they are specific to each unit of health outcome used.	
Cost of illness	Cost of illness analysis is a way of measuring medical and other costs resulting from a specific disease or condition. Estimates total costs incurred because of a disease or condition.	(Hodgson and Cai 2001) Medical care expenditures for hypertension, its complications, and its comorbidities.
	– Costs of medical resources to treat disease; costs of non-medical resources to treat disease; and loss in productivity.	
Budget impact	A budget impact analysis (BIA) is an economic assessment that estimates the financial consequences of adopting a new intervention. A budget impact analysis is usually performed in addition to cost-effectiveness analysis. Cost-effectiveness analysis evaluates whether an intervention provides value relative to an existing intervention. A budget impact analysis evaluates whether the high-value intervention is affordable.	(Trueman et al. 2001) Developing guidance for budget impact analysis.

[a] CUA = cost-utility analysis.

common HEOR types with their definitions and objectives. Interest in the field of HEOR has grown exponentially as governments and other payers grapple with how to provide the best possible health outcomes at affordable costs.

The focus of much of the HEOR is the emphasis on assessing health interventions in the context of actual quality of life measures (as opposed to clinical endpoints as one would determine in a clinical trial). Figure 17.2 provides a simple visual representation of the gains in quality of life as measured by QALYs (quality-adjusted life-year) with the intervention of a healthcare option and program. Such graphical displays offer an easy-to-understand visual communication that is easy for policymakers and healthcare providers to understand and support.

An important organization for supporting HEOR is NICE (National Institute for Health and Clinical Effectiveness). NICE aspires to provide authoritative, robust, and reliable guidance on current best practices in HOER. It has a remit to produce national guidance on individual technologies, appraisal management of specific conditions, clinical guidance, and clinical audit.

One mechanism to get connected to this field of research and discipline is through its many professional societies, which provide both current content of the field as it evolves and networking opportunities for those engaged in various aspects of the discipline. Of the many professional societies supporting health economics in some capacity, the premier organizations include the following: The American Society of Health Economists (ASHEcon; https://www.ashecon.org), The International Health Economics Association (*i*HEA; https://www.healtheconomics.org), and The Professional Society For Health Economics and Outcomes Research (ISPOR; https://www.ispor.org). An added

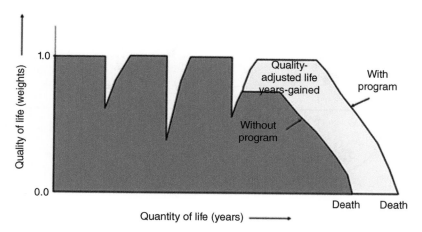

Figure 17.2 HECON assessment illustrating QALYs gained by health intervention.

benefit to such professional societies is the ability to attract students, faculty, and industry scientists to the same venue so that educational and research opportunities have an opportunity to develop from both a scientific and financial perspective.

Industry Investment in Health Economics Research

The pressure for healthcare systems to provide more resource-intensive healthcare and newer, more costly therapies is significant, despite limited healthcare budgets. It is not surprising, then, that demonstration that a new therapy is effective is no longer sufficient to ensure that it can be used in practice within publicly funded healthcare systems. The impact of the therapy on healthcare costs is also important and considered by decision-makers, who must decide whether scarce resources should be invested in providing a new therapy. The impact of therapy on both clinical benefits and costs can be estimated simultaneously using economic evaluation, the strengths and limitations of which are discussed. When planning a clinical trial, important economic outcomes can often be collected alongside the clinical outcome data, enabling consideration of the impact of the therapy on overall resource use, thus enabling performance of an economic evaluation, if appropriate. Figure 17.3 shows the often-cited lifecycle timeline and illustrates time-dependent milestones (e.g. the break-even point) that a company uses to forecast future product success upon as well as strategize about lifecycle management opportunities. With the growing costs of developing new drugs, increasing

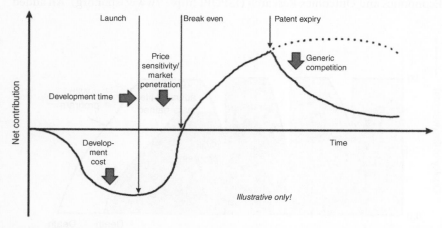

Figure 17.3 Schematic representing the impact of health economics of pharmaceutical product lifecycle and lifecycle management – external pressures result in shortened product life cycles; HEOR allows companies to manage expectations.

competition, and shortening times to peak sales, the importance of getting a drug launch right has never been greater. As payer influence grows and new therapies target smaller patient populations with complex needs, developing and executing an appropriate launch strategy becomes increasingly difficult. Health economic analyses allow companies to benchmark their performance against competitors and recent product launches factoring in current and future healthcare trends.

Examples of Industry Sponsored HECON Research

Decision-makers around the world are faced with limited budgets for funding health research, such as new clinical trials, and funding health technologies, such as new drugs. Resources spent on funding new research may otherwise be spent directly on patient care, and so it is desirable to consider the cost-effective use of limited resources. The methods of health economics facilitate resource allocation decisions by evaluating the maximum health gained for the resources spent.

As medication expenditures rise, payers are increasingly looking for evidence of the economic value of new medications during formulary decision-making. Payers in one survey ranked healthcare resource utilization (HCRU) data second behind post-marketing clinical efficacy and safety in terms of usefulness for decision making. To meet this demand for early evidence of value, manufacturers often generate economic models using assumptions, expert opinion, or literature to derive the models' HCRU inputs. Many sectors including academic, regulatory, the pharmaceutical industry, and healthcare providers including insurance companies represent key stakeholders conducting HEOR trials through few examples exist where these communities do so in a collaborative manner.

There has long been discussion as to whether commercial sponsorship of clinical studies produces a conflict of interest, and this sentiment certainly carries forward to the subject of health economics research (Hartman et al. 2003). In general, it can be assumed that companies avoid conducting head-to-head economic trials, particularly when they are unlikely to reveal the superiority of a new treatment or drug. In comparison with studies sponsored by nonprofit organizations, industry-sponsored studies were 1.9 times more likely to be cost-minimization analyses and 2.5 times less likely to be cost-effectiveness analyses. The reason for these relationships becomes apparent when the definition of cost-minimization analysis is considered. Such analyses involve comparisons of equal-effective alternatives based on net costs, with the aim of determining the less costly option. If it can be demonstrated in a cost-minimization analysis that, with identical clinical outcomes, the sponsor's drug reduces costs of treatment compared to a competitor's drug, the sponsor can readily expand market share. Drug companies can use such successful trials to market their products.

In comparison to studies sponsored by nonprofit organizations, the industry-sponsored studies were 3.5 times more likely to involve drugs and 25 times less likely to involve the evaluation of diagnostic screening methods. For industry, in contrast with nonprofit organizations, the economic evaluation of screening methods is usually of limited financial interest. However, since a medical treatment exhibits a reduction of its marginal utility with increasing exploitation, the economic evaluation of screening procedures is of particular interest. Only by means of health economics assessment is it possible to provide the preparatory groundwork for the decision on who should be screened, how often, and at what cost.

Cost-effectiveness plays an important role at two points in the health technology assessment (HTA) process: to determine whether it is cost-effective to fund a new health technology and whether it is cost-effective to fund a piece of research. For the first of these questions, cost-effectiveness analysis compares the costs and benefits (often measured using quality-adjusted life years (QALYs)) of competing health technologies to identify which represent value for money (Jo 2014; Manns 2009). For the second type of evaluation, value of information analysis (VOIA) assesses whether it is worthwhile collecting further information, such as performing further clinical trials, to reduce decision uncertainty (Rheinberger et al. 2016; Sempa et al. 2012). Amongst other inputs, a reliable health economic analysis requires an accurate estimate of the treatment effect and associated confidence interval, often provided by a clinical trial.

One of the frontier areas for HEOR is the design and conduct of prospective clinical trials designed to investigate the heal economic benefits of new treatments relative to the standard of care. Adaptive designs are one innovative approach to conducting a clinical trial. Unlike a traditional fixed sample size design, data are examined as the trial progresses to inform modifications to the trial. This can potentially save time and resources, as well as prevent patients from being needlessly randomized (Flight et al. 2020). The number of trials using adaptive methods has increased from 11 per 10 000 registered trials between 2001 and 2005 to 38 per 10 000 registered trials between 2012 and 2013 (Flight et al. 2020).

Adaptive designs and their implementation are commonly based on demonstrating clinical effectiveness. Despite its importance, cost-effectiveness is often a secondary consideration. It is currently unclear what impact the use of an adaptive design has on a health economic analysis. Additionally, opportunities are potentially being missed to incorporate health economics into the design and analysis of adaptive clinical trials. The use of health economics in the design and analysis of adaptive clinical trials has the potential to increase the efficiency of health technology assessments worldwide.

References

Babigumira, J.B., Castelnuovo, B., Lamorde, M. et al. (2009a). Potential impact of task-shifting on costs of antiretroviral therapy and physician supply in Uganda. *BMC Health Services Research* 9: 192. https://doi.org/10.1186/1472-6963-9-192.

Babigumira, J.B., Sethi, A.K., Smyth, K.A., and Singer, M.E. (2009b). Cost effectiveness of facility-based care, home-based care and mobile clinics for provision of antiretroviral therapy in Uganda. *Pharmacoeconomics* 27 (11): 963–973. https://doi.org/10.2165/11318230-000000000-00000. PMID: 19888795; PMCID: PMC3305803.

Brent, R.J. (2009). A cost-benefit analysis of female primary education as a means of reducing HIV/AIDS in Tanzania. *Applied Economics* 41 (14): 1731–1743.

Brouwer, W., van Baal, P., van Exel, J., and Versteegh, M. (2019). When is it too expensive? Cost-effectiveness thresholds and health care decision-making. *The European Journal of Health Economics* 20: 175–180. https://doi.org/10.1007/s10198-018-1000-4.

Flight, L., Julious, S., Brennan, A. et al. (2020). How can health economics be used in the design and analysis of adaptive clinical trials? A qualitative analysis. *Trials* 21: 252. https://doi.org/10.1186/s13063-020-4137-2.

Hartmann, M., Knoth, H., Schulz, D., and Knoth, S. (2003 Oct 20). Industry-sponsored economic studies in oncology vs studies sponsored by nonprofit organisations. *British Journal of Cancer* 89 (8): 1405–1408. https://doi.org/10.1038/sj.bjc.6601308. PMID: 14562007; PMCID: PMC2394350.

Hodgson, T.A. and Cai, L. (2001). Medical care expenditures for hypertension, its complications, and its comorbidities. *Medical Care* 39 (6): 599–615.

Huynen, M.M., Martens, P., and Hilderink, H.B. (2005). The health impacts of globalisation: a conceptual framework. *Global Health* 1: 14. https://doi.org/10.1186/1744-8603-1-14.

Jakovljevic, M.M. and Ogura, S. (2016). Health economics at the crossroads of centuries – From the past to the future. *Frontiers in Public Health* 4: 115. https://doi.org/10.3389/fpubh.2016.00115. PMID: 27376055; PMCID: PMC4899886.

Manns, B.J. (2009). The role of health economics within clinical research. *Methods in Molecular Biology* 473: 235–250. https://doi.org/10.1007/978-1-59745-385-1_14. PMID: 19160742.

Jo, C. (2014). Cost-of-illness studies: concepts, scopes, and methods. *Clin Mol Hepatol* 20 (4): 327–337. https://doi.org/10.3350/cmh.2014.20.4.327. (accesses 24 December 2014). PMID: 25548737; PMCID: PMC4278062.

Rheinberger, C.M., Herrera-Araujo, D., and Hammitt, J.K. (2016). The value of disease prevention vs treatment. *Journal of Health Economics* 50: 247–255. https://doi.org/10.1016/j.jhealeco.2016.08.005. (accessed 29 August 2016). PMID: 27616486.

Sempa, J., Ssennono, M., Kuznik, A. et al. (2012). Cost-effectiveness of early initiation of first-line combination antiretroviral therapy in Uganda. *BMC Public Health* 12: 736. https://doi.org/10.1186/1471-2458-12-736.

Trueman, P., Drummond, M., and Hutton, J. (2001). Developing guidance for budget impact analysis. *Pharmacoeconomics* 19: 609–621. https://doi.org/10.2165/00019053-200119060-00001.

Chapter Self-Assessments: Check Your Knowledge

Questions:
- Define and explain HEOR?
- Explain how health economics is used to promote health?
- How do professional societies support health economics?
- Discuss how adaptive trial designs benefit HEOR?

Answers:
- Health Economics is an applied field of study that pursues the systematic and rigorous examination of the problems faced in promoting health for all. By applying economic theories of consumer, producer, and social choice, health economics aims to understand the behavior of individuals, healthcare providers, public and private organizations, and governments in decision-making. The benefits and costs of medical interventions can vary dramatically, and the benefits can be economic, clinical, both, or may include hard-to-measure costs or benefits the patient experiences directly. Health economics and outcomes research (HEOR) can help healthcare decision-makers – including clinicians, governments, payers, health ministries, patients, and more – to adequately compare and choose among the available options.
- Health economics is used to promote health through the study of healthcare providers, hospitals and clinics, managed care, and public health promotion activities. Health economists apply the theories of production, efficiency, disparities, competition, and regulation to better inform the public and private sector on the most efficient, or cost-effective, and equitable course of action. Such research can include the economic evaluation of new technologies, as well as the study of appropriate prices, anti-trust policy, optimal public and private investment, and strategic behavior. Health economics can also be used to evaluate how certain social problems, such as market failure and inequitable allocation of resources, can impact on the health of a community or population.

- One mechanism to get connected to this field of research and discipline is through its many professional societies, which provide both current content of the field as it evolves and networking opportunities for those engaged in various aspects of the discipline. An added benefit to such professional societies is the ability to attract students, faculty, and industry scientists to the same venue so that educational and research opportunities have an opportunity to develop from both a scientific and financial perspective.

- One of the frontier areas for HEOR is the design and conduct of prospective clinical trials designed to investigate the heal economic benefits of new treatments relative to the standard of care. Adaptive designs are one innovative approach to conducting a clinical trial. Unlike a traditional fixed sample size design, data are examined as the trial progresses to inform modifications to the trial. This can potentially save time and resources, as well as prevent patients from being needlessly randomized. Adaptive designs and their implementation are commonly based on demonstrating clinical effectiveness. Despite its importance, cost-effectiveness is often a secondary consideration. It is currently unclear what impact the use of an adaptive design has on a health economic analysis.

Quiz:

1 The challenge of financial sustainability facing modern-day health systems remains primarily attributable to _____. Choose the best answer
   - **A** population aging, infectious diseases, immigration, under-developed supply chains, and innovation in medicine
   - **B** population aging, prosperity diseases, large scale migrations, rapid urbanization, and technological innovation in medicine
   - **C** poor leadership, population aging, and infectious diseases
   - **D** None of the above

2 As the Western world's healthcare systems often focus on the treatment of disease, it is a common criticism that preventative approaches are underfunded compared to treatment. A standard practice in health economics that disadvantages prevention approaches discounts the value of _____ (select the correct answer)
   - **A** future lives
   - **B** QALYs
   - **C** Burden of disease (BOD)
   - **D** None of the above

3 Which common health economic analysis does the following description describe? It places monetary values on inputs (costs) and outcomes, thereby allowing comparison of projects (or interventions or investments) across the economy. It also allows the assessment of intrinsic value i.e. if benefits exceed costs, the intervention is worth doing (ignoring deadweight loss from taxation and fiscal constraints).

   A Cost-minimization analysis
   B Cost-threshold analysis
   C Cost-benefit analysis
   D Budget impact analysis

4 Which common health economic analysis does the following description describe? It is used when outcomes are equal or assumed to be equal (owing to outcomes being roughly identical). Historically recommended for economic evaluations of trials showing no statistical significance in effectiveness and conducts separate and sequential hypothesis tests on costs and effects to determine whether incremental cost-effectiveness is necessary.

   A Cost-minimization analysis
   B Cost-threshold analysis
   C Cost-benefit analysis
   D Budget impact analysis

18

Current State of Affairs

Attrition Rates and Evolving Corporate Strategies

Eileen (Doyle)Castranova

Certara, Inc., Princeton, NJ, USA

The Cost of Drug Development

Most of the costs of new drug development are related to the costs of failed projects (Paul et al. 2010). Drug development is characterized by high attrition rates, large capital expenditures, and long timelines.

If any given asset has a 1 in 10 000 chance of becoming a drug, how does one shepherd the "right" one to market? To assess this, we need to think about where we are good at choosing the compound that progresses to the next phase, and in the cases where we are poor at choosing, figure out how to choose more wisely.

In this chapter, you will learn about the factors that affect attrition rates and considerations for making successful decisions to choose which assets (drug candidates) progress through clinical studies and which programs should be terminated. The milestones of success in each phase will be discussed, as will the elements of successful transitions from phase to phase. Proposed causes for decreased efficiency amidst scientific advances will be explained, as will mechanisms for increasing efficiency. Finally, recent frontiers (evolving corporate strategy) will be introduced.

Probability of Successful Drug Approval by Clinical Study Phase Transition

In a 2016 report, the Tufts Center for the Study of Drug Development estimates the overall likelihood of approval (LOA) for drugs entering humans in Phase 1 to ultimately be approved for marketing is 9.6%. This composite score is based on the probability of success in each clinical phase. Said another way, it is based on the probability of successfully advancing to the next phase of development.

Fundamentals of Drug Development, First Edition. Edited by Jeffrey S. Barrett.
© 2022 John Wiley & Sons, Inc. Published 2022 by John Wiley & Sons, Inc.
Companion website: www.wiley.com/go/Barrett/FundamentalsDrugDevelopment

To successfully progress through drug development, a candidate needs to meet the milestones of each phase. The drug candidate must be safe to administer in humans and have a reasonable chance of reaching clinically effective concentrations (Phase 1), it must show efficacy in a small population of people with the disease or condition of interest (while maintaining a favorable safety profile) (Phase 2), and this safety and efficacy in early clinical development must be borne out in pivotal (typically Phase 3) trials in a larger segment of the population to be treated. Additionally, the candidate must be able to meet commercial, marketing and payer reimbursement goals, and still meet an unmet medical need at the time of launch (often eight to nine years after clinical development begins).

Phase 1 to Phase 2

The first human experience typically happens in Phase 1, where the drug is often administered to healthy volunteers. Phase 1 is focused on defining the therapeutic window in healthy volunteers and determining the dose range to be studied in Phase 2. The big question in Phase 1 is, "is it safe?" Using healthy study participants, researchers determine the mode of action, safety, and side effects of the drug candidate. Phase 1 studies are the smallest studies. A small Phase 1 study can enroll as few as nine subjects; typically Phase 1 studies enroll 12–50 healthy volunteers.

The first patient experience typically happens in Phase 2. Phase 2 is focused on defining the therapeutic window in patients and determining dose selection for Phase 3. The big question in Phase 2 is, "is it effective?" Researchers use individuals who have the disease in question to determine if the drug is effective against that disease or condition. Phase 2 studies typically enroll 50–200 subjects.

A Phase 1 program can be considered successful if safety in humans is established and a reasonable dose range (to consider proof of concept and proof of mechanism in Phase 2 trials) has been defined. The ability to safely dose individuals with drug candidates and achieve drug exposure that should lead to drug efficacy is key. Other characteristics of successful Phase 1 programs are those in which an assay is developed and suitable for patient trials; the formulation is reasonable to pursue the final market image; biomarkers suitable for Phase 2 have been defined; measures of efficacy have been selected; and there are no major predictable issues with respect to dosing (such as an excessively short half-life relative to the indication, a significant food effect or a significant drug-drug interaction).

A poor transition to Phase 2, or (said another way), transitioning a drug candidate from Phase 1 to Phase 2 without achieving most of the characteristics of a successful Phase 1 program can lead to disaster in Phase 2 (a more public and expensive phase to fail).

Phase 2 to Phase 3

Phase 2 trials are screening trials to determine whether to study a new treatment in larger Phase 3 trials. Phase 2a is exploratory, non-pivotal studies that have clinical efficacy, pharmacodynamics, or biological activity as a primary endpoint. The trial can be conducted in patients or healthy volunteers. Phase 2a trials are used to provide proof of concept, proof of efficacy, or proof of mechanism. They can be mechanistic studies, dose-ranging studies, or pilot studies. Phase 2b trials are the definitive dose range finding study in patients with efficacy as the primary endpoint. Exceptionally, Phase 2 studies can be used as pivotal trials if the drug is intended to treat life-threatening or severely debilitating illnesses (as in oncology indications). The endpoints of Phase 2 trials must be associated with clinical benefits. Toxicity must be understood (and often correlated with drug candidate dose and/or exposure). Phase 2 trials are designed to show statistical differences between the effect of the drug candidate, often via comparison with placebo or the standard of care treatment (if there is one). Proof of concept involves providing evidence that a candidate drug might be effective for a disease. Criterial for continuing to investigate the candidate drug include safety, tolerability, pharmacokinetics, duration of action, efficacy, patient acceptability, and commercial viability.

The patient experience that determines approval occurs in Phase 3. The patient population determines the indication, labeling, and commercial strategy of the drug candidate. Phase 3 is focused on determining safety and efficacy in the target population(s) in a statistically significant manner. Often enrolling several hundred to several thousand participants (depending on indication), researchers work to understand if the drug candidate is safe and effective in large groups of diverse people. One gains a more thorough understanding of the drug's effectiveness, benefits, and the range of possible adverse events. Variability of drug exposure is also studied in Phase 3.

A Phase 2 program can be considered successful if "active" or efficacious doses can be discriminated (from each other and from placebo), drug activity can be projected relative to clinical outcomes, and the patient population of interest is viable (i.e. treatable, and accessible). Milestones of Phase 2 include defining the therapeutic window relative to the standard of care (and the marketplace), assessing the variability in patient response to determine overall response rate relative to "enriched" or sub-populations, and coming up with strategies for dose adjustments relative to the requirements for the mainstream (post-study) patient population. There needs to be good evidence that clinical benefit can be established and that the drug will be superior to placebo (or an active comparator/standard of care).

Clinical trials fail for many reasons. The top five include inadequate study design, improper dose selection, non-optimal assessment schedules, inappropriate efficacy metrics/markers, and issues with how data are analyzed.

A poor transition to Phase 3, or (said another way), transitioning a drug candidate from Phase 2 to Phase 3 without achieving most of the characteristics of a successful Phase 2 program, can lead to disaster in Phase 3 (again, a more public and expensive phase to fail).

In a 2019 Genetic Engineering and Biotechnology News report, Philippidis describes 13 clinical trial failures from the previous year. The leading cause was a failure to meet a Phase 3 endpoint. In one case, Incyte's epacadostat, an IDO1 inhibitor in trials for the treatment of unresectable or metastatic melanoma, which was being administered in combination with Merck & Co.'s Keytruda® (pembrolizumab), failed the Phase 3 ECHO-301/KEYNOTE-252 trial. Two pivotal lung cancer trials of the epacadostat + pembrolizumab combination were converted to Phase 2 studies.

In 2017 the FDA published "22 case studies where Phase 2 and Phase 3 trials had divergent results." Fourteen case studies described Phase 3 trials that demonstrated a lack of efficacy in a promising experimental therapy; one described a Phase 3 trial that demonstrated a lack of safety in a promising experimental therapy, and seven described Phase 3 trials that demonstrated a lack of efficacy and safety in promising experimental therapies.

Despite statistically significant results in reducing the symptoms of Schizophrenia in Phase 2, in Phase 3 trials Roche's bitopertin failed to improve the negative symptoms of Schizophrenia. Differences in the patient population between Phase 2 and Phase 3 (inclusion/exclusion criteria), as well as the length of study (eight weeks in Phase 2 and 24 weeks in Phase 3), could be major contributors to this failure. Failure could potentially have been avoided by more careful study design consideration (using a more representative patient population in Phase 2 as what was planned in Phase 3 or matching the Phase 3 population to the Phase 2 population) or greater scrutiny of the dose-response in Phase 2.

In Phase 2 per-protocol population (Umbricht et al. 2014), eight weeks of treatment with bitopertin was associated with a significant reduction of negative symptoms in the 10 mg/d cohort and the 30 mg/d cohort, but not the 60 mg/d cohort. Only the 10 mg/d cohort showed a significantly higher response rate and a trend toward improved functioning when compared with placebo.

In the Phase 3 program (Bugarski-Kirola et al. 2017), three (SunLyte [WN25308], DayLyte [WN25309], and FlashLyte [NN25310]) Phase 3, multicenter, randomized, 24 weeks, double-blind, parallel-group, placebo-controlled studies evaluated the efficacy and safety of adjunctive bitopertin in stable patients with persistent, predominant negative symptoms of Schizophrenia treated with antipsychotics. SunLyte met the prespecified criteria for lack of efficacy and was declared futile. The primary efficacy endpoint was mean change from baseline in Positive and Negative Syndrome Scale negative symptom factor score at week 24. At week 24, mean change from baseline showed improvement in all treatment arms but no statistically significant separation from placebo in Positive and Negative Syndrome Scale negative symptom factor score and all other endpoints.

The high attrition rate in Phase 2 is expected, as this is the time that definitive efficacy is tested in a segment of the target population for the first time. Still, Phase 2 transition success varies by disease. Phase 2 success rates range from a high of 56.6% in the field of hematology to a low of 23.7% in psychiatry.

Thus far, numerically, the biggest problem in drug development is still the attrition rate between Phase 2 and Phase 3. We have shown that despite our technological and scientific advances, we cannot translate early efficacy in a small segment of the to-be-treated population to confirmed efficacy in pivotal Phase 3 trials. For some companies, however, that is changing.

Examples of Midphase Success

Pfizer reported a jump in clinical trial success rate, from 9% overall in 2019 to 21% overall in 2020. Historically, Pfizer's clinical research success rate has been on the lower end of the average. In 2015, for example, Pfizer calculated that 5% of its assets entering first-in-human studies went on to gain marketing approval. The industry average was 11%.

The 2020 success is attributed to a reduction in mid-phase attrition. From 2010-2015, Pfizer had a 5-year average success rate in Phase 2 to 3 transition of 15%. Recently, they reported their Phase 2 success rate as 52%, comparing this to the industry average of 29%.

In addition to running better-informed trials and making stronger go/no-go decisions between Phase 2 and Phase 3, portfolio reprioritization and therapeutic area selection play a role in success rate calculations. In recent years Pfizer exited neuroscience research and development, a notoriously difficult therapeutic area with lower-than-average success rates. Reattribution of assets and movement to more successful ventures can increase the metrics of end-to-end success as much as better science can.

Decreasing Efficiency Amidst Scientific Advances

The Innovation-Stagnation Challenge

In 2004, the FDA released a white paper – Challenges and Opportunities report. In it, the authors note that basic science discoveries are not leading to more effective, affordable, and safe medicines for patients. Product development has become more challenging, inefficient, and therefore costly. The number of new drug applications

(NDAs) and biologics license applications (BLAs) have declined year-over-year. They point out the trend of innovators concentrating on product development of assets with the potential for high market return while developing products for important public health needs (less-common diseases, diseases prevalent in parts of the world with lesser resources, drug candidates aimed at prevention rather than treatment, and individualized therapies) remain under-explored.

The hypothesis of the article is that the applied sciences needed for drug development have not kept pace with the tremendous advances in basic sciences. The solution, they propose, is new tools to get better answers about how the safety and effectiveness of new products can be demonstrated.

Many companies accepted the challenge and started developing such tools. *In silico* modeling boomed in the period after the Critical Path Initiative was launched, and that boom continues today. Carrying out biological experiments entirely in a computer has the potential to speed the rate of discovery (and development) while reducing the need for expensive and time-consuming lab work (and clinical trials).

In Silico Experimentation in Drug Discovery

In discovery, drug candidates can be produced and screened based on physical properties. For example, using protein docking algorithms, researchers identified enzyme inhibitors associated with cancer. Fifty percent of the molecules were later shown to be active inhibitors in vitro (Ludwig Institute 2010; Röhrig et al. 2010). The use of in silico screening based on proposed protein-ligand interactions can be achieved in a fraction of the time of a traditional high throughput screening (HTS), with much greater success rates (the expected hit rate of HTS is 1%).

In addition to compound screening, computer models of cellular behaviors have been established. In 2007, researchers developed an in-silico model of tuberculosis. The in-silico system allowed for simulations of cellular growth rates, so phenomena of interest could be observed in minutes instead of months (University of Surrey 2007).

In Silico Experimentation in Drug Development

Model-based drug design expands off the benchtop and into the clinic as well. Techniques such as physiologically-based pharmacokinetic (PBPK) modeling, quantitative structure-activity relationships (QSAR), and molecular dynamics simulations can be integrated with techniques like cell culture analogs *in vitro*, with the potential to obviate the need for (or at least decrease the number of) animal experiments necessary to support drug safety in humans. However, despite almost a decade of advances, science was not keeping pace with rising costs.

Eroom's Law

In 2012 an opinion paper was published in Nat Rev Drug Disc describing the inverse trend between scientific advances and the efficiency of drug research and development. The moniker, "Eroom's Law," is "Moore's Law" backward (a description of the exponential increase in technology improvement over time). The paper highlighted the decline in the number of new drugs approved per billion US dollars spent on research and development, with a half-life of nine years. Taking inflation into consideration, the researchers claimed an 80-fold decrease in R&D efficiency since 1950. The number of new drugs introduced per year has been broadly flat over the period since the 1950s, and costs have grown steadily (Munos 2010).

Four factors were considered to be primary causes of decreasing drug research and development efficiency: the "Better than the Beatles" problem, the Cautious Regulator problem, the "throw money at it" tendency, and the "basic research-brute force bias." Each of these four issues is described in more detail below.

The "better than the Beatles" Problem

The limited marketing protection for novel therapeutics via patent mean blockbusters are quickly overtaken by generics. A ceaselessly improving catalog of choices (at different costs and prices) increases the complexity of the development process for new drugs and raises the evidence requirement necessary for approval, use (i.e. prescription), and reimbursement. This phenomenon encourages R&D in difficult to treat therapeutic areas and deters R&D in areas that have treatment (but which might benefit from so-called "me too" drugs) – statin example. A crowded therapeutic area reduces the economic value of future drugs, even if such drugs are superior (in efficacy, for example, as opposed to cost).

Pammolli et al. (2011) have provided a quantitative illustration of the "better than the Beatles" problem. Their analysis compared R&D projects started between 1990 and 1999 with those started between 2000 and 2004.

Attrition rates rose during the latter period. However, the increase could be largely explained by a shift in the mix of R&D projects from commercially crowded therapeutic areas in which historic drug approval probabilities were high (for example, genitourinary drugs and sex hormones) to less crowded areas with lower historical approval probabilities (for example, antineoplastics and immunomodulatory agents). Innovation in pharmaceuticals is a cumulative process, and markets in which the POS is high are those in which effective compounds are already available. Payers discourage incremental innovation and investments in follow-on drugs in already established therapeutic classes, mostly using reference pricing schemes and bids designed to maximize the intensity of price competition among different molecules (Pammolli et al. 2011). Consequently, R&D investments tend to focus on new therapeutic targets, which are characterized by high uncertainty and difficulty, but lower expected post-launch competition (Ma and Zemmel 2002).

The Cautious Regulator Problem

As more drugs become available, better reporting infrastructures and quicker information flow become available, and the public has an instant and constant opinion, regulators have developed lower risk tolerance. Higher (or at least more) standards are set, and associated costs of R&D increase. The general public's opinion (in the United States) of the pharmaceutical industry hit a record low in 2019, when a Gallup poll found that the pharmaceutical industry is the most poorly regarded industry, ranking last on a list of 25 industries that Gallup tests annually. Safety of the public is paramount. When public perception is that corporations cannot be trusted to do the right thing, regulators are called upon to pick up the slack.

Once a regulation is in place, overwhelming evidence must be presented in order to not conform to it, even if omitting that step could be achieved without causing significant risk to drug safety. For example, the Ames test for mutagenicity adds little to the assessment of drug safety, but a positive finding kills some drug candidates. There are large capital costs associated with asking to not conform to a requirement. However, when doing so is warranted, it is worth it, saving time, money, patient stress, and getting unmet medical needs met more quickly (e.g. concentration-QT studies instead of thorough QT studies). In the last 60 years, there has been only one rise in R&D efficiency until lately. In the late 1990s, efficiency rose. This is attributed to the clearing of the regulatory backlog at the FDA following the implementation of the 1992 Prescription Drug User Fee Act (PDUFA), with a small contribution from the rapid development and approval of several HIV drugs, hastened by input from community action and patient support group efforts.

Additionally, our audit-based approach to regulatory documentation leads to increased documentation time, documentation specialization, and increased cost to prepare a filing.

In general, regulators are more risk-tolerant when treatment options are fewer, as the risk/benefit ratio tilts in favor of taking on more risk for higher benefit. Conversely, when several treatment options exist, the regulatory bar for other drugs in the same indication is raised.

The "throw money at it" Tendency

This tendency describes adding human and capital resources to research and development, which has led to a rise in R&D spending until recent years. This tendency has prevailed because it has worked: for most of the past 60 years, companies saw a favorable return on investments. Being second or third to market instead of first has large financial costs, so investing additional resources in order to be first to launch is warranted.

The "basic research-brute force" Bias

This bias is the tendency to overestimate the ability of advances in basic research, such as molecular biology and empiric screening methods (for example, high throughput screening of compound libraries), which are standard in discovery and preclinical research, to increase the probability that a molecule will be safe and effective in humans. Classical R&D is a search, filter, and selection process. While scientific advances have increased the breadth of potential targets and technologies such as medicinal chemistry to modify and synthesize more drug-like molecules, the quality of selection was largely ignored. The use of pathway analysis for target selection, the use of transgenic mice for target validation, and the use of knowledge of druggability metrics such as absorption, distribution, metabolism, excretion, and toxicology characteristics of early targets must be employed to generate rational-not numerous-drug candidates.

As of the early 2000s, much of pharmaceutical R&D was based on the idea that high-affinity binding to a single biological target linked to disease will lead to medical benefits in humans (Hopkins et al. 2006). However, many drugs do not act on a single target, and many pathways have redundancy or alternatives such that blocking a single signal does not disrupt the complex network of signal transduction. Meanwhile, more first-in-class small molecule drugs approved between 1999 and 2008 were discovered using phenotypic assays than using target-based assays (Swinney and Anthony 2011). It is possible that target-based approaches are efficient in identifying new candidates for already validated targets yet inefficient in the search for innovative drugs. Automation, systemization, and process generation have worked in other industries. It seemed reasonable that screening millions of leads against a genomics-derived target could streamline the R&D process (Scannell et al. 2012).

Increasing Efficiency: Focus on the Biggest Problems First

Breaking Eroom's Law

Starting in 2010, the trend reversed: the number of drugs per billion US dollars in R&D spending began to rise (Ringel et al. 2020). By 2018, a net of 0.7 new molecular entity launches per billion per year were achieved. Success rates increased, especially in late-stage R&D, driving the turnaround in productivity. This turnaround in the timing and amount of attrition could only be achieved as a result one or more of three main factors: better information, better use of information, and a changing threshold for regulatory approval (Ringel et al. 2020).

Better Information

Emphasis has been placed on the use of genetic data to understand disease processes. Focus has increased on validated targets (by genome-wide association studies). Targeting genetically validated candidates could lead to approvals in smaller populations, given that many issues typically treated as a single disease based on clinical presentation have heterogeneous underlying mechanisms. Rare diseases, which have been a particular focus of pharmaceutical R&D over the past few years, are often caused by specific genetic variants. Understanding this and targeting sub-populations has led to an increase in drug candidates approved for the treatment of rare diseases.

Better Use of Information

Better decision-making processes approach optimal decision-making. Cognitive biases set the threshold for terminating programs too high. Overvaluing optionality, optimism bias, loss aversion.

Changes to the Threshold for Approval

As already discussed, regulators are more risk-tolerant when treatment options are fewer. Greater focus on the R&D of underserved diseases lacking specific therapies leads to a greater desire to meet the unmet medical need. In particularly severe diseases, a less favorable drug profile (adverse events, low efficacy, etc.) will be accepted by regulators, as the other option is worse (e.g. the approval of eteplirsen for the treatment of Duchenne muscular dystrophy).

Delivering the Right Dose to the Right Patient at the Right Time

The primary increase in cost is during the clinical phases of development. Therefore, anything that can be done to decrease the cost of clinical trials (at any phase) will increase the efficiency of drug development.

Improving Clinical Trial Efficiency

A study from the Tufts Center for the Study of Drug Development estimated the cost to bring a drug to market at $2.6 billion (DiMasi et al. 2016). Factors that contribute to these high costs include increased clinical trial complexity, larger clinical trial size, higher cost of goods, greater focus on targeting chronic and degenerative diseases, changes in study design to gather information via health technology, and testing on comparator drugs to accommodate payer demands for comparative efficacy data.

Clinical trial costs can be broken down into per-study cost, per-site cost, per-patient cost, site overhead cost, and all additional costs not captured. Per-study costs include data collection, management, analysis and quality control, and the cost per institutional review board approval (including amendments). Per-site costs are the sum of site recruitment, site retention, administrative staff, and site monitoring costs. Per-patient costs include patient recruitment and retention, staff (such as physician, registered nurse, or clinical research associate), the cost of administering the procedure itself, and central lab costs. A 2014 study by the US Department of Health and Human Services presented costs by therapeutic area and by clinical trial phase. Phase 1 costs ranged from $1.4 million in the endocrinology therapeutic area (TA) to $6.6 million in the immunomodulation TA. Phase 2 costs ranged from $7 million in cardiovascular to $19.6 million in hematology. Phase 3 costs ranged from $11.5 million in dermatology to $52.9 million in pain and anesthesia. Post-marketing (Phase 4 studies) ranged from $6.8 million in the genitourinary TA to $72.9 million in the respiratory therapeutic area.

In 2018, Moore, Zhang, and Anderson published estimates of pivotal trial costs for novel therapeutic agents approved by the FDA between 2015 and 2016 (Moore et al. 2018). In their study of 59 novel therapeutics, the estimated median cost of pivotal efficacy trials was $19 million. At the extremes, there was a 100-fold cost difference. The analysis considered studies with fewer than 15 patients to more than 8000 patients, depending on therapeutic area.

While streamlining clinical studies is one possible cost-saving measure, model-based analysis is another.

In Silico Medicine

The idea of simulating a clinical trial in silico has been around for over 20 years (Bonate 2020). In recent years, new approaches in modeling and simulation have provided insights into biomedicine. Now, modeling and simulation are an integral part of achieving marketing authorization for therapeutics in many regions. These advances inform more intelligent clinical trials and often substitute for animal or human experimentation. In silico clinical trials use patient-specific models to create virtual cohorts for testing the safety and efficacy of drug candidates.

Linked pharmacometric-pharmacoeconomic modeling and simulation explores the contingency of market access and pharmaceutical pricing on the ability to demonstrate comparative effectiveness and cost-effectiveness. Predictions of the economic potential of drug candidates in development inform decisions across the product life cycle, and, in this way, are integral to portfolio management (Hill-McManus et al. 2020).

Model-Based Drug Development

The concept of model-based (MBDD) or model-informed (MIDD) drug development was born to address the low productivity and escalating cost of drug development. The FDA "critical path" document (ref) defined MBDD as the development and application of pharmaco-statistical models of drug efficacy and safety from preclinical and clinical data to improve drug development knowledge management and decision-making. Since the early 1990s, scientists in the field have stressed the importance of integrating the concepts of toxicokinetics, pharmacokinetics, and pharmacodynamics in drug development (Peck et al. 1992) and treating drug development as a cycle of learning and confirmation (Sheiner 1997), with each phase informing the next phase, and feeding back to the previous phase. The rationale for MBDD is a model's ability to integrate prior information when analyzing and interpreting the results of a current study. Models allow the integration of data from different studies based on understanding of the drug and the disease (Lalonde et al. 2007). Knowledge about a compound is continuously updated to inform decision-making for the next study (or development phase). The ability to pool information across trials, doses, and even compounds with the same mechanism of action leads to an increase in actionable information. Models developed using all the pertinent clinical data can be used to simulate (predict) the results of future trials. These predictions, when designed appropriately, take into consideration what we know about the compound and quantify what we do not know about the compound (confidence in the estimates of the parameters; variability).

MBDD can be broken down into six components: PK-PD and disease/placebo models; competitor information and meta-analysis; design and trial execution models; data-analytic models; quantitative decision criteria; and trial performance metrics.

PK-PD

Pharmacokinetic-pharmacodynamic (PK-PD) models describe the temporal relationship between drug administration and drug action (exposure and response). PK-PD modeling allows for the characterization of the time course of drug response – efficacy and toxicity. Characterizing differences in formulations, routes of administration, dosing regimens, and differences in patient populations can all be achieved with PK-PD modeling. Dose selection is one of the most challenging aspects of drug development. Choosing the "right dose" and defending the posology (dose, route, schedule, patient population) is one of the most challenging parts of drug discovery. Failure to explore and defend dose selection adequately is likely to blame for many Phase 2 to Phase 3 phase transition failures.

Meta-Analysis

The commercializability of a drug candidate depends on the competitive landscape. Whether or not the candidate is first-in-class, comparisons to the standard of care with respect to effect size, time to response, and durability of response must be made. Short of running a head-to-head trial against every possible competitor, the best way to understand the pharmacodynamics of a competitor is the curation of published results and meta-analysis of those data. From the published (often only summary-level) data, a representative subset of studies with adequate information must be identified in order to pool the data. Integration of the different data sources and different levels of information (and uncertainty) requires a strong grasp of statistics and a thorough definition of assumptions.

Design Considerations and Trial Execution Models

Adaptive designs allow for more efficient, agile decision-making, leading to their popularity among pharmaceutical companies and regulators. Such "enhanced" designs focus on learning from data as they are collected and making (pre-planned) changes to the study design when those milestones are met. Examples of adaptive designs include assessments for futility/success after a certain number of subjects; dropping dose levels that do not meet a certain (predefined) threshold; and changing the placebo or standard of care arm as time passes (e.g. to adjust for changing standards of care in a disease such as SARS-CoV-2).

Data-Analytic Models

These are the statistical models at the modelers' disposal. The question to be answered determines the model selection. The best choice may be a simple pairwise analysis of covariance (e.g. effect size determination), which requires few assumptions, or it may be best addressed with a regression-based approach. Similarly, the complexity needed depends on a detailed understanding of the question. For example, endpoint analysis (fixed time point) to estimate the dose-response relationship may require a simple (linear or nonlinear) regression model, while a longitudinal analysis will require more complex computations (and potentially more assumptions).

Quantitative Decision Criteria and Trial Performance Metrics

The ability to know the treatment effect of a drug (specifically relative to a comparator) is key in development. Clinical trials provide data to support the estimation of the treatment effect. Quantitatively, models of the distribution of the treatment effect can be used to generate and then refine the current state of knowledge about the true drug effect. Selecting quantitative decision criteria as cut points for go/no go decisions expand the utility of clinical trials, enabling decision making.

Examples of the use of clinical trial simulation to support decision making, designing an adaptive study, and MBDD's impact on regulatory decision making can be found in recently published accounts (Lalonde et al. 2007).

Recent Frontiers and Continued Call to Action

The past 40 years have seen a maturation of pharmacometrics as a discipline. Regulators routinely expect a comprehensive pharmacometrics package as part of any regulatory submission. Portions of the analyses have become automated and relatively push-button. In his 2020 perspective paper, Barrett reminds us that it is time to expect more from pharmacometrics (Barrett 2020).

One way to expand pharmacometrics utility is through collaboration with other disciplines. Candidate selection can be enhanced by combining pharmacometrics techniques with systems pharmacology; in pediatric extrapolation, with systems pharmacology and artificial intelligence/machine learning (AI/ML); and in product differentiation with AI/ML integration with real-world data analytics.

Creation of applications such as R Shiny tools, which provide data visualization of real-time simulations (for collaboration with non-modelers and other subject matter experts), bring the power of modeling and simulation to the project team in clear, immediately relevant terms. Recent work by colleagues at Certara, Inc. on the COVID-19 Pharmacology Resource Center, including an in-silico compound screening dashboard (Dodds et al. 2021) and the CODEx COVID-19 outcomes database (Tomazini et al. 2020) (containing summary-level endpoint data from 167 studies reported in 156 references at the time of this publication) allow for the integration of multiple modeling platforms with real world evidence data.

Integration of Real-World Evidence/Big Data

Coincident with poor attrition rates over the past decades, the healthcare industry has been experiencing rapid growth in data sources such as electronic health records, insurance claims data, patient registries, surveys, medical devices, imaging, genomics, and more that capture vast amounts of patient health and medical information. This real-world data (RWD) can provide valuable health information in the context of patients' day-to-day lives including real clinical practice. With this added clinical context, RWD becomes real-world evidence (RWE) that can be used to evaluate the epidemiology and burden of disease, comorbidities, treatment patterns, adherence, and outcomes of different treatments. RWE can therefore be used to model clinical studies, inform hypotheses, and thus improve the likelihood of approval and successful treatment launch. Not only can these applications contribute to compressing clinical trial timelines and drive down treatment costs, but RWE can also serve as a complement to evidence gathered from randomized control trials (RCTs), which continue to be the trusted standard for assessing biopharmaceutical drug safety and efficacy.

One issue for drug sponsors hoping to benefit from RWD is to develop an appropriate strategy for leveraging RWD sources. Many of these sources also fall into the category of "big data" practically meaning that the size is large (constantly moving target as of 2012 ranging from a few dozen terabytes to many zettabytes of data) and the format is often complex (can include unstructured, semi-structured and structured data). Such strategies involve decisions on whether to purchase such data from data providers or pay a fee for services such as an annual license. Strategies vary based on cost, internal expertise investment, and scope of work, of course, but most big Pharma companies recognize that they will make a substantial investment here in the hope of improving attrition rates by leveraging big data based on the perceived value of RWD and RWE.

FDA Guidance and Appreciation Evolving

Global regulatory authorities are also expanding their appreciation for this topic. Advances in the availability of RWD sources – such as electronic health records, registries, medical claims, pharmacy data, and feedback from wearables and mobile technology – have increased the potential to generate robust RWE, to support FDA regulatory decisions. RWE is the clinical evidence regarding the usage, and benefits and risks, of a medical product derived from the analysis of RWD. The real-life clinical performance of a medical product might be more clearly demonstrated through RWD/RWE because a controlled clinical trial often cannot evaluate all applications of a product in clinical practice across the full range of potential users (FDA Guidance 2017).

Three central considerations drive FDA's evaluation of RWE: whether the RWD are fit for use; whether the trial or study design used to generate RWE can provide adequate scientific evidence to answer or help answer the regulatory question; and whether the study conduct meets FDA regulatory requirements. As part of the agency's RWE efforts, the US Food and Drug Administration recently announced four grant awards (RFA-FD-20-020) to examine the use of RWD to generate RWE in regulatory decision-making. Through this awards program, the agency seeks to encourage innovative approaches to further explore the use of RWD while ensuring that scientific evidence supporting marketing approvals meets FDA's high evidentiary standards. As directed by the Twenty-first Century Cures Act, FDA is exploring the potential use of RWD and RWE to support the approval of new drug indications or post-approval study requirements for approved drugs. In December 2018, FDA published a strategic RWE Framework in support of this goal (FDA Guidance 2017). The grant awards are:

- Enhancing evidence generation by linking RCTs to RWD
- Applying novel statistical approaches to develop a decision framework for hybrid randomized controlled trial designs which combine internal control arms with patients' data from real-world data source
- Advancing standards and methodologies to generate real-world evidence from real-world data through a neonatal pilot project
- Transforming Real-world evidence with Unstructured and Structured data to advance Tailored therapy (TRUST)

RCTs are no longer the sole source of data to inform guidelines, regulatory, and policy decisions. RWD, collected from registries, electronic health records, insurance claims, pharmacy records, social media, and sensor outputs from devices form RWE, which can supplement evidence from RCTs. Benefits of using RWE include less time and cost to produce meaningful data; the ability to capture additional information, including social determinants of health that can impact health outcomes; detection of uncommon adverse events; and the potential to apply machine learning and artificial intelligence to the delivery of health care.

The Twenty-first Century Cures Act, signed into law on 13 December 2016 by the FDA, aims to encourage the pharma industry to adopt new technologies into drug development and approval. The Cures Act helped to provide faster approval for regenerative medicine advanced therapy such as cell therapy, gene therapy, and therapeutic tissue. The Cures Act, recently supported by The National

Coordinator for Health Information Technology (ONC) and Centers for Medicare Medicaid Services (CMS), also helps physicians, hospitals, insurers, and information networks exchange data faster.

References

Barrett, J.S. (2020). Time to expect more from pharmacometrics. *Clinical Pharmacology and Therapeutics* 108 (6): 1129–1131. https://doi.org/10.1002/cpt.1914. (accessed 19 June 2020). PMID: 32562273; PMCID: PMC7687126.

Bonate, P.L. (2020). On the shoulders of giants. . .. *Journal of Pharmacokinetics and Pharmacodynamics* 47 (1): 1. https://doi.org/10.1007/s10928-020-09674-4. PMID: 31960232.

Dragana Bugarski-Kirola, T. Blaettlerb, C. Arango et al. *Bitopertin in negative symptoms of Schizophrenia – Results from the phase III FlashLyte and DayLyte studies.* 2017;82(1):8–16. https://doi.org/10.1016/j.biopsych.2016.11.014. (accessed 15 December 2016).

DiMasi, G., Grabowskib, H.G., and Hansenc, R.W. (2016). Innovation in the pharmaceutical industry: new estimates of R&D costs. *Journal of Health Economics* 47: 20–33.

Dodds, M., Xiong, Y., Mouksassi, S. et al. (2021). Model-informed drug repurposing: a pharmacometric approach to novel pathogen preparedness, response and retrospection. *British Journal of Clinical Pharmacology* 1–10: https://doi.org/10.1111/bcp.14760.

Examination of Clinical Trial Costs and Barriers for Drug Development (2014). US Department of Health and Human Services, https://aspe.hhs.gov/report/examination-clinical-trial-costs-and-barriers-drug-development (accessed 20 February 2021).

Guidance for Industry and Food and Drug Administration Staff (2017). Use of Real-World Evidence to Support Regulatory Decision-Making for Medical Devices. U.S. Department of Health and Human Services Food and Drug Administration, Center for Devices and Radiological Health and the Center for Biologics Evaluation and Research. https://www.fda.gov/regulatory-information/search-fda-guidance-documents/use-real-world-evidence-support-regulatory-decision-making-medical-devices.

Daniel Hill-McManus, Scott Marshall, Jing Liu, Richard J. Willke, Dyfrig A. Hughes (2020). Linked Pharmacometric-Pharmacoeconomic modeling and simulation in clinical drug development. *Clinical Pharmacology and Therapeutics.* https://doi.org/10.1002/cpt.2051 (accessed 16 September 2020).

Hopkins, A.L., Mason, J.S., and Overington, J.P. (2006). Can we rationally design promiscuous drugs? *Current Opinion in Structural Biology* 16: 127–136.

Lalonde, R.L., Kowalski, K.G., Hutmacher, M.M. et al. (2007). Model-based drug development. *Clinical Pharmacology and Therapeutics* 82: 21–32.

Ludwig Institute for Cancer Research (2010). New computational tool for cancer treatment. *ScienceDaily.* (accessed 12 February 2021).

Ma, P. and Zemmel, R. (2002). Value of novelty? *Nature Reviews Drug Discovery* 1: 571–572. https://news.gallup.com/poll/266060/big-pharma-sinks-bottom-industry-rankings.aspx.

Moore, Z., Zhang, H., Anderson, G., and Alexander, G.C. (2018). Pivotal trial costs for novel therapeutic agents approved by the FDA between 2015–2016. *JAMA Internal Medicine* 178 (11): 1451–1457.

Munos, B. (2010). Lessons from 60 years of pharmaceutical innovation. *Nature Reviews Drug Discovery* 8: 959–968.

Pammolli, F., Magazzini, L., and Riccaboni, M. (2011). The productivity crisis in pharmaceutical R&D. *Nature Rev. Drug Discov.* 10: 428–438.

Paul, S.M., Mytelka, D.S., Dunwiddie, C.T. et al. (2010). How to improve R&D productivity: the pharmaceutical industry's grand challenge. *Nature Reviews Drug Discovery* 9: 203–214.

Peck, C.C., Barr, W.H., Benet, L.Z. et al. (1992). Opportunities for integration of pharmacokinetics, pharmacodynamics, and toxicokinetics in rational drug development. *Clinical Pharmacology and Therapeutics* 51 (4): 465–473. https://doi.org/10.1038/clpt.1992.47. PMID: 1563216.

Philippidis, A. (2019). Unlucky 13: Top Clinical Trial Failures of 2018. Biopharmas pursue costly studies despite data showing low success rates. https://www.genengnews.com/category/magazine/march-2019-vol-39-no-3.

Ringel, M.S., Scannell, J.W., Baedeker, M., and Schulze, U. (2020). Breaking Eroom's Law. *Nature Reviews Drug Discovery* 19 (12): 833–834. https://doi.org/10.1038/d41573-020-00059-3. PMID: 32300238.

Röhrig, U.F., Awad, L., Grosdidier, A.L. et al. (2010). Rational design of Indoleamine 2,3-Dioxygenase inhibitors. *Journal of Medicinal Chemistry* **53** (3): 1172–1189.

Scannell, J.W., Blanckley, A., Boldon, H., and Warrington, B. (2012). Diagnosing the decline in pharmaceutical R&D efficiency. *Nature Reviews Drug Discovery* 11 (3): 191–200. https://doi.org/10.1038/nrd3681. PMID: 22378269.

Sheiner, L.B. (1997). Learning versus confirming in clinical drug development. *Clinical Pharmacology and Therapeutics* 61 (3): 275–291. https://doi.org/10.1016/S0009-9236(97)90160-0. PMID: 9084453.

Swinney, D.C. and Anthony, J. (2011). How were new medicines discovered? *Nature Reviews Drug Discovery* 10: 507–519.

Tomazini, B.M., Maia, I.S., Cavalcanti, A.B. et al. (2020). Effect of dexamethasone on days alive and ventilator-free in patients with moderate or severe acute respiratory distress syndrome and COVID-19: the CoDEX randomized clinical trial. *JAMA* 324 (13): 1307–1316. https://doi.org/10.1001/jama.2020.17021.

Umbricht, A., Alberati, D., Martin-Facklam, M. et al. (2014). Effect of Bitopertin, a Glycine Reuptake inhibitor, on negative symptoms of Schizophrenia: a randomized, double-blind, proof-of-concept study. *JAMA Psychiatry* 71 (6): 637–646. https://doi.org/10.1001/jamapsychiatry.2014.163.

University of Surrey (2007). In Silico cell for TB drug discovery. *ScienceDaily*. (accessed 12 February 2021).

Chapter Self-Assessments: Check Your Knowledge

Questions:

- Provide a high-level of the goals and transition milestones across drug development stages and examine how they contribute to the overall likelihood of approval (LOA)?
- What was the main conclusion of the 2004 FDA white paper – Challenges and Opportunities report?
- In the last 60 years, there has been only one rise in R&D efficiency until recently. Describe the like event and explanation?
- Explain how in silico pharmacoeconomic modeling helps drug development and can be utilized as a tool to improve attrition rates.

Answers:

- To successfully progress through drug development, a candidate needs to meet the milestones of each phase. The drug candidate must be safe to administer in humans and have a reasonable chance of reaching clinically effective concentrations (Phase 1), it must show efficacy in a small population of people with the disease or condition of interest (while maintaining a favorable safety profile) (Phase 2), and this safety and efficacy in early clinical development must be borne out in pivotal (typically Phase 3) trials in a larger segment of the population to be treated. Additionally, the candidate must be able to meet commercial, marketing, and payer reimbursement goals, and still meet an unmet medical need at the time of launch (often eight to nine years after clinical development begins).
- FDA's position in the white paper was that there was a growing trend of innovators concentrating on product development of assets with the potential for high market return while developing products for important public health needs (less common diseases, diseases prevalent in parts of the world with lesser resources, drug candidates aimed at prevention rather than treatment, and individualized therapies) remain under-explored.
- In the late 1990s, R&D efficiency rose for a short time. This was attributed to the clearing of the regulatory backlog at the FDA following the implementation of the 1992 Prescription Drug User Fee Act (PDUFA), with a small contribution from the rapid development and approval of several HIV drugs, hastened by input from community action and patient support group efforts.

- Linked pharmacometric-pharmacoeconomic modeling and simulation explores the contingency of market access and pharmaceutical pricing on the ability to demonstrate comparative effectiveness and cost-effectiveness. Predictions of the economic potential of drug candidates in development inform decisions across the product life cycle and, in this way, are integral to portfolio management and ultimately better decisions that can improve attrition rates.

Quiz:

1 In a 2016 report, the Tufts Center for the Study of Drug Development estimates the overall likelihood of approval (LOA) for drugs entering humans in Phase 1 to ultimately be approved for marketing is _____ %. Choose the best answer
   A <2
   B 9.6
   C 10–15
   D 25

2 The moniker, "Eroom's Law," is _____ (select the correct answer)
   A "Moore's Law" spelled backwards
   B A principle defined by Jacon Eroom that describes the probability of success across drug development
   C A principle that estimates the decrease in R&D efficiency
   D None of the above

3 A study from the Tufts Center for the Study of Drug Development estimated the cost to bring a drug to market at $2.6 billion (DiMasi et al. 2016). Factors that contribute to these high costs include all but which of the items below (choose the best answer)
   A Increased clinical trial complexity
   B Larger clinical trial size
   C Higher cost of goods
   D Greater focus on targeting chronic and degenerative diseases
   E Changes in study design to gather information via health technology
   F All are correct

4 <u>All but which</u> are considered components of Model-based drug development?
   A PK-PD and disease/placebo models
   B Competitor information and meta-analysis
   C Market size assessment and landscape analysis
   D Design and trial execution models
   E Data-analytic models
   F Quantitative decision criteria and trial performance metrics

19

Medical Devices

Jenny Zhuang[1], BS and John(Chengfeng) Zhuang[2], PhD

[1] *Veeva Systems, Pleasanton, CA, USA*
[2] *VAsieris Pharmaceuticals, Shanghai, China*

The medical device industry is large and rapidly growing, representing hundreds of billions of dollars in sales each year. It is also diverse, with millions of different medical devices available on the market worldwide. These devices vary in complexity from tongue depressors to implantable cardiac pacemakers/defibrillators. Some of the larger medical device manufacturers include Johnson & Johnson, Medtronic, Boston Scientific, and Baxter. Compared to the drug industry, there are more small- to mid-size manufacturers in the medical device industry.

Similar to drugs, medical devices are highly regulated to protect the well-being of users and patients. The regulations governing medical devices vary from one country to another, but all follow a similar approach, i.e. risk-based classification and marketing requirements. This chapter introduces the United States regulations of medical devices. Readers who are interested in learning more about the regulations of medical devices by other countries or regions are encouraged to consult the websites of the health authorities in those countries or regions.

Definition of Medical Device

In the US, medical devices are regulated by the Food and Drug Administration (FDA). Within the FDA, the Center for Devices and Radiological Health (CDRH) regulates medical devices and radiation-emitting products. CDRH is charged with the evaluation of the safety and effectiveness of medical devices before and after reaching the market to ensure patients and health care providers have timely and continued access to the medical devices they need.

The first question often asked is: what is a medical device? In the US, Section 201(h) of the Federal Food, Drug, and Cosmetic Act (FD&C Act) provides the definition. The term "medical device" means: an instrument, apparatus,

Fundamentals of Drug Development, First Edition. Edited by Jeffrey S. Barrett.
© 2022 John Wiley & Sons, Inc. Published 2022 by John Wiley & Sons, Inc.
Companion website: www.wiley.com/go/Barrett/FundamentalsDrugDevelopment

implement, machine, contrivance, implant, in vitro reagent, or other similar or related article, including any component, part, or accessory, which is: (1) recognized in the official National Formulary, or the United States Pharmacopeia, or any supplement to them; (2) intended for use in the diagnosis of disease or other conditions, or in the cure, mitigation, treatment, or prevention of disease, in man or other animals, or (3) intended to affect the structure or any function of the body of man or other animals; and which does not achieve its primary intended purposes through chemical action within or on the body of man or other animals; and which is not dependent upon being metabolized for the achievement of its primary intended purposes.

Therefore, a key distinction between a device and a drug is that a device achieves its primary intended purpose through physical or mechanical action, whereas a drug through chemical or biochemical action.

The device regulations in the US are codified in 21 Code of Federal Regulations (CFR): Parts 800-1050. Parts 800-861 are general device requirements, whereas Parts 862-1050 are device-specific requirements. In addition, 21 CFR: Parts 1-99 are general medical requirements that also apply to medical devices.

History of Medical Device Regulations

Prior to the twentieth century, food, drugs, and devices were largely unregulated in the United States. This lack of regulation led to unsafe or unsanitary conditions in food processing plants as well as the use of toxic or unknown ingredients in both food and drugs. In the early 1900s, public exposure of these practices and concern about their impact on human health necessitated regulations and resulted in the establishment of the Pure Food and Drug Act of 1906. The Pure Food and Drug Act introduced regulations on the labeling of drugs and food and was enforced by the US Bureau of Chemistry, which later became the FDA in 1930.

FD&C Act, passed in 1938, replaced the Pure Food and Drug Act and established more stringent laws on the production of food and drugs; manufacturers were now required to provide proof of safety for new drugs prior to placing them on the market, and the FDA was given more authority to inspect production facilities and remove or recall unsafe products. In addition, medical devices and cosmetics were regulated for the first time. However, at that time, medical devices were relatively simple and low-risk instruments, such as scalpels and syringes, and from a regulatory control perspective, there was no distinction between drugs and devices.

In the post-World War II era, rapid growth in technology brought about more complex and advanced medical devices, such as heart-lung and x-ray machines. It became clear that medical devices had a greater range of risk and complexity, and therefore the level of regulatory scrutiny and control needed for devices should differ accordingly. The Medical Device Amendments to the FD&C Act in 1976 established three medical device classes, which are described in the following section.

Classification of Medical Devices

The US device classification system was established by the 1976 Medical Device Amendments to the FD&C Act to enable the regulation of devices in accordance with their risks and benefits. It recognized the wide variation in complexity and potential for harm among devices and defined the degree of regulatory control needed to ensure the safety and effectiveness of the device. The classification depends on the intended use of the device and the degree of risk posed by the device. There are three classes of devices based on risk level: Class I (lowest risk), Class II (moderate risk), and Class III (highest risk). Each Class of device is subject to different regulatory control and submission requirements, as summarized in Table 19.1 below.

Class I devices are those that pose minimal potential for harm to patients and are often simpler in design. Examples of Class I devices include elastic bandages, examination gloves, hand-held surgical instruments, lead shields, and toothbrushes. Class I devices are only subject to general controls.

General controls include establishment registration, medical device listing, manufactured in accordance with Quality System regulation (21CFR Part 820), labeled in accordance with 21 CFR Part 801 or 809, and submission of a premarket notification (also known as a 510(k), see below for more details). However, over 90% of Class I devices are exempt from premarket notification requirements. Many Class I devices are also exempt from cGMP requirements (except for 21CFR 820.180 Records and 820.198 Complaint Files). Applicable exemptions are stated in the device-specific regulations (21 CFR 862-892).

Class II devices that pose higher risk than Class I but not as high as Class III. This Class includes those devices for which general controls alone are

Table 19.1 Device classification, regulatory control, and submission requirements

Class and risk level	General controls	Special controls	Common submission
Class I – lowest risk	x		Exempt
Class II – moderate risk	x	x	510(k)
Class III – highest risk	x		Premarket approval (PMA)

insufficient to ensure safety and effectiveness, and methods exist to provide additional evidence of the device's safety and effectiveness. Examples include powered wheelchairs, infusion pumps, surgical drapes, orthopedic implants, and imaging systems. Class II devices must comply with general controls plus special controls.

Special controls are specific to Class II devices and usually for well-established device types. Special controls might include special labeling requirements, limitation for indications, contraindications, warnings; mandatory performance standards; testing requirements; performance criteria; and post-market surveillance. The specific special control requirements for a device can be found in the "(b) Classification" section of the device regulation (e.g. 21 CFR 876.5860(b) for high permeability hemodialysis system).

Class III devices are devices for which insufficient information exists to ensure safety and effectiveness solely through general or special controls. These devices usually support or sustain human life, are of substantial importance in preventing impairment of human health, or present a potential, unreasonable risk of illness or injury. Examples of Class III devices include extended wear contact lenses, replacement heart valves, and high-energy defibrillators. Class III devices must comply with general controls and may also require a premarket approval (PMA) submission and approval (see next).

Submission Types and Requirements

Most Class I devices are exempt from submission requirements since they pose a low risk to patients, and their safety and effectiveness can be established by general controls. On the other hand, most Class II devices require submission of pre-market notification, also known as a 510(k), and Class III devices submission of PMA. In fact, 510(k) and PMA are the two most common types of US submissions for medical devices. The following section explains these two submission types in greater detail.

A 510(k) is a premarket submission made by device manufacturers to notify the FDA of their intent to market a medical device at least 90 days in advance. It is called a 510(k) because Section 510(k) of the FD&C Act specifies this submission requirement. The 510(k) procedure is further described in 21 CFR 807 Subpart E.

A key and distinguishing component of a 510(k) submission is to provide a rationale for substantial equivalence. The purpose of this component is to present sufficient information for FDA to be able to determine that the device to be marketed is substantially equivalent (SE) to other, similar, legally marketed devices (i.e. predicate devices). The regulation defines "legally marketed," or predicate, devices as follows:

- Pre-amendment device – a device that was marketed prior to 28 May 1976, or
- A device that has been reclassified from Class III to Class II or Class I, or
- A device that has been found to be substantially equivalent to such a device through the 510(k) process, or
- A device that was granted marketing authorization via the De Novo classification process (see below)

A device is substantially equivalent to the predicate if it has the same intended use and technological characteristics as the predicate device. Alternatively, a device is substantially equivalent to the predicate if it has the same intended use as the predicate device and has different technological characteristics, but the information submitted to FDA does not raise new questions of safety and effectiveness due to the difference in technological characteristics, and the submitted information demonstrates that the device is as safe and effective as the predicate device. To facilitate review, a comparison table is typically generated to describe the similarity and differences between the new device and the predicate device. Table 19.2 provides an example of such a comparison table.

It is important to note that substantial equivalence does not mean that the new and predicate devices need to be identical. However, the new and predicate devices should have the same intended use, and any differences in technological characteristics (such as material in the example in Table 19.2) should not raise questions of safety and effectiveness. The 510(k) submission allows the FDA to review the scientific methods used to evaluate differences in technological characteristics and performance data. This performance data can include clinical data and non-clinical bench performance data (e.g. engineering performance testing, sterility, electromagnetic compatibility, software validation, biocompatibility evaluation), among other data. The device manufacturer may only market the new device after it receives an order from FDA declaring the new device is SE to the predicate. This process is sometimes referred to as FDA "clearance" of the device

Table 19.2 Example comparison table for illustrating SE designation

Attribute	Your new hip	Predicate hip (K06XXXX)
Intended Use	To replace hip. . .	To replace hip. . .
Design	Fluted, coated, sizes from 10–18	Fluted, coated, sizes from 12–18
Materials	CoCr	Titanium
Performance	10 million cycles	10 million cycles
Sterility	Radiation 10^{-6} SAL	Radiation 10^{-6} SAL
Packaging	Tyvek pouch	Tyvek pouch

to be marketed. Unlike a new drug application (NDA) or PMA, the term "clearance" rather than "approval" is used because, in a 510(k) submission, FDA does not evaluate scientific evidence that *proves* the device to be safe and effective.

Due to the level of risk associated with Class III devices, FDA has determined that general and special controls alone are insufficient to ensure the safety and effectiveness of Class III devices. These devices require a PMA application under section 515 of the FD&C Act in order to obtain permission to market the product in the US. PMA is the FDA process of scientific and regulatory review to evaluate the safety and effectiveness of Class III medical devices. FDA must determine that PMA contains sufficient valid scientific evidence to ensure the device is safe and effective for its intended use(s). A PMA for a device is equivalent to an NDA for a drug. The regulation governing premarket approval is located in Title 21 Code of Federal Regulations (CFR) Part 814.

Valid scientific evidence is evidence from well-controlled investigations, partially controlled studies, studies and objective trials without matched controls, well-documented case histories conducted by qualified experts, and reports of significant human experience with a marketed device, from which it can fairly and responsibly be concluded by qualified experts that there is a reasonable assurance of the safety and effectiveness of a device under its conditions of use.

Device safety is defined in 21 CFR §860.7(d)(1) as: "There is a reasonable assurance that a device is safe when it can be determined, based on valid scientific evidence, that the probable benefits to health from the use of the device for its intended uses and conditions of use, when accompanied by adequate directions and warnings against unsafe use, outweigh any probable risks." An effective device is defined in 21 CFR §860.7(e) as: "when it can be determined based on valid scientific evidence, that is a significant portion of the target population, the use of the device for its intended uses and conditions of use, when accompanied by adequate directions for use and warnings against unsafe use, will provide clinically significant results." Table 19.3 illustrates the similarities and differences between a 510(k) and a PMA.

Table 19.3 Comparisons between 510(k) and PMA

510(k)	PMA
Most Class II devices; a few Class I devices	Most Class III devices
Demonstrate substantial equivalence (clinical trials seldom needed)	Prove safety and efficacy (Clinical trials usually needed)
Device is "cleared" to market	Device is "approved" by FDA
User fee is thousands of dollars	User fee is hundreds of thousand dollars
Typical review time 30–90 d	Typical review time 6–12 mo

Table 19.4 Examples of medical devices and regulatory approval pathways.

Device	Device class	Regulatory approval pathway
Band-Aid® flexible fabric bandages	I	General controls
Oral-B® Pro 1000 electric toothbrush	I	General controls
3M™ Littmann® classic stethoscope	I	General controls
The Tether™ – Vertebral body tethering system	Not classified	HDE
FilmArray global fever panel	II	De Novo classification, 510(k)
First Response® early result pregnancy test	II	510(k)
Fitbit ECG app	II	510(k), IDE
WaveLight Allegretto wave excimer laser system	III	PMA, IDE
ZOLL LifeVest wearable defibrillator	III	PMA, IDE

There are other premarket submission types for medical devices, such as investigational device exemption (IDE), De Novo classification request, and humanitarian device exemption (HDE). An IDE is similar to a drug IND and is required to conduct clinical evaluation of a device in human subjects. A De Novo classification request is a regulatory procedure for FDA to classify a novel device, or an existing device with new intended use or with different technological characteristics that raise different questions of safety and effectiveness, for which there is no predicate device for determination of substantial equivalence. Upon receipt of a de novo classification request, FDA makes a risk-based evaluation for classification of the device into Class I or II. After that, the newly classified device can be used as a predicate device for future devices. Finally, an HDE is a premarket submission for a Humanitarian Use Device (HUD). A HUD is a medical device intended to benefit no more than 8000 individuals in the United States per year. An HDE is exempt from the effectiveness requirements; it only needs to provide reasonable assurance of safety and probable benefit. Table 19.4 below provides real-world examples of several medical devices and their varying pathways for regulatory approval.

Combination Products

Certain products are made up of multiple drugs, devices, and/or biologics. Cardiac drug-eluting stents, for example, are comprised of a wire mesh scaffold (device) that actively releases a drug at the stent implantation site. Products that are comprised of two or more regulated components are defined as combination products in the US. For the purpose of jurisdiction within the FDA, a combination product is designated either as a drug, device, or biologic depending on the primary mode of action (PMOA). FDA's Office of Combination Products (OCP) was established to facilitate this jurisdiction. A company may obtain a formal determination of a combination product's PMOA to the OCP, which designates an FDA center – Center for Drug Evaluation and Research (CDER), Center for Biologics Evaluation and Research (CBER), or CDRH – for oversight over the combination product development and approval process.

Comparison of Device and Drug Development

Drugs and medical devices differ not only in their regulatory approval pathways but also in their development. There are significant distinctions between drug and medical device design and manufacturing considerations as well as clinical trials.

Clinical trials for pharmaceuticals are structured in multiple phases (Phase 1, Phase 2, Phase 3, Phase 4), which are performed after the preclinical phase, where the drug is tested in animals to collect pharmacology, toxicology, and ADME information. These phases evaluate the safety and efficacy of the investigational drug in human subjects. For devices, clinical trials are typically conducted in pilot, pivotal, and post-market phases, which test not only the safety and efficacy of the device, but also the usability of the device which feeds input into the device design and improvement. Therefore, for device trials, there may be ongoing overlap between clinical trials and device development and design activities. Table 19.5 outlines additional differences between drug and device clinical trials.

Table 19.5 Differences between drug and medical device clinical trials.

	Drugs	Medical devices
Phases	Phase 1, 2, 3, 4	Pilot, Pivotal, Post-Market
Average population size	Larger (100s–1000s)	Smaller (10s–100s, 2000 max)
Average trial duration	Years	Months
Randomization	Common	Varies
Control	Placebo	Competing technology or none
Average cost	$10-100+ MM per phase	Pilot, < $5 MM Pivotal, $10–50 MM Post-Market, varies

From a design/development and manufacturing perspective, both drugs and medical devices must adhere to Good Manufacturing Practice (GMP) requirements to ensure products are manufactured with adherence to quality standards that ensure their safety and efficacy. These quality standards apply to the entire product lifecycle, from design/development and testing to manufacturing and commercialization.

For medical devices, the Quality System Regulation (QSR), codified in 21 CFR 820, provides detailed FDA regulations for implementing GMP for medical device manufacturers. This includes requirements for design control (e.g. device design and development) as well as risk management. Depending on the device type and Class, only certain parts of the regulation may be pertinent for a particular device.

Drug FDA GMP requirements are detailed in 21 CFR 210 and 21 CFR 211. 21 CFR 210 outlines requirements for manufacturing, facilities, and controls for the manufacture, processing, packing, and holding of drugs, whereas 21 CFR 211 outlines requirements for finished drug products. Although the regulations for device and drug GMP differ, they both aim to achieve the same goal - to ensure products are designed and manufactured with quality to ensure their safety and effectiveness.

Summary

The regulation of medical devices in the United States began in 1938 with the Federal FD&C; device regulatory framework allows more flexibility for the diverse nature of medical devices as opposed to new chemical entities. Medical devices are classified as Class I, II, and III based on risk level. Regulatory requirements are based on "Risk" and "Intended Use" of the device, and clinical trials are only required for the highest risk medical devices. Common premarket submissions to FDA include 510(k) and PMA. Some devices do not require any submission to FDA prior to marketing. Drugs and medical devices vary greatly in how they are developed, such as manufacturing and conducting clinical trials.

Chapter Self-Assessments: Check Your Knowledge

Questions:
- Who regulates medical devices in the United States?
- Define medical devices and the distinction between a device and a drug?
- Discuss premarket notification requirements for medical devices?
- Describe the rationale for a 510(k) submission for demonstrating substantial equivalence?

Answers:

- In the US, medical devices are regulated by the Food and Drug Administration (FDA). Within the FDA, the Center for Devices and Radiological Health (CDRH) regulates medical devices and radiation-emitting products. CDRH is charged with the evaluation of the safety and effectiveness of medical devices before and after reaching the market to ensure patients and healthcare providers have timely and continued access to the medical devices they need.

- Section 201(h) of Federal Food, Drug, and Cosmetic Act (FD&C Act) provides the definition of a "medical device" as an instrument, apparatus, implement, machine, contrivance, implant, in vitro reagent, or other similar or related article, including any component, part, or accessory, which is: (1) recognized in the official National Formulary, or the United States Pharmacopeia, or any supplement to them; (2) intended for use in the diagnosis of disease or other conditions, or in the cure, mitigation, treatment, or prevention of disease, in man or other animals, or (3) intended to affect the structure or any function of the body of man or other animals; and which does not achieve its primary intended purposes through chemical action within or on the body of man or other animals; and which is not dependent upon being metabolized for the achievement of its primary intended purposes. A key distinction between a device and a drug is that a device achieves its primary intended purpose through physical or mechanical action, whereas a drug through chemical or biochemical action.

- Over 90% of Class I devices are exempt from premarket notification requirements. Many Class I devices are also exempt from cGMP requirements (except for 21CFR 820.180 Records and 820.198 Complaint Files). Applicable exemptions are stated in the device-specific regulations (21 CFR 862-892). Class II devices that pose higher risk than Class I but not as high as Class III. This class includes those devices for which general controls alone are insufficient to ensure safety and effectiveness, and methods exist to provide additional evidence of the device's safety and effectiveness.

- The purpose of this component is to present sufficient information for FDA to be able to determine that the device to be marketed is Substantially Equivalent (SE) to other, similar, legally marketed devices (i.e. predicate devices). The regulation defines "legally marketed," or predicate, devices as follows: Pre-amendment device - a device that was marketed prior to May 28, 1976, or a device that has been reclassified from Class III to Class II or Class I, or a device that has been found to be substantially equivalent to such a device through the 510(k) process, or a device that was granted marketing authorization via the de novo classification process. A device is substantially equivalent to the predicate if it has the same intended use and technological characteristics as the predicate device. Alternatively, a device is substantially equivalent to the predicate if it has the same intended use as the predicate device and has different technological characteristics, but the information submitted to FDA does not raise new questions of safety and effectiveness due to the difference in technological characteristics, and the submitted information demonstrates that the device is as safe and effective as the predicate device.

Quiz:

1 True or False. The Food, Drug, and Cosmetic Act (FD&C), passed in 1938, replaced the Pure Food and Drug Act and established more stringent laws on the production of food and drugs, and included medical devices. (Answer False)

2 True or False. There are three classes of devices based on complexity level: Class I (lowest complexity), Class II (moderate complexity), and Class III (highest complexity). (Answer False)

3 Examples of Class I devices include all but which of the following: (choose the correct answer) (answer = d)
   A elastic bandages
   B examination gloves
   C hand-held surgical instruments
   D suppositories
   E toothbrushes

4 Examples of Class II devices include all but which of the following: choose the correct answer) (answer = a)
   A suppositories
   B powered wheelchairs
   C infusion pumps
   D surgical drapes
   E orthopedic implants
   F imaging systems.

Useful Resources

US FDA website: www.fda.gov
"Medical Devices" tab
Navigate the Medical Devices Section
Medical Device: Comprehensive Regulatory Assistance

20

Supply Chain
Donna Humski

Johnson & Johson Pharmaceutical Company

The approval of a new drug opens the doors to improving the lives of people. Of course, since the early concept phases of this drug, the goal has been to be able to broadly supply this drug to patients everywhere. Throughout the development process designing for manufacturability was a priority. Compliant raw materials have been carefully selected and built into the formulation. The compounding process is robust, repeatable, and scalable. Now it's time to stand up a supply chain that will deliver this drug to patients and deliver profit. This chapter will define the core functions of the supply chain and their critical relationship to each other. Case studies will explore the decisions required to design an optimal supply network to bring the newly approved drug to the patients it is intended to treat.

What Is the Supply Chain?

The supply chain is a series of actions that must be completed to produce and distribute goods. These actions are integrally linked, which is the basis for the reference to a chain. The best product can grow or fail in the market because of the supply chain regardless of the strength of the science behind the product. A reliable supply chain will win favor with customers and bring promotional opportunities for a product at the expense of competitors.

Lack of reliability can result in financial penalties such as re-slotting fees to put products back on the shelf, credits for loss of revenue taken against the invoice, and revocation of advertising spots. Typically assessed on a per-unit basis, these fees can total millions of dollars and can be applied month after month. In the retail space, warehouse clubs only carry the top product in a category. If the supply chain cannot reliably meet demand, the product is removed from the shelf, and another put in its place. The manufacturer must then work to get back on the shelf, possibly paying slotting fees and providing unique offerings. In chain retail, the store manager will replace the missing branded product with the chain's generic alternative. Customers may see the generic as the only option. An unreliable supply chain results in ongoing lost sales but patient frustration and reputation damage.

In the prescription drug market, a lack of reliability can result in patient harm. Patients unable to purchase anti-rejection medications for transplanted organs, anti-depression medications to balance their actions, and even high blood pressure medications can suffer physical repercussions from a lack of supply for a drug on which they depend. Recognizing that the supply chain is critical to both the patient and the successful life of the product reinforces the importance of the design decisions supply chain leaders must make.

Supply chains come in many different sizes and shapes. Networks can be as simple as a single manufacturing and distribution sites to multiple globally located sites performing individual steps of product construction and distribution. The supply chain should be customized for the product to drive differentiated service and profitability outcomes. The wrong configuration will drive high costs, frequent disruptions, and compliance issues if not matched to the product requirements and patient needs. On the other hand, a supply chain can be a significant competitive advantage in the market if a resilient and reliable match to the marketplace expectations. First, let's examine the components or functions of the supply chain.

Traditionally, PLAN, SOURCE, MAKE, and DELIVER represent the core functions of the supply chain. Each of these functions is dependent on, linked to, the others (see Figure 20.1). For example, PLAN requires insight from Deliver to understand what customers desire. SOURCE requires a plan to know

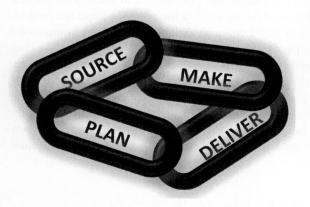

Figure 20.1 Core functions and relationships of the pharmaceutical supply chain.

what to purchase. PLAN also requires data from MAKE to align on how much can be produced. MAKE requires raw materials from suppliers and direction from PLAN. DELIVER requires a reliable schedule from MAKE to commit to customer orders. Each of the links must be robust enough to keep the chain intact through daily challenges and crises. Figure 20.1 depicts the circular connection between the functions. Change in any one link will pull the other links in the same direction. Digging deeper into each function will show how the chain works.

PLAN

PLAN is the conductor of the supply chain orchestra, the coach of the team, or the air traffic controller. A planning team creates a forecast from historical distribution statistics combined with insight from the sales team on new customers or promotions. The forecast minus existing inventory is used to determine requirements for SOURCE and MAKE. This is often called the *operating* plan. The operating plan is further refined to drive a time-phased production schedule based upon raw material availability and manufacturing capacity. PLAN will also provide the DELIVER function with a projection for storage and transportation. PLAN controls the amount of inventory maintained at each supply chain step. Inventory is critical to order fulfillment and ultimately profit. A planner must have strong analytics skills, broad supply chain experience to be able to challenge peers to ensure that the forecast is realistic and drive adherence to the production schedule and inventory plan.

An ideal operating plan is depicted below. Target inventory is the amount of inventory required to cover future sales plus a buffer for errors in forecasting or producing defects, among other unanticipated issues reflected as a min-max range. There are many target inventory calculation methods. Production planning software, such as SAP and Oracle, have models embedded in the software. Manual calculation is possible but complicated. All models will require data such as forecast accuracy, production lead time and frequency, defect rate, and desired order fulfillment or service level. If a drug is a custom made for a patient, then no inventory is made in advance and total production lead time is the governing factor.

Figure 20.2 shows an example of an operating plan. This company is producing a drug with frequent production runs each month. Production is planned when the on-hand inventory drops below the minimum target inventory level. Each month's sales demand is covered by the inventory on-hand from *prior* months' production (operating plan). Inventory must be available prior to the start of the sales month. Note that promotions are being planned for three future months.

	Month 1	Month 2	Month 3	Month 4	Month 5	Month 6
Inventory On-Hand	100	125	125	115	115	115
Base Forecasted Sales	−50	−50	−50	−75	−75	−50
Promotional Demand		−25	−10		−25	
Operations plan	+75	+75	+50	+75	+100	+50
Ending Inventory	125	125	115	115	115	115

Figure 20.2 Operating plan example.

	Month 1	Month 2	Month 3	Month 4	Month 5	Month 6
Inventory On-Hand	100	100	75	65	40	−10
Base Forecasted Sales	−50	−50	−50	−75	−75	−50
Promotional Demand		−25	−10		−25	
Operating plan (Lot size = 25, max 2 lots per month)	+50	+50	+50	+50	+50	+50
Ending Inventory	100	75	65	40	−10	−10

Figure 20.3 Operating plan with constrained equipment capacity.

Production is adjusted to meet those anticipated increased sales and maintain the inventory target level.

In this example, the target for inventory on-hand is expected to be between 100 and 125 units at the end of each month based upon a monthly sales forecast, 30-day production lead time, a lot size of 25 units, and a maximum of four lots per month.

Figure 20.3 depicts a service challenge scenario that often occurs. MAKE informs PLAN that only 50 units can be made during Months 2 through 6 due to a required equipment upgrade. Now, using the operating plan depicted in Figure 20.2, inventory ending inventory dips below the projected sales demand resulting in at least two months of deficit.

PLAN now has a few options to protect against the inventory shortage.

- PLAN can align with SALES to delay the Month 5 promotion to Months 8 or 9. This may be unacceptable as the supply chain is constraining the market; however, it may be the only alternative.
- PLAN can ask MAKE to postpone the equipment upgrade until Month 3 while increasing production in Months 1 and 2 to cover the deficit. The equipment upgrade may be necessary to keep the equipment functioning, and a delay could be damaging. SOURCE must be consulted to obtain the necessary increased amounts of raw materials.

Neither of these options meets all of the functions' goals. A compromise will be required here. If this demand pattern will continue, an increase in total equipment capacity may be necessary. What would you do?

SOURCE

In the example above, SOURCE action is required as PLAN makes a change to the operating plan. SOURCE is often also known as procurement or purchasing. The goal of the sourcing organization is to ensure the continuity of raw material supply at the best value, whether production increases or decreases. Source teams may form long-term supply relationships with key partners or spot buy in the market depending on the raw material. The forecast generated by the planning function is the basis for purchasing contracts; however, the operating plan is the driver of actual purchases.

SOURCE professionals must be able to discern value vs. price. As an example, a low-priced material with a long lead time may be less valuable than a higher priced material from a supplier with a quick turnaround. Given the example in Figure 20.2, a local supplier with available capacity will be critical if MAKE is to increase Month 1 and 2 production output. High-quality raw material is more valuable than a less costly raw material from a questionable supply source. Material defects will slow production, cause scrap, and potentially result in lost manufacturing time. Value, not price, should be the focus of the sourcing process.

MAKE

MAKE refers to the organization that physically manufactures or assembles the product. Manufacturing involves the act of producing and the support of engineering, equipment mechanics, crew supervisors, and operators among other professionals. MAKE takes direction from PLAN on how much is to be produced and when. MAKE is dependent on SOURCE for compliant and available raw materials. MAKE also partners closely with Research and Development to design a product and manufacturing process that is repeatable, reliable, and cost-effective. In the healthcare manufacturing industry, routine audits by government agencies such as the Food and Drug Administration (FDA), Pharmaceuticals and Medical Devices Agency (PMDA), Technischer Überwachungsverein (TUV), European Medicines Agency (EMA), etc. will be done on the manufacturing site, equipment, documentation, and processes. MAKE partners with both Regulatory and Quality to ensure constant compliance to current Good Manufacturing Practices (cGMPs).

DELIVER

DELIVER represents a broad group of functions including customer management, order management, order fulfillment, storage, Customs management, and transportation. This is the organization that directly connects with the customer. In the best supply chains, this customer connection provides critical insight that will improve the accuracy of the sales forecast. That forecast ignites the entire supply chain process.

Storage can be very complicated for healthcare products. Regulatory controlled substances, temperature-controlled substances, hazardous or explosive materials, and the like all require specialized storage and transportation conditions. Distribution centers and warehouses are subject to regulatory body audits. In addition to maintaining proper product conditions, traceability at the unit level, building security, cleanliness, proper management of material expiry and destruction are all requirements.

Customer order management must be performed in compliance with financial policies for the country in which the product is sold. In the United States, compliance with Sarbanes Oxley regulations requires that shipments must be aligned with customer orders. This is particularly challenging at financial period closes when company revenue is in the spotlight. Important, also, is that customs management of imported and exported products is typically owned by the DELIVER function. Mismanagement of Customs requirements has potentially large financial implications if noncompliant. The manufacturer can be fined on both the import and export side and lose the right to ship to a given country. In the other direction, the manufacturer could be missing advantageous tax benefits if Customs regulations are not managed properly.

Although this function sounds like trucking, DELIVER is much more than storage and transport. Customer relationship, regulatory compliance, customs, and financial compliance are also critical components of this function. In today's global market, easy order placement and accurate delivery are customer expectations, not optional. A healthcare supply chain with a well-designed, efficient, and compliant DELIVER function will be able to hold lower inventories, avoid fines, and exceed customer expectations.

QUALITY

Why isn't QUALITY a supply chain function or a link in the chain? QUALITY is the function that interprets regulations into policies and guidelines to direct the supply chain. QUALITY is not an inspector at the end of production but rather the author of the requirement at each step that when completed properly will result in a compliant product. As defects and unanticipated results can be more frequent than expected, QUALITY sets the criteria for acceptable and unacceptable actions and outcomes. In the healthcare industry, QUALITY is established as an independent organization. This is specifically to ensure that compliance decisions, audits, and assessments are not influenced by the supply chain. QUALITY will work closely with SOURCE to determine the acceptability of a supplier. QUALITY partners with MAKE to ensure robust procedure execution and compliant goods. QUALITY also collaborates with DELIVER to manage customer allocations for controlled substances and adherence to distribution standards. In all these actions, QUALITY must have the ability to independently assess and determine that the product and process are compliant.

Now with an understanding of the functions, let's examine the supply chain configuration options and how to determine which option is best for the product, the marketplace, and the patient.

Patients Are Not Patient

The ultimate objective of the supply chain is often misunderstood. Manufacturing products is one outcome. Delivering goods is another outcome. Improving the cost of materials is also considered a measure of supply chain success. All of these are important but secondary to the true objective – product availability. Patients are not patient.

In the healthcare market, having the product reliably available where and when the patient requires treatment is an unquestionable expectation. A patient cannot benefit from the science and design of the formulation if they cannot purchase and use the product. Once diagnosed, a patient will want or need to begin treatment immediately. Patients are not patient.

Fulfilling patient demand, also known as customer service, is typically measured in orders fulfilled at the time of demand. Routinely reflected as a percentage of total units or dollars shipped to the total ordered in the agreed upon lead time, customer service shows the effectiveness of the end-to-end supply chain. Understanding that fulfilling the patient demand for a product is truly the ultimate measure of success for a supply chain shapes the configuration decisions that a company's leaders can make. Where and how to manufacture and distribute the product are decisions that must be thoughtfully made as plants, equipment, and processes are expensive, regulated, and may not be able to be changed quickly.

Patients are not patient. However, anticipating when the patient will require treatment is often difficult. New drugs especially experience significant demand variability during the introductory phase. This means that the chance of under forecasting the need for this new drug and disappointing the patient is likely. Unfortunately, so is over forecasting and over producing, resulting in having an excess product that expires before the sale. The supply chain must be configured to be agile to produce enough products to meet unpredictable patient demand when it occurs. Decisions must also consider protecting the profitability of the drug. In a nutshell, profitability is a result of selling a product. Without a sale, no profit is made. The common cold provides an excellent example of why the fulfilling patient need is the ultimate objective.

Consider the example of a patient who recognizes that she is experiencing a cold. In fact, she remembers that her children and their friends at school have similar symptoms and is confident in her self-diagnosis. She stops at the pharmacy on her drive home to purchase cold medication. She would like to purchase Brand A cold medication as it is well known as safe and very effective. In the pharmacy, she cannot find Brand A in stock. She is in need of treatment now and purchases Brand B cold medication instead. After a few days, she is healthy and returns to work. She sees that a colleague has the same symptoms. She recommends Brand B to her colleague. Brand A has now lost two sales. This is why product availability is the ultimate objective of the supply chain. Patients are not patient.

Configuring the Supply Chain... One Site or Multiple? To Manufacture or to Buy?

With the understanding that reliable product availability is the primary objective, the construct of the supply chain becomes clear. The configuration of manufacturing and distribution sites is the foundation of the supply chain. The first decision that must be made is whether this product requires one or multiple manufacturing sites. There are many factors to evaluate when determining the number and location of sites. The construct of the manufacturing network must be based upon the understanding of the location of your patient and the size and variability of patient demand. A supply chain set up for the needs of the product will be able to not only satisfy the patient but optimize cost and, in turn, drive profit. Are your patients located in one country or region? Are your patients spread around the globe? Is this treatment a regiment in nature or trauma-related – predictable or extremely variable? Given this, the number of manufacturing sites must be calculated to optimize customer service to the patient. This is not a routine formula. The answer lies in the balancing of factors such as scalable equipment capacity, country regulations, compliance reputation, proximity to shipping lanes that reach the patient, labor availability and skill level, and total cost, among other business criteria.

Single or Multiple Sites

With no silver bullet formula, supply chain leaders will evaluate using a single manufacturing site or multiple manufacturing sites. While either choice can be successful, there are key advantages and disadvantages of choice. Matching the drug and patient need is the key criteria.

Single or sole sources of supply are advantageous for complex or proprietary products. A sole source of supply is when there is only one possible site of manufacturing. A single source of supply is when only one of many capable sites is selected to manufacture. The single supply site allows for skilled focus on an intricate production process or the protection of a patented compounding, compression, coating, packaging, or testing processes. A single or sole source can offer tighter quality oversight. However, a single or sole supply site can also provide challenges reaching global customers, higher costs, or managing forecast variation.

Multiple sources of supply can provide the capacity necessary for growth. Manufacturers located near the patient population reduce the lead time from production to the patient and can improve product availability. Cost competition is also a benefit of multiple sources of supply. However, managing complex processes or making changes can be complicated by the number of supply sites. As explained in the pandemic example later in this chapter, multiple sites can be advantageous in minimizing supply disruption in unexpected situations.

What Manufacturing Network Would You Set Up?

To illustrate the point, let's assume we have a newly approved drug that has been approved to mitigate hair loss in men and women. Patients over the age of 12 may use this drug with a prescription from their doctor. This treatment is not typically associated with a trauma event; however, patients will not want to pause treatment. The drug has been designed using a combination of globally available raw materials; however, raw material 12 346 must meet tighter than standard criteria for the efficacy of this formulation. The drug is now approved in the US and Canada. Approvals in other regions of the world are expected over the next 18–24 months. Additionally, the raw materials must be carefully mixed in a specifically timed sequence for the right chemical reaction to occur. This requires custom handling equipment at the manufacturer and highly skilled employees. The drug is also approved in two strengths for mild and extreme cases.

Since your patient population is predominantly located in one physical region and the process requires special attention, finding one manufacturing site with ample capacity is a recommended starting point. This single-site should perfect the process batch after batch. When future global regulatory approvals are secured, additional manufacturing sites could be added to facilitate meeting customer demand.

Customized Drugs

The future of health care is trending toward custom-made products. A product that is custom or unique to a patient requires the ability to manage at an order by order level. Typically, customized products require specialized equipment or processes. This often starts with one qualified manufacturing site coupled with strong distribution capabilities. Expanding to a network with multiple qualified manufacturing sites will come in time as the approvals, and use of the product grows.

Eyeglasses are a simpler example of a patient custom product. While you may visit an optometrist near your home, your glasses will likely be produced at a qualified, centralized manufacturing operation. This centralized site has the necessary equipment and experts to deliver high-quality eyeglasses to meet the patient's prescription and a lead time promise.

Drugs for Broad-Scale Treatment

If a drug is designed to treat a broad group of patients like an antibiotic, the construct of the manufacturing network should be based on the capacity to efficiently meet the volume needs of the patient population. The use of multiple manufacturing sites can provide capacity for growth, backup in the event of an emergency, and a comparison of costs and efficiency.

Think of the many over-the-counter pharmaceuticals available in the market today. Manufacturing occurs in the region where the patient is located to ensure product availability to the broad user base. The patient has many choices of branded and generic varieties to meet their needs. Successful manufacturing and distribution networks are constructed to support seasonal demands at many pharmacy or retail outlets around the globe.

In Figure 20.4, the advantages and disadvantages of one vs. multiple manufacturing sites are illustrated. Remember, serving the patient best must be the driving decision factor.

Make vs. Buy – Another Decision Point

As only compliant and approved manufacturers can be considered for the manufacturing network, now the choice to make or buy comes into play. To *make* a product refers to controlling the manufacturing capability typically by using company-owned, aka internal, assets and teams to produce. To *buy* is to hire an external source or third-party contracted manufacturing to produce the product.

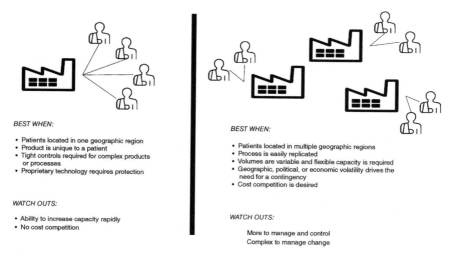

BEST WHEN:

- Patients located in one geographic region
- Product is unique to a patient
- Tight controls required for complex products or processes
- Proprietary technology requires protection

WATCH OUTS:

- Ability to increase capacity rapidly
- No cost competition

BEST WHEN:

- Patients located in multiple geographic regions
- Process is easily replicated
- Volumes are variable and flexible capacity is required
- Geographic, political, or economic volatility drives the need for a contingency
- Cost competition is desired

WATCH OUTS:

- More to manage and control
- Complex to manage change

Figure 20.4 Schematic comparing single and multiple manufacturing site options.

Contract Manufacturers

Often external manufacturing, *buy,* is more flexible and agile than internal manufacturing capabilities. New products have no historical sales data on which to build a demand forecast. A significant degree of variability is to be expected. Contract manufacturers are often designed to increase and decrease production quickly. Contractors utilize quick-change equipment and deploy flexible staffing models to rapidly scale up and down. This agility is particularly valuable when patient demand takes an unexpected turn. The choice to use external manufacturing must be further refined to determine who will own sourcing and planning decisions. The buyer can provide a forecast and purchase completed, which is known as a pure buy. The buyer can also decide to retain the responsibility to purchase raw materials and own process oversight. This is known as a managed buy.

The alternative to buying the product is to *make* it using the company's internal manufacturing capabilities. Internal manufacturing can be less agile due to the high utilization of equipment for profitability reasons and permanent staffing. However, internal manufacturing offers detailed production oversight, which may be critical to protect proprietary processes or materials or ensure high quality on complex products.

As long as compliance and regulatory criteria are met, either make or buy are viable options. A supply chain network may be built with both make and buy manufacturing deployed simultaneously. For example, internal make can be deployed as the primary manufacturing site with an external third-party used for surge or contingency capacity.

Once the manufacturing network configuration is determined, the other links in the supply chain should be confirmed. The distribution network design is particularly important in meeting patient needs.

Configuring the Distribution Network to Have the Right Product, at the Right Place, at the Right Time

While the manufacturing network decision is the first configuration point, the distribution design decision is also critical to complete an optimal supply chain network for the product. Remember that DELIVER is defined as all the actions from processing a customer order to storing inventory to financial management, to transporting the drugs from point to point. This includes the management of customs filings when borders are crossed. The DELIVER function can be centralized to manage a distribution network around the globe. The number of distribution sites is a critical decision like that of manufacturing sites.

Distribution in the healthcare business requires significantly more process controls than distributing most other retail products. Each unit, tablet, bottle, syringe, or dosage must be accounted in all steps of the distribution process. Traceability from the manufacturing site to the pharmacy, retailer, or patient is required to ensure proper use of the substance and to recall the product if necessary. Specialized drug distribution capabilities are not available at all distribution centers. Large-scale distribution providers, UPS, DHL, Federal Express, Kuehne & Nagel, etc., have healthcare-specific logistics services. These solutions include highly skilled staff, secure facilities, as well as systemically and compliantly maintaining unit traceability, temperature-controlled storage and shipments, and security for controlled substances.

Just as with manufacturing sites, Deliver is also a question of one or multiple. Again, going back to the impatient patients, the number and location of distribution points are based on the customer service promise. Patients expect to purchase the product as needed. If patients are located in many geographic areas, multiple distribution points are equally required. If manufacturing sites are located in many regions, multiple distribution nodes or points not only provide timely order fulfillment but also provide a contingency plan should crisis arise.

For broad-scale use products, distribution sites are most effective when located near major transportation hubs with the ability to reach the patient or doctor within a few days. For customized or unique products, shipment within hours may be required. These situations may require a distribution site located at or near the manufacturing site. For global products, the country or region transportation infrastructure must be carefully considered. Even today, delivery in some countries is via bicycle, train, or wagon. Crossing multiple borders in one shipment can be problematic. Temperature impacts must be considered. As with manufacturing sites, a balanced decision inclusive of many factors is necessary. There is no one formula that will produce the right answer.

The distribution network is also susceptible to many of the same risks as manufacturing sites. Environmental, political, and economic issues provide serious obstacles to reaching patients. In 2010, Eyjafallajokull, a volcano in Iceland, erupted from April to June. While this eruption was not featured on the nightly news, the smoke grounded and re-routed air transport causing unanticipated and long-term delays for products flying between Europe and Asia and the Americas. The delays resulted in manufacturing sites shut down as they were starved for raw materials and patients unable to obtain products. Some products expired before they were finally able to reach their destination. Likewise, political and economic instability in Venezuela closed one of the largest shipping ports in South America starting in 2010 with continued impacts even a decade later. These and many other examples highlight the risks of a single node distribution network vs. multiple sites. Single sites offer the ability to compile and redirect inventory when demand shifts; however, multiple distribution sites offer proximity to patients.

Close Patient Connection – Using a Distributor

Direct distribution to individual patients is efficient for custom or unique patient products; however, most patients will obtain a drug through a pharmacy or retail outlet. One unique trait in the healthcare industry vs. electronics or other industries is the use of a distributor to link between the manufacturing company distribution and the pharmacy. A distributor, as shown in Figure 20.5, provides the pharmacy the opportunity to efficiently order multiple products from many manufacturers at one time. This reduces the administrative time spent managing purchase orders at the pharmacy allowing the pharmacists to dedicate time to their patients. Distributors like McKesson and Cardinal Health are the AMAZON of the drug industry. Having a relationship with distributors benefits the manufacturing company as well. A distributor is a large customer who's scale drives efficiency in the order fulfillment process. In other words, the distributor consolidates what would be smaller shipments into one efficient order for a manufacturer. The distributor also holds inventory to ensure higher product availability for their pharmacies and retailers, which in turn improves availability for the impatient patient. Remember that inventory is held at each transition point or node in the network. Distributors do add more inventory to the network but also bring considerable efficiency at the same time.

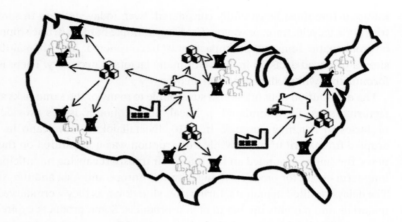

LEGEND

🏭	Manufacturing Sites	📦	Distributor
🏠	Manufacturer Distribution Center	♟	Pharmacy/Retailer

Figure 20.5   Distribution network options.

With the Deliver network design decision, the supply chain network is set. Why not look at raw material supplier locations? The location of raw material suppliers is important as the availability of these materials is impacted by economic, environmental, political, and other factors. However, raw material suppliers are likely to change more frequently than manufacturing and distribution sites. For this reason, having strong sourcing relationships and thoughtful inventory calculations are recommended to mitigate supplier risk to continuity of supply.

Product Availability and Cost

This chapter has been focused on the supply chain network configuration choices. Realistically, though, no supply chain conversation can be complete without understanding cost. Cost not profitability. The profitability of a product is based on many factors such as taxes, royalties, marketing spends, price elasticity, and other costs. The supply chain is a cost in this profitability equation. Supply chain costs include:

- Raw materials
- Transportation, normal modes, and expedited

- Storage facilities
- The cost of detection and destruction of defective products
- Plant, equipment, and personnel
- Order management
- The cash invested in inventory that could be otherwise used

If the supply chain network design does not meet the patient's consumption of the product, profitability issues are guaranteed. Optimizing the cost in one individual link of the supply chain can lead to incremental cost across the network. If SOURCE negotiates a reduced cost but the lead time to obtain raw materials is too long, an investment in inventory will be needed to offset the lead time. Buying inventory ties up cash and is a risk. If the raw material is purchased and not consumed before expiry, it cannot be used. SOURCE is not alone in cost risks. If the manufacturing process yields a high percentage of defects, this also offsets profitability. If poor forecasting results in a lack of product, then expedited freight at an incremental cost will be required to satisfy the patient demand. Running air conditioning units beyond their useful life at the storage facilities can seem to be less expensive than replacing them but may result in temperature violations that render the product unsalable. While intended to optimize cost, these are all obviously detrimental to profitability goals.

Poor supply chain choices and outcomes drive incremental costs that cannot be recovered in the price of the product charged to the patient. Afterall, the patient should not pay for the inefficient or ineffective results. Looking back at our cold product purchase example, assuming products are similarly effective, if the patient cannot purchase a product, she will purchase an alternative. If the product is overpriced compared to the competition, she will likely purchase an alternative. Without a sale, no profit can be made, inventory will accumulate, cash will be tied up, and the additional costs will not be recouped. Profitability is undeniably dependent on the supply chain and, ultimately, product availability.

Case Study: The Impact of a Global Pandemic on the Supply Chain

Whatever supply chain configuration you choose, Murphy's Law typically holds true; disruption will occur. Let's examine how a global pandemic tests the optimal supply chain configuration challenging customer service and ultimately profitability for a company.

In 2020 a global pandemic, COVID-19, impacted patients on nearly all continents. Supply chain resilience was challenged for months with long-term impacts for many products. As the COVID-19 virus infection spread, supply chain organizations experienced disruptions that spanned from raw material supply to employee availability to transportation deficits to retail closures. Employees were not able to work, resulting in raw material production disruptions and shortages. What raw materials were produced often encountered difficulty being delivered to manufacturing plants due to a lack of flights and ground transportation. Manufacturing employees were also infected, and output decreased. The incoming raw materials now needed to be sanitized before use. Manufacturers wrote new procedures to handle raw material receipts that delayed availability. The finished products then experienced similar transportation challenges on the way to the market. Retailers and pharmacies had limited operations, which made obtaining products difficult for patients. Resilient supply chains experienced minimal disruption. Having multiple approved sources for raw materials improved the continuity of supply to the manufacturing operation. Having multiple manufacturing sites also improved the continuity of supply. Manufacturing and distribution sites located close to patients simplified, making the product available. While supply chains do not encounter pandemics routinely, economic, political, and environmental issues do. This has caused supply chain leaders to pause and re-examine the construct of their networks.

"The pandemic has exposed one of the major weaknesses of many supply chains: the inability to react to sudden, large-scale disruptions. This lack of resiliency has been especially notable in the supply chains of the life sciences, health care, and food industries. The resulting turmoil has generated calls for companies that had offshore production to Asia (and China, in particular) to bring it back home. But this approach is no panacea. For one thing, given the huge size of the Chinese market, most global companies will need to keep a presence there to serve it. For another, since China is now a dominant, if not sole, source, for thousands of items, reducing dependence on it in many cases will take considerable investment and time.

For industries that are essential to the country, such as pharmaceuticals and health care, there needs to be government involvement to ensure that supply chains are resilient. There is a precedent: "In the wake of the 2008 financial crisis, the US government and European Union instituted a stress test for banks to guarantee that major institutions whose failure could cause the entire financial system to collapse had the wherewithal to survive a future crisis. Based on our experience in supply chain risk, we suggest that similarly critical supply chains should be required to pass stress tests." (Simchi-Levi and Simchi-Levi 2020)

Designing for continuity of supply, product availability, and resilience upfront pays benefits day-to-day and in a pandemic. For those companies who were not resiliently configured, estimating the cost and lead time to return to a more normal supply cadence is unknown. Patients and consumers alike may have moved on to alternative products permanently. Patients, and consumers, are not patient.

Hopefully, global pandemics are not frequent events in the future. However, constructing the appropriate, resilient supply chain is. The right number of manufacturing and distribution sites for this new drug will determine the supply chain resiliency. When the design is based on serving the patient, the network is poised for growth and success. Afterall, patients are not patients even during a pandemic.

The Best Supply Chains

Great supply chains are custom built to reliably, resiliently, and profitably meet the needs of their patients. These supply chains are agile; adapting and thriving when challenged with opportunities or crises. These businesses invest not only in cost reductions and tactics but in advancement in capabilities that allow them to reliably satisfy the needs of the impatient patient in the most uncertain times. The Gartner Group evaluates major company supply chains on an annual basis. The criteria are largely focused on the evaluation of a supply chain strategy and execution against that strategy. Agility, strategy deployment, and the adoption of Digital Technology are the themes dominating the performance of the top health care supply chains.

"Agility" has become a much-coveted capability, even as the debate of how to define the term persists. Gartner analysts Thomas O'Connor and Pierfrancesco Manenti define agile supply chains as those that are "able to quickly respond to changes in both customer demand and supply sources, without incurring extra costs or penalties." (See "The Agile Supply Chain Imperative: Defining How Companies Can Sense and Respond to Change.") This is an attractive trait, as it allows supply chains to insulate customers from supply disruptions and take advantage of demand spikes.

As healthcare is a care-driven industry, naturally, digital applications are centered on ways to improve this care, usually patient experience, and connected or "smart" devices. The descriptions of this year's leading supply chains found later in this report provide good examples on the breadth of digital projects being implemented by organizations. Leaders in healthcare supply chains realize that just like the visible portion of the iceberg is the smallest part, so is strategy development. The real effort is required in deploying the strategy across a complex organization, ensuring that every individual in the organization is aware of their expected contribution to the initiative. As a result, leading healthcare supply chains invest heavily in communication, education and governance – putting tools in place that propagate the components of the strategy, address questions and concerns and monitor progress toward goals.

In the health care industry, Gartner has endorsed the top performers, Johnson and Johnson, Cleveland Clinic, CVS Health, and Banner Health. McKesson and

(Gartner press release 2019). Interesting that only many of these healthcare supply chains are hospital networks vs. drug manufacturers. Remember that hospital networks purchase and distribute healthcare products across large patient populations. These top performers have consistently outstanding financials, compliance, and strategic, long-term focused supply chain leadership.

Conclusion

Your newly approved drug is now dependent upon a supply chain to reach its market. Through this chapter, you've seen that supply chains are not one-size-fits-all. The careful selection of manufacturing and distribution capabilities to match the drug and the patient needs provides the foundation for growth and profitability. A supply chain designed and then managed properly to ensure product availability for patients is a competitive advantage in driving trust and loyalty to this new drug.

Success is achieved when the four functions of the supply chain – PLAN, SOURCE, MAKE, and DELIVER – optimize the potential of the design. The four functions must collaborate to fuel the manufacturing and distribution operations with accurate operating plans, compliant and available raw materials, robust process design and execution, compliance to regulations, and outstanding customer management to put this new drug into the hands of the patients. The benefits of focusing on adaptability and value, not cost, are critical. The supply chain must be poised for constant change and ready for unpredictable disruption. What can go wrong will and the patient is not patient.

Through this chapter, you have seen the options to customize the construct of the supply chain to prioritize compliantly serving the patient while withstanding even the impacts of a global pandemic and ultimately generating profitability. How are you designing your supply chain?

References

Gartner Press Release (2019).Gartner Announces Rankings of the 2019 Supply Chain Top 25. Phoenix, Ariz. https://www.gartner.com/en/newsroom/press-releases/2019-05-16-gartner-announces-rankings-of-the-2019-supply-chain-t (accessed 28 July 2020).

Simchi-Levi D and Simchi-Levi S (2020). Building Resilient Supply Chains Won't be Easy, Harvard Business Review. https://hbr.org/2020/06/building-resilient-supply-chains-wont-be-easy (accessed 28 July 2020).

Chapter Self-Assessments: Check Your Knowledge

Questions:

- There are four key functions in the supply chain. No function can operate without impacting the other three. Discuss how SOURCE is impacted by PLAN, MAKE and DELIVER?
- Describe the supply chain configuration you would establish for a drug that is customized to the specific patient?
- Why is customer service more important than profitability in supply chain design?
- Why is QUALITY an independent function in healthcare supply chains?
- How can the supply chain prepare for disruptions?

Answers:

- The Sourcing function is responsible to ensure continuity of supply of raw materials and semi-finished goods to the manufacturing sites. Sourcing is dependent on annual and short-term forecasts from PLAN to ensure that supplier capacity is adequate to not only meet the known needs but to flex to cover both increased and decreased demands. The Sourcing function must collaborate with the DELIVER function in order to get the raw materials and semi-finished goods to the sites. Rather than allow multiple small deliveries to barrage a manufacturing site dock, Sourcing can partner with DELIVER to arrange efficient delivery cycles, just-in-time or Kanban deliveries on set schedules, or consolidated deliveries to optimize dock time. Finally, the MAKE function will work with SOURCE when defected raw materials are supplied or to enact a change to the specification of raw material to optimize the manufacturing process. Extra Credit: Source and Quality are also integrated in supplier management, supplier quality reviews, and supplier audits.
- In the case of a drug customized to a specific patient, the supply chain configuration should be a manufacturing site located near the patient that has DELIVER capabilities. The manufacturing site should double as a distribution node in order to eliminate shipping damage and ensure speed and quality of delivery to unique customers. This is a special case where having the end-to-end supply chain in one operation is extremely efficient and beneficial. Batch sizes are very small. The manufacturing capabilities must be scaled down to avoid waste. Order management must have the ability to accurately receive and process orders from individuals – including payment receipts and shipment tracking. Example: genetic-specific treatments designed for one patient. Medical devices such as contact lenses also fit this archetype of supply chain design.

- Companies who prioritize meeting the needs of their customers build long-term loyalty and credibility with their patients. Continuity of supply drives favorability with retailers and pharmacies as well as patients. If a company makes the decision to lengthen delivery lead times or require a patient to purchase in bulk or increases delivery charges, a patient may be able to go elsewhere for their treatment or choose another treatment option. Additionally, insurance providers may decide to remove the company as an approved supplier. The profitability gain must be tested against the patient's ability to absorb the change. Ideally, a supply chain is responsible to routinely improving production costs to increase profitability without impact to customer service. There are many examples of choosing less costly raw materials that are of lesser quality. These can create inefficiency in manufacturing and result in lack of supply and increased costs/decreased profitability. Another common pitfall is to shift sourcing to other regions of the world with lower wages. If the tradeoff is piece price for significantly longer lead times, this will result in holding more inventory, potential quality issues, and less ability to flex with demand changes.
- In the healthcare supply chain, Quality reports under separate management from the PLAN, SOURCE, MAKE, and DELIVER teams. This allows Quality to have an unaffiliated, agnostic ability to identify defects, audit processes, and determine the ultimate acceptance of the drug product. Quality is responsible to keep manufacturing and suppliers compliant to regulations and requirements. Quality must be able to audit processes and sites at will. The supply chain functions, and Quality will collaborate, but Quality must make decisions independent of cost, service, and capacity.
- There are many acceptable answers to this question. They all start with having an aligned business continuity plan. This will require investment in dual supplier contracts, regional and global backup sites, having capabilities at the ready – such as transportation for emergency import/export. Having unutilized but available capacity is one of the most difficult investments for a supply chain. Manufacturing sites are measured on efficiency or cost per unit. When capacity is idle, the cost per unit rises to compensate for unused equipment or crewing. Therefore, most plants load at over 85% of their equipment capacity to payback the investment. When a disruption surfaces, the lack of capacity in manufacturing or material suppliers will slow or stymy response time. Some companies chose to hold inventory in lieu of additional capacity. Regional regulatory differences and languages can make holding the right inventory complicated. The inventory theory does not work when products expire or change quickly. Inventory is a non-liquid cash expense that is often a target for reduction in corporations. Obsolescence and quality defects can be large risks when using inventory as a flexibility plan. Companies who produce lifesaving drugs often have multiple layers of continuity plans containing both inventory and excess

capacity strategies. Companies must align to a strategy or risk being lulled by temporary stability into reducing inventory protection or filling up reserve capacity. The COVID19 pandemic proved these points. In some cases, demand surges consumed all inventory, and manufacturing capacity could not keep up with growth. In other cases, the inventory was located in different parts of the globe and could not be transported to the right region due to lack of flights and sailings with the added complexity of political constraints in place. In the pandemic, even if the equipment was available, at times, workers were not. Any business continuity plan must consider political unrest, natural disaster, competitive challenges, and now, pandemics as potential scenarios. Net, there are many right answers. Additional capacity, dual suppliers, multiple sites approved to produce the same product or inventory all come with significant investment and effort. The company must choose which will provide the right amount of insurance. Every company must have a risk assessment and a continuity plan that is aligned between commercial and supply chain executives and prioritizes customer service.

Quiz:

1   True or False. The supply chain is a series of actions that must be completed to produce and distribute goods. True.

2   What are the four core functions of the supply chain? Choose the best answer d
  A PLAN, SOURCE, SUPPLY, and DISTRIBUTE
  B PLAN, SCOPE, MAKE, and DELIVER
  C PLAN, SCOPE, SUPPLY, and DISTRIBUTE
  D PLAN, SOURCE, MAKE, and DELIVER

3   True or False. Target inventory is the amount of inventory required to cover future sales plus a buffer for errors in forecasting or producing defects, among other unanticipated issues reflected as a min-max range. True

4   Manufacturers located near the patient population reduce the lead time from _____ to patient and can improve product _____. Cost _____ is also a benefit of multiple sources of supply. Choose the best answer to complete the sentence. a
  A production, availability, competition
  B distribution, performance, containment
  C manufacture, acceptance, adjustment
  D production, performance, acceptance

21

Sales, Marketing and Advertising
Jeffrey S. Barrett

Aridhia Digital Research Environment

Introduction

Modern drug development is a highly regulated process, as we have discussed in previous chapters in this text. Prior to the introduction of regulation to the industry, sales and marketing existed. Even prior to investment in research and development (R&D), selling of amulets, pardons and charms were included with medicinal remedies lumped together for the benefit of the seller and at least hopefully some of the would-be patients in the early days of what would become the pharmaceutical industry as was discussed in Chapter 1. At present, the global market for pharmaceuticals is $900 billion, and this figure is fully expected to exceed $1.1 trillion USD in the next few years (DiMasi et al. 2016). In fact, recent studies show the industry is growing at a rate of 5%, which is just behind the two other major healthcare segments – medical services and equipment. Drug affordability and disease prevalence continue to drive this rate; while government policies and regulations can impede or slow down this growth, their impact on the bottom line is likely small. This demand is not likely to slow down anytime soon, and many trends attest to this statement. Likewise, in addition to the pressures of discovering new medicines, there is the necessity of marketing, advertising, and selling the approved products in a company's portfolio to maximize profits and pay for R&D expenditures (both successes and failures).

The pharmaceutical industry, which researches, develops, produces, and markets prescription drugs in the United States, is the most heavily regulated of all industries when it comes to the advertising and promotion of its products. Through its Drug Marketing, Advertising, and Communications Divisions, the Food and Drug Administration (FDA) regulates all advertising and promotional activities for prescription drugs, including statements made to physicians and pharmacists by pharmaceutical sales representatives. Advertising of over-the-counter (OTC) drugs, which is not regulated by the FDA, is under the jurisdiction of the Federal Trade Commission (FTC).

Before a new prescription drug is approved for marketing, the FDA and the pharmaceutical company sponsoring the NDA must agree on the "full prescribing information" that will accompany the product, and that must be included in all advertisements, brochures, promotional pieces, and samples. This full prescribing information must include, in the correct order, the following information about the drug: its trade name, its assigned name, the strength of its dosage form, a caution statement (stating that a prescription is required), a description of its active ingredient, the clinical pharmacology of the drug, indications for its usage, contraindications for usage, precautions, adverse reactions, instructions on what to do in case of overdosage, dosage and administration, and how the drug is supplied.

This chapter will describe the process by which a pharmaceutical company markets and sells newly approved medicines in addition to how it promotes their products through various advertising campaigns. The scrutiny on these approaches imposed by various regulatory authorities and governmental oversight will be described in addition to the legal framework that allows such promotion. Penalties for firms going outside the prescribed boundaries will also be discussed. Finally, details regarding marketing strategies and sales forecasting will also be described.

Definitions and Context for Pharmaceutical Sales

Sales of Pharmaceutical products, which may include medicines, or surgical devices, consumables of any form, machines, and equipment used in surgeries, can be generally referred to as pharmaceutical sales. The target audience is doctors of any kind, pharmacists or chemists, and/or purchase in-charge agents in hospitals or pharmacies. The pharmaceutical sale is vastly different from regular sales of any kind, right from the product to the customer to the process of selling. Of all sales career opportunities, pharmaceutical sales is one of the most lucrative and high-paying and one of the most challenging jobs requiring a lot of education and training on the salesperson's part. Like every sale, there is a buyer and a seller. In this case, the buyer depends on the product of the manufacturer. For simplicity, we can consider all buyer kinds. The "buyer" for medicines of any kind could include Chemists/Pharmacists, Distributors, and Hospital Pharmacies. Buyers for medical devices, instruments, implants could include Doctors, Purchase Officers, and similar institutional representatives. It is more commonplace in both settings to also engage procurement representatives who have the role of negotiating favorable terms and pricing. They also ensure supplier quality, efficiency, and timeliness.

The general process of pharmaceutical sales followed by most of the pharma companies involves the definition of the target (potential customers), engagement of the sales process and the sales call. A proper target is passed on to the sales team, which will consist of value business to be gotten from the area. The target is considered after taking into consideration last year's target(s), market potential, and industry growth. Proper targeting is necessary for the profitability of the company. Once the targets have been defined, the salesperson can then work on customers and segregate according to the potential of the customers. Then the salesperson arranges visits to the relevant customers. In the case of medicines, the customers would be doctors and chemists, while in the case of medical devices, it would be a doctor himself and purchase in charge. The salesperson visits doctors by scheduling an appointment. The visit can be carried out as often as necessary. In the case of medicines, the objective of a sales call would be to influence the decision of the doctor and make him/her the desired brand of medicines. For that, the salesperson ensures that there are relevant studies supporting the product and that proper documentation in the form of publications, publicly available technical reports, and other freely available technical documentation. The studies are discussed with the doctor, and a few samples of the products are kept with him/her for doctors to test for themselves. In the second visit, glitches if any on the doctor's part is counted with examples and counter studies from the salesperson.

The second visit is the determining visit to convert the doctor. If there are any more doubts in the minds of the doctor, the salespersons' visits increase, but the objective remains the same that is to convince the doctor to write the desired brand of medicine. Once the doctor is convinced, the salesperson then visits the pharmacist or chemist shops to check the availability of the product. The pharmacist or chemist then gives the requirement for the product to the salesperson, who then arranges it from the proper supply chain access point (e.g. warehouse). On the third visit, the salesperson checked the stock with the pharmacist/chemist. If the stock is unconsumed, that means the doctor is not writing the brand in which case the salesperson increases the visit to the doctor with more armamentarium and studies to influence his prescription. Decision influencers may be used in the stage. In this capacity, a decision influencer refers to people who the decision-maker allows, or invites, to have a say or play a role in the decision-making process. They may not be able to say "yes" or make the final call, but they can definitely deliver a "no" that could impact the outcome both positively and negatively.

Pharmaceutical companies have invested a lot of money in incentivizing (gifting) doctors; however, this is not a step that is supported by any medical council all over the world. Gifting, in cash or kind, or any act done to influence the writing of medicines by a doctor, is prohibited, and if caught in the process, the salesperson will lose his job, the doctor loses his practicing license, and the pharmaceutical companies are fined heavily. If the doctor is convinced, he starts prescribing the product. While the convincing takes place at the clinic, the sale happens at the pharmacist's or chemist's shops. The salesperson keeps on replenishing the stock at the chemist and collects payments. In the case of the Hospital pharmacies, a similar procedure is followed. Relatively recent legislation has sought to increase the transparency around such incentivization, especially to remove any potential to bias the clinical prescriber. The Sunshine Act requires manufacturers of drugs, medical devices, biological and medical supplies covered by the three federal health care programs Medicare, Medicaid, and State Children's Health Insurance Program (SCHIP) to collect and track all financial relationships with physicians and teaching hospitals and to report these data to the Centers for Medicare and Medicaid Services (CMS). The goal of the law is to increase the transparency of financial relationships between health care providers and pharmaceutical manufacturers and to uncover potential conflicts of interest (S.301 Congressional dialogue 2009). The bill allows states to enact "additional requirements," as six states already had industry-pay disclosure laws (O'Reilly 2009). Industry personnel now take rigorous care to avoid any suggestion that their practices are biased, and internal SOP training rigorously makes this point to all employees where this issue is a concern. In 2013, the American Medical Association offered physicians training to understand the Sunshine Act (Lazarus 2013); training on the Sunshine Act is also considered a part of a physician's continuing education curricula as well.

In the case of surgical products, the process changes slightly. The pharmaceutical salesperson visits the doctor by taking his appointment. Unlike the sales of medicines, in this case, there is an in-depth discussion of the salesperson and the doctor about the procedure where that particular product is going to be used. Here the salesperson acts more as a solution provider to the doctor where he educates the doctor about the procedure and where and how his product can be used. The salesperson then asks the doctor for an appointment to shadow him or assist him or observe his surgical technique in Operation Room. Post permission from the doctor, the salesperson visits and observes the techniques of the doctor and then positions his products accordingly. The sales call in this technique is not only limited to the out-patient doctor (OPD) visits but also extends to the operating rooms (ORs). The salesperson then offers a sample of his product to the doctor for use in OR and assists the doctor in the surgery, if need be. Following that visit, the salesperson then takes feedback of the product from the doctor and tries to get an order. If the doctor likes the product, he will place an order with the salesperson and the disease caused if the doctor does not like the product, a salesperson continuous his visits clarifying the doubts until the doctor is convinced to buy the product.

In some cases, the salesperson must-visit purchase officers. This is true, especially in the case of multispecialty hospitals. The salesperson, after influencing the doctor in the multispecialty hospital, then talks to the purchase officers to place an order. If the doctor is convinced, he places the order with the purchasing officer, who in turn places the order with the salesperson, and the sale is closed. In the case of medical equipment, there is an additional step of demonstration wherein the doctor takes a demo of the machine in the surgery if required. After taking the demo, it is up to the doctor whether to purchase or not. The salesperson can follow up with the doctor.

A product/market development strategy concerns developing new products or modifying existing products, and offering those products to current or new markets. These strategies typically surface when there is little opportunity for growth in an organization's existing market. The four most common strategies in this category derive from the Ansoff Matrix (see Figure 21.1): market penetration (growing sales of an existing product in existing markets), market development (launching an existing product in a new market), product development (introducing an existing product into a new market) and diversification (introducing a new product into a new market).

Pharmaceutical products or services will be promoted in accordance with the Ansoff Matrix almost every time and can dictate the marketing strategy adopted. For example, so many partnerships and mergers occur in this industry, where pharmaceutical organizations combine their resources and leverage their strengths to increase market share in this manner. Should a pharmaceutical organization want to sell more products in current markets, it might decide to invest more in its marketing budget.

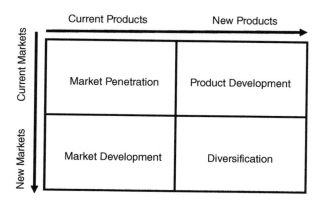

Figure 21.1 Ansoff Matrix illustrating a product/market development strategy concerning the development of new products or modification of existing products and offering those products to current or new markets.

Pharmaceutical Sales and Marketing: Historical and Current Practices

With the average American spending $1200 USD on drugs a year in 2020, marketing is a top priority for the major players in the pharmaceutical industry (https://www.bloomberg.com/quicktake/drug-prices). With so much spending involved, most companies understand the great role and importance of marketing in pharma. Traditionally, the advertising and promotion of pharmaceutical products were directed primarily to physicians, with some limited advertising and promotion being directed to pharmacists. With the expiration of patents on some major drugs in the 1980s and 1990s, generic versions of the drugs became available from competing manufacturers. The generic drugs were priced lower than the brand-name products, so pharmacists got laws passed allowing them to substitute generic products for the brand-name products. This gave pharmacists more control over which generic company's products to purchase and dispense. Advertising and promotion to pharmacists increased. When committees, usually composed of pharmacists, became very important in deciding which drugs could, or could not, be prescribed or reimbursed under third-party payment programs (Medicaid, HMOs, and other insurance programs), advertising and promotion were also directed to the decision-makers in those organizations. More recently, advertising is also being directed to the consumer.

In the mid-1980s, two pharmaceutical companies began direct-to-consumer (DTC) advertising. Pfizer led the way with its healthcare series of ads to the general public. Merrell Dow was next, using DTC ads to inform the public that physicians had a new treatment to help smokers who wanted to stop smoking. When the company's new, nonsedating antihistamine became available, it used DTC ads to tell allergy sufferers that physicians now had a new treatment for allergies. The ads did not mention the name of the products; rather, they asked patients with specific problems or symptoms to see their physician.

The next phase of DTC advertising led to ads in magazines and newspapers that mentioned the name of the product and its indication for use. The advertising of prescription drugs on television or radio remained greatly restricted at this time since it was not possible to include the necessary summary of prescribing information on the air. Because of this limitation, the ads on television or radio had to focus on either the name of the product or the indication for the product. To promote Nicorette, a nicotine-containing gum designed to help smokers stop smoking, Merrell Dow advertised it on television with the message that Nicorette was now available at pharmacies but only by prescription and under a doctor's supervision. According to FDA rules at that time, Merrell Dow could not say that Nicorette was useful in helping smokers who wanted to stop smoking since it had included the name of the product in the commercial. When a company has the only – or the major – product in the market, this approach can be very effective because it increases awareness among patients that a new treatment is available and influences them to see their doctors.

In 1997, the FDA changed the regulations regarding DTC advertising of prescription products on television and radio. It now allows both the name of the product and indications for it to be advertised, as long as the main precautions or warnings are given in the commercial. This has led to many prescription products being advertised on television, such as Rogaine, Claritin, Allegra, Viagra, Pravachol, Prilosec, and others. Nicorette, by this time, had been cleared by the FDA to be sold over the counter and, since it no longer required a prescription, the product was no longer governed by FDA rules but rather by FTC regulations. By 2020 the current role of the pharmaceutical industry's sales and marketing workforce will be replaced by a new model as the industry shifts from a mass-market to a target-market approach to increase revenue.

One of the primary activities of the commercial part of every pharmaceutical company is the development of a marketing strategy and the forecasting of sales of current and future compounds in development that represent a company's portfolio of products. A pharmaceutical marketing strategy or a range of strategic marketing options is initiated by asking the sales and marketing groups to collaborate on answering the following questions in accordance with their marketing/organizational objectives:

- How will our marketing activities help make sales?
- What market trends are emerging that we need to respond to?
- What position will we strive to achieve?
- Which pharma market segments will be targeted with which propositions?
- What communication strategies will be used to support customer acquisition?
- What experience will we look to create for our audiences?
- How can we differ from our competitors?

Some of the answers to these questions will be embedded into a compound's target product profile (TPP) for development compounds (see earlier chapters) and into the formal marketing strategy for approved products.

With respect to actual forecasting of pharmaceutical sales (regardless of development stage or commercial availability), five main steps define the process:

1) Estimating base population considering different dynamics such as changing demographics
2) Epidemiology data forecasting: quantifying incidence or prevalence-based data
3) Estimating the number of patients that are (correctly) diagnosed with the disease
4) Estimating the number of patients treated (or could be treated) with a certain drug therapy
5) Estimate the total pool of treated patients divided into relevant segments based on age, gender, the severity of disease, etc.

There are various methods that can be employed to accomplish sales forecasting including conjoint analysis, scenario planning, and historical analogy with the incorporation of trending events to further refine the projections. The choice of methodology often depends on the market status (new product on the market or not) and the availability of data. In any event, forecasts are made with some frequency and updated periodically. Depending on these projections, a company may abandon a development compound, entire therapeutic area or make an acquisition of some kind (compound(s) or entire company) so their input is highly valued within the context of internal decision making.

Advertising

The US pharmaceutical industry spent $6.1 billion on advertising prescription drugs directly to consumers in 2017. Since 1962 these ads have been regulated by the Food and Drug Administration (FDA) to ensure that they are not false or misleading. The United States and New Zealand are the only two countries where DTC advertising of prescription drugs is legal. In the 1700s and 1800s, drug compounds (also called "patent medicines") were advertised in newspapers in ways that were often exaggerated or deceptive. For example, Lydia E. Pinkham's Vegetable Compound was mass advertised starting in 1876 and purported to "cure entirely the worst form of Female Complaints, all ovarian troubles, Inflammation and Ulceration, Falling and Displacements, and the consequent Spinal Weakness, and is particularly adapted to the Change of Life," in addition to curing headaches, depression, indigestion, insomnia, and other ailments. By the twentieth century, drugs were separated into two classes: (1) "ethical drugs" that were listed in the United States Pharmacopoeia (USP) by the American Medical Association (AMA) including morphine, quinine, and aspirin; and (2) patent medications that were mostly made of water with a bit of alcohol or opium and other unknown ingredients, which were advertised without regulation (including Lydia E. Pinkham's Vegetable Compound, Hamlin's Wizard Oil, Kick-a-poo Indian Sagwa, and Warner's Safe Cure for Diabetes to name a few). By the early 1900s, patent medication ads accounted for nearly half of the total ad revenue for newspapers. Physicians wrote prescriptions for drugs, but prescriptions were not required to obtain drugs from physicians, pharmacists, or people like Lydia Pinkham.

A historical perspective on pharmaceutical advertising has been provided by Julie Donohue (Donohue 2006) and fundamentally concludes that the recent pharmaceutical promotion of prescription drugs to consumers was made possible by the rise of consumer-oriented medicine following the social movements for patients' and consumers' rights. Previously we discussed the sales aspects is somewhat generic terms and focused on the unique aspects of pharmaceutical sales.

Here too, the distinction between the terms patient and consumer may often appear blurred, but the differences between the legal rights of patients and consumers are clear. Although entitlement to information is central to both the patients' and consumers' rights movements, the goals of providing this information are different. Others (Mariner 1998) contend that the main tool of consumer protection laws is the disclosure of information in order to level the playing field between buyers and sellers. The rights of patients then are developed outside the context of commercial markets, independently of health insurance, and without regard to the existence or source of payment for health care. Ordinarily a patient is in a relationship with a physician or other health care professional.

Direct-to-consumer pharmaceutical advertising (DTCPA) has grown rapidly during the past several decades and is now the most prominent type of health communication that the public encounters (Ventola 2011). DTCPA can be defined as an effort (usually via popular media) made by a pharmaceutical company to promote its prescription products directly to patients (Mogull 2008; U.S. Department of Health and Human Services 2017). The United States and New Zealand are the only countries that allow DTCPA that includes product claims (see Table 21.1 for FDA requirements on DTC advertising). Most other countries don't allow DTCPA at all; however, Canada does allow ads that mention either the product or the indication, but not both. The pharmaceutical industry and lobby groups have tried unsuccessfully to overturn bans against DTCPA in Canada and other

Table 21.1 Types of DTC advertisements with corresponding FDA requirements.

Type of Ad	Regulatory requirement
Product claim Ad: Names a drug and the indication(s); makes claims regarding safety and efficacy	• Product claims are made, so "fair balance" doe apply, and risks are required to be included in a summary. Or for broadcast ads only: • Risks must be included in a "major statement" and "adequate provision" for access to a "brief summary" is provided.
Reminder Ad: Names a drug, dosage form, and possibly cost but not its uses	• No product claims are made, so "fair balance" doesn't apply, and mention of risk in "brief summary, "adequate provision, or "major statement" is not required. • FDA does not allow this type of ad for drugs with serious risks (e.g. black box warning)
Help-seeking Ad: Describes a disease or condition but does not mention a specific drug that treats it.	• No product is mentioned, nor any claims made, so "fair balance" does not apply; inclusion of risks in "brief summary," major statement," or "adequate provision" is not required.

countries or regions, such as in the European Union (EU). Notably, in 2008, 22 of the 27 EU member states voted against proposed legislation that would have allowed even limited "information to patients" to be provided.

The FDA regulates DTCPA, but critics say that the rules are too relaxed and inadequately enforced.4–6 Although only limited data exist, research suggests that DTCPA is both beneficial and detrimental to public health. The number of arguments that favor or oppose DTCPA is fairly evenly balanced, and viewpoints presented by both sides can be supported with evidence. Although there have been calls to ban or severely curtail consumer drug advertising, remedies to maximize the benefits and minimize the risks of DTCPA are more frequently suggested.

Legislation and Regulation

In the United States, prescription drug advertising and promotion are monitored by FDA's Office of Prescription Drug Promotion (OPDP). OPDP reviews prescription drug advertising and promotional labeling to ensure the information contained in these materials is not false or misleading. The mission of OPDP is "to protect the public health by assuring prescription drug information is truthful, balanced, and accurately communicated. This is accomplished through a comprehensive surveillance, enforcement, and education program, and by fostering better communication of labeling and promotional information to both health care professionals and consumers (OPDP, FDA 2017)." In the United States, the Federal Food, Drug, and Cosmetic Act (FDCA) and Title 21 of the Code of Federal Regulations Part 202 (21 CFR Part 202 2017) primarily govern prescription drug advertising and promotion. Together, the FDCA and 21 CFR Part 202 regulate how pharmaceutical companies can promote prescription drugs to both health care professionals and consumers. In addition to laws and regulations, FDA has issued draft and final Guidance documents, which help provide their current thinking on several topics related to prescription drug promotion.

Failure to adhere to certain requirements of the FDCA may deem a drug to be misbranded. Per FDA regulations, all prescription drug promotion must adhere to the following expectations: be consistent with the FDA-approved Prescribing Information (PI); be truthful and non-misleading; contain a fair balance of product benefits and risks; and include material information. One of the fundamental requirements of prescription drug promotion in the United States is the requirement that companies promote only uses that are "on label" or consistent with the FDA-approved PI. The term "off label" generally refers to the promotion of a product for uses that are inconsistent with the FDA-approved labeling or PI. This can relate to the promotion of uses, dosing/administration, or patient populations that are not FDA-approved. Another requirement is that promotional materials must not be false or misleading. There are many ways that promotional materials can be deemed false or misleading, including promotion that somehow characterizes a drug to be more effective or safe than what is supported by the FDA-approved label. All promotional materials that include efficacy/benefit claims must provide a fair balance between benefit information and information on risks associated with the product. Fair balance not only refers to the inclusion of the safety information but also to the overall prominence of its presentation as compared to the benefit presentation. Promotional materials that do not present the benefits and risks in a fair and balanced manner could be deemed as false or misleading.

Prescription drug promotion may also be considered false and misleading if there is a failure to include material information. Material information refers to anything that may be critical for a patient or health care provider to know before or while using a drug that would help ensure safe and effective use of the product. As you can sense, much of the language around this terminology is framed in legal terminology where strict definition and thus interpretation is an expectation and the basis upon which companies are held accountable under the law Mogull and Balzhiser (2015).

Surveillance and Consequences of False and Misleading Promotion

OPDP reviews promotional materials submitted at the time of initial dissemination on Form FDA 2253, as well as through routine surveillance and the Bad Ad Program. Subpart H drugs designated for accelerated approval follow a different process outside the scope of this article. FDA's Bad Ad Program is an outreach program designed to help healthcare providers recognize potentially false or misleading prescription drug promotion. The Bad Ad Program seeks to raise awareness among health care professionals regarding false and misleading promotions and provides a venue to report violations or misleading promotional materials to the OPDP. If it (OPDP) finds materials to be violative, they can issue two different types of letters. The first and less serious type is called an Untitled Letter, also known as Notice of Violation (NOV), which usually requires that a company cease the offending promotions. The second type is called a Warning Letter and can result in a company also being required to do corrective advertising, which can be laborious and costly. Both types of letters identify the various violations OPDP has found in the promotional materials, which often include false or misleading risk presentation, false or misleading benefit presentation, and lack of adequate directions for use. FDA provides training and continuing education on the Bad Ad program through its website (https://www.fda.gov/drugs/office-prescription-drug-promotion/bad-ad-program).

Table 21.2   A comparison of regulatory advertising oversight across select countries.

Country	Surveillance agency	Requires presentation of risk information	Prohibits false and misleading advertising	Prohibits off-label promotion	Direct to consumer advertising permitted
USA	FDA OPDP	✓	✓	✓	✓
Japan	MHLW and PMDA	✓	✓	✓	X
Brazil	ANVISA	✓	✓	✓	X
Chile	ANAMED	✓	✓	✓	X
Columbia	INVIMA	✓	✓	✓	X

FDA = Food and Drug Administration; OPDP = Office of Prescription Drug Promotion; MHLW = Ministry of Health, Labor and Welfare; PMDA = Pharmaceuticals and Medical Devices Agency; ANVISA = Agencia National de Vigilancia Sanitaria; ANAMED = Agencia Nacional de Medicamentos; INVIMA = El Instituto Nacional de Vigilancia de Medicamentos y Alimentos.

In 2016, OPDP issued a total of 11 letters, eight Untitled Letters, and three Warning Letters. Of the 11 letters, four related to investigational new drugs that had not yet been approved by FDA. The violative promotional materials at issue in four of these letters were detected via Form FDA 2253, while seven were captured through routine monitoring and surveillance. A primary violation noted in over half of the 2016 OPDP letters was that the promotional materials contained a false or misleading risk presentation or omitted risk information. Other violations included omission of material facts, false or misleading benefit presentations, failure to submit under Form FDA-2253, and lack of adequate directions for use.

In addition to laws and regulations, there are voluntary industry organizations such as the Pharmaceutical Research and Manufacturers of America (PhRMA) and Biotechnology Innovation Organization (BIO) where member companies may adopt codes and guidelines (Ministry of Health 2017; PhRMA 2017; PMDA 2017). Similarly, surveillance oversight is provided by many regulatory organizations around the world, so there is the ability to provide an almost global oversight. Table 21.2 provides a representative comparison of regulatory oversight of advertising across various regulatory authorities.

Summary

The profitability of every pharmaceutical company is directly dependent on their ability to effectively market, advertise and sell their products. While there are common and best practices that guide the commercial groups that support these endeavors, there is also an implicit expectation that their employees will be actively involved with the R&D effort, have a seat at the table at early-stage project team meetings, and contribute greatly to the evolving target product profile particularly on the latest evaluation of the marketplace. These groups must keep abreast of external health economic studies and conduct their own evaluations that guide the marketing strategies and sales forecasting. Good commercial performance influences both the return on investment but portfolio confidence and ultimately the share price of the company.

References

21 CFR 202.1 (2017). https://www.accessdata.fda.gov/scripts/cdrh/cfdocs/cfCFR/CFRSearch.cfm?fr=202.1 (accessed 31 July 2017).

DiMasi, J.A., Grabowski, H.G., and Hansen, R.W. (2016). Innovation in the pharmaceutical industry: new estimates of R&D costs. *Journal of Health Economics* 47: 20–33.

Donohue, J. (2006). A History of drug advertising: the evolving roles of consumers and consumer protection. *Milbank Quarterly* 84 (4): 659–699. https://doi.org/10.1111/j.1468-0009.2006.00464.x PMCID: PMC2690298, PMID: 17096638.

Lazarus, Jeremy (2013). The Physician Payment Sunshine Act is here: Are you ready kevinmd.com. https://www.kevinmd.com/blog/2013/04/physician-payment-sunshine-act-ready.html (accessed 27 March 2021).

Mariner, W.K. (1998). Standards of care and standard form contracts: distinguishing patient rights and consumer rights in managed care. *Journal of Contemporary Health Law and Policy* 15: 1–55.

Ministry of Health, Labour and Welfare (2017). Ministry of Health, Labour and Welfare. http://www.mhlw.go.jp/english/index.html (accessed 4 June 2017).

Mogull, S.A. (2008). Chronology of direct-to-consumer advertising regulation in the United States. *AMWA Journal* 23: 3.

Mogull, S.A. and Balzhiser, D. (2015). Pharmaceutical companies are writing the script for health consumerism. *Communication Design Quarterly Review* 3 (4): 35–49. https://doi.org/10.1145/2826972.2826976. ISSN 2166-1200.

O'Reilly, Kevin B. (2009). 'Sunshine' bill sets $100 trigger for disclosing drug industry pay to doctors. *American Medical News*. https://amednews.com/article/20090223/profession/302239977/4 (accessed 27 March 2021).

OPDP (2017). Center for Drug Evaluation and Research. About the Center for Drug Evaluation and Research – The Office of Prescription Drug Promotion (OPDP). U S Food and Drug Administration Home Page. https://www.fda.gov/aboutfda/centersoffices/officeofmedicalproductsandtobacco/cder/ucm090142.htm. (accessed 4 June 2017).

Pharmaceutical Research and Manufacturers of America (2017). PhRMA. http://www.phrma.org (accessed 4 June 2017).

PMDA (2017) Pharmaceuticals and Medical Devices Agency. Pharmaceuticals and Medical Devices Agency. http://www.pmda.go.jp/english (accessed 4 June 2017).

S.301 (2009). Physician Payments Sunshine Act of 2009 - 111th Congress (2009-2010). Library of Congress (Congress.gov). (accessed 27 March 2021).

U.S. Department of Health and Human Services (2017). Consumer Updates – Keeping Watch Over Direct-to-Consumer Ads. U S Food and Drug Administration Home Page. https://www.fda.gov/ForConsumers/ConsumerUpdates/ucm107170.htm (accessed 27 June 2017).

Ventola, C.L. (2011). Direct-to-consumer pharmaceutical advertising: therapeutic or toxic? *Pharmacy and Therapeutics* 36 (10): 669–674.

Chapter Self-Assessments: Check Your Knowledge

Questions:

- Who regulates advertising and promotion of products by the pharmaceutical industry?
- What is included in the full prescribing information to be provided by the sponsor to FDA?
- Explain the Sunshine Act and why its relevant to pharmaceutical sales?
- Explain what a pharmaceutical marketing strategy is and what questions it seeks to address?

Answers:

- Through its Drug Marketing, Advertising, and Communications Divisions, the Food and Drug Administration (FDA) regulates all advertising and promotional activities for prescription drugs, including statements made to physicians and pharmacists by pharmaceutical sales representatives. Advertising of over-the-counter (OTC) drugs, which is not regulated by the FDA, is under the jurisdiction of the Federal Trade Commission (FTC).
- Full prescribing information must include, in the correct order, the following information about the drug: its trade name, its assigned name, the strength of its dosage form, a caution statement (stating that a prescription is required), a description of its active ingredient, the clinical pharmacology of the drug, indications for its usage, contraindications for usage, precautions, adverse reactions, instructions on what to do in case of overdosage, dosage and administration, and how the drug is supplied.
- The Sunshine Act requires manufacturers of drugs, medical devices, biological and medical supplies covered by the three federal healthcare programs Medicare, Medicaid, and State Children's Health Insurance Program (SCHIP), to collect and track all financial relationships with physicians and teaching hospitals and to report these data to the Centers for Medicare and Medicaid Services (CMS). The goal of the law is to increase the transparency of financial

relationships between healthcare providers and pharmaceutical manufacturers and to uncover potential conflicts of interest.

- A pharmaceutical marketing strategy or a range of strategic marketing options is initiated by asking the sales and marketing groups to collaborate on answering the following questions in accordance with their marketing/organizational objectives:
  - How will the marketing activities help make sales?
  - What market trends are emerging that the company needs to respond to?
  - What position will the company strive to achieve?
  - Which pharma market segments will be targeted with which propositions?
  - What communication strategies will be used to support customer acquisition?
  - What experience will the company look to create for their audiences?
  - How can the company differ from its competitors?

Quiz:

1 True or False. Before a new prescription drug is approved for marketing, the FDA and the pharmaceutical company sponsoring the NDA must agree on the "full prescribing information" that will accompany the product, and that must be included in all advertisements, brochures, promotional pieces, and samples.

2 The "buyer" for medicines of any kind could include **all but which** of the following:
   A Chemists/Pharmacists
   B Distributors
   C Third-party payers
   D Hospital Pharmacies.

3 The four most common pharmaceutical marketing strategies include **all but which** of the following:
   A market penetration
   B survey landscape development
   C market development
   D product development
   E diversification

4 What are the only two countries where direct-to-consumer (DTC) advertising of prescription drugs is legal?
   A United States and New Zealand
   B United States and Canada
   C United States and Mexico
   D United States and United Kingdom

22

Generic Drugs and the Generic Industry
Robert G. Bell, PhD

Drug and Biotechnology Development, LLC, Clearwater, FL, USA

Introduction

What do you think of when you hear the word "generic"? It may conjure up the notion of shopping at a grocery store and seeing the plain boxes marked corn flakes rather than the fancy branded colorful boxes that are advertised on television. Even the definition by Merriam-Webster (Generic 2021) tends to convey a lack of significance and quality: relating to or characteristic of a whole group or class; not being or having a particular brand name, and having no particularly distinctive quality or application. However, this terminology does not accurately describe or define generic drugs. Generic drugs are made with the same active ingredient and are therapeutically equivalent to the brand name drug (21 CFR § 314.3(b) n.d.), usually at a fraction of the cost.

In 2012, 84% of prescriptions in the United States were filled with generic drugs; in 2014 (Boehm et al. 2013), the use of generic drugs in the United States led to US$254 billion in healthcare savings (Generic Drug Savings in the U.S (PDF) 2015). In addition to the cost savings, over 85% of all drugs dispensed in the United States in 2018 were generic, and only 10% were branded drugs (Matej Mikulic. Branded vs. generic U.S. drug prescriptions dispensed 2005–2018 2019).

Generic drugs have been around for about one hundred years since generic versions of aspirin hit the shelves. The drug registration regulations have changed considerably since then, and the "modern" generic regulation process occurred in 1984 with the passage of the Drug Price Competition and Patent Term Restoration Act, commonly known as the Hatch-Waxman Act (Drug Price Competition and Patent Term Restoration Act of 1984 1984). The new law made it much easier and less expensive to bring new generic drugs to market by eliminating lengthy human trials and using scientifically sound methodologies to prove that the generic drug had the same active ingredients and that they performed in the body the same way as the brand-name drug. The Hatch-Waxman Act also included provisions that increased the amount of time the branded company could hold patents and exclusivities on a new drug and granted 180-day exclusivities for certain generic applicants. The Hatch-Waxman Act provided a win-win scenario for both the brand and generic industries, and within a year, the FDA received hundreds of abbreviated new drug applications (ANDA) for new generic drugs (U.S. Food and Drug Administration 2009; Food and Drug Administration 2009).

All prescription drugs are approved by the United States Food and Drug Administration (USFDA) under the Federal Food, Drug, and Cosmetic Act (FFDCA). New drugs are codified under FFDCA section 505 (b)(1) and 505 (b)(2), whereas generic drugs are codified under FFDCA 505 (j). The FDA ensures that all drugs, both new and generic drugs, are rigorously tested prior to approval into the US marketplace. New drugs are examined for safety, especially side effects (adverse events) and efficacy. Generic drugs are examined for therapeutic equivalence to the new drug they wish to copy. The approval of a new drug can take up to ten or more years and can cost hundreds of millions of dollars to develop. Generic drugs can take up to five years to develop and can cost up to several millions of dollars to develop.

Although not considered generic, subsequent entry biological drugs, otherwise known as biosimilars, have a more comprehensive pathway under the FFDCA and the Public Health Service Act (PHSA) required for approval (section 351 (k)). Additionally, medical devices are approved under the FFDCA Medical Device Regulation Act or Medical Device Amendments of 1976, which defines the requirements for premarket approval and subsequent substantial equivalent devices (section 510) for medical device approval. This chapter will not discuss biosimilars or medical devices, only generic drugs. The interested reader will find information regarding biosimilars at www.fda.gov/biosimilars and for medical devices at www.fda.gov/medicaldevice. Medical devices are also covered in this text in Chapter 19 as well. The objectives of this chapter are to familiarize the reader with the generic drug approval process and the quality attributes associated with generic drug products. So, what is a generic drug, and what are the differences between generic and new (branded) drug products?

New vs. Generic Drugs

Since the early part of the last century, new drugs have been approved by the submission of a new drug application (NDA) before it can be commercially sold in the United States. A new drug is usually a medication that has not been used before in clinical practice to treat a disease or condition. The NDA application is the documented evidence through which the drug sponsors have proven their drug is safe and efficacious for its' intended use to request the FDA approve this new pharmaceutical product for sale and marketing in the United States. The data is gathered by studying the drug in animal (preclinical testing) and human studies (Phases 1–3) through the use of an investigational new drug (IND) application which becomes part of the NDA submission. A new drug may be an innovative new compound that is classified as a new molecular entity (NME), or it may be related to a previously approved product. The FDA drug approval process is a multi-step process that takes up to ten or more years and costs up to a billion (or more) dollars (Prasad and Mailankody 2017).

A generic drug is a product that compares to the branded drug (or reference listed drug product (RLD)) which is usually the first approved branded drug product) in dosage form, route of administration, strength, quality, safety, and performance characteristics. The generic drug must have the same intended use as listed in the drug label as the branded RLD. Generic drugs typically do not have to undergo pre-clinical animal drug testing or human clinical trials (Phase 1, 2, or 3 human studies) only demonstrate therapeutic equivalence and submit an ANDA. Therapeutic equivalence consists of pharmaceutical equivalence and bioequivalence. The concept of therapeutic equivalence applies only to drug products containing identical active ingredient(s) and does not encompass a comparison of different therapeutic agents used for the same condition. See Table 22.1 for a comparison of new drug requirements vs. generic drug requirements.

The US generic drug market reached a value of $US115.2 billion in 2019 and grew at a compound annual growth rate of 11.7% during 2014–2019 (US Generic Drug Market n.d.). This is mainly due to the loss of patent protection from brand drugs which then allow for generic drugs to be developed and approved. As depicted in Table 22.1, generic manufacturers also pay user fees under the Generic Drug User Fee Amendments of 2012 and reauthorized in 2017. These generic user fees allow for additional FDA resources to provide for a timely review of the ANDA. In addition, the US government has introduced several programs for offering incentives to physicians and pharmacists to promote generic substitution (Sarpatwari et al. 2015; State Policy Options To Reduce Prescription Drug Spending n.d.), further saving the consumers, the patients and government, a significant amount of money through substitution of therapeutically equivalent generic drug products.

The major manufacturers of generic drug products include both US and international companies (21 Largest Generic Drug Companies n.d.). Currently the largest global generic pharmaceutical company is Teva Pharmaceutical Industries Limited (Teva), based in Israel. Other large generic manufacturers include Sandoz, a division of the Novartis Group based in Munich, Mylan Pharmaceuticals, Inc. (who recently merged with Pfizer-Upjohn to form Viatris) based in Pennsylvania, Sun Pharmaceuticals, Lupin Pharmaceuticals and Dr. Reddy's Laboratories

Table 22.1 New drug vs. generic drug registration requirements.

Stage/Activities	Brand drugs	Generic drugs
Drug discovery	Yes High throughput screening of thousands of drugs. Intellectual property/Patents	No
Drug development	Yes Absorption, distribution, metabolism, excretion. Mechanism of action. Dosing and dose design. Side effects and adverse events. Toxicology. Drug-drug and drug-food interactions. Comparative effectiveness.	No
Preclinical research nonclinical pharmacology and toxicology	Yes In vitro and In vivo animal studies	No
Human clinical studies	Yes Phase 1, 2, and 3 Human Studies	No
Bioavailability and bioequivalence	Yes	Yes
Chemistry	Yes	Yes
Manufacturing	Yes	Yes
Testing and stability	Yes	Yes
Labeling	Yes	Yes
Regulatory inspections	Yes	Yes
Market exclusivity	Yes	No (However, first approved generic receives 180 d exclusivity)
Patents	Yes	No, but possible
Post marketing studies	Yes	No
Post marketing safety surveillance	Yes	Yes
User fees	Yes	Yes

Limited which are based in India, Actavis, Inc. (formerly known as Purepac Pharmaceutical Co., Inc.) and Par Pharmaceuticals based out of New Jersey, Hospira Inc. (acquired by Pfizer) based in Illinois, and Endo Pharmaceuticals based in Ireland.

Furthermore, an "authorized generic" drug may be marketed by the brand name drug company, or another company with the brand company's permission. An authorized generic drug is a term commonly used to describe an approved brand name drug that is marketed without the brand name on its label - it is the exact same drug product as the branded product but does not have the brand name on its label. As you can see from the above, generic manufacturers are global and in some instances, owned by branded drug companies and produce their own authorized generic drug product.

Generic Drug Requirements

The FDA's publication, *Approved Drug Products with Therapeutic Equivalence Evaluations* (commonly known as the *Orange Book* (Approved Drug Products with Therapeutic Equivalence Evaluations 2020)), identifies drug products, both new and generic, approved on the basis of safety and effectiveness by the FDA under the FFDCA. The main criterion for the inclusion of any product is that the product is the subject of an application with approval that has not been withdrawn for safety or efficacy reasons. As described in the FDA's *Orange Book*, the FDA classifies a generic drug as therapeutically equivalent those RLD (brand drug product) that meet the following general criteria:

- **They are approved as safe and effective to the comparative branded drug.** Efficacy is the ability for the drug to produce the desired effect, such as lowering high blood pressure or an antibiotic resolving an infection. Drug safety evaluates information relating to side effects of drugs during the clinical trials and post-approval phase. Safety and efficacy go hand in hand – the pharmaceutical drug producers must continually monitor health outcomes and report to the FDA any and all evidence of possible "adverse events," or negative side effects that some patients experience once the medicine is made available to the general patient population. The FDA evaluates new drugs during its' investigational development before they can be commercialized to ensure that prescription and over-the-counter drugs, both brand name and generic, work correctly and that the health benefits outweigh the known risks of the drug. The FDA's review of new drug applications assures the new drug is efficacious and prevents unsafe drugs from entering the market. This rigorous evaluation by the FDA of new drugs and knowledge gained during the marketing exclusivity period of the new drug paves the way for understanding the safety and efficacy needed to develop a generic drug.

- **They are pharmaceutical equivalents**, meaning that they contain identical amounts of the identical active drug ingredient in the identical dosage form (e.g. oral tablet, capsule, solution, etc.) and route of administration (e.g. intravenous, oral, transdermal, nasal spray, etc.), and meet compendial such as the United States Pharmacopeia (USP) or other applicable standards such as FDA guidance regarding strength, quality, purity, and identity. The generic version usually differs from its brand-name counterpart in size, color, shape, and flavor. For instance, using the narrow therapeutic index drug Coumadin® 5 mg tablets as the branded reference product, generic 5 mg warfarin tablets must contain the appropriate amount of warfarin sodium to result in 5 mg of warfarin, the appropriate excipient amounts of lactose, starch, magnesium state, and FD&C yellow No. 6 aluminum lake and comply with the USP specifications (Warfarin Sodium Tablet Monograph 2020) for identification (high-performance liquid chromatography (HPLC) and infrared spectroscopy), assay (95–105% by HPLC), dissolution (80% in 30 minutes), content uniformity (The tablets comply if not more than one individual content is outside the limits of 85–115% of the average content and none is outside the limits of 75–125% of the average content), organic impurities are ≤0.5% and the product is preserved in tight, light-resistant containers and stored at controlled room temperature. Barr Laboratories, Inc. (Pomona, NY) received approval of generic warfarin in 1997 and conducted contentment uniformity on over 4 million warfarin tablets (Mengler and Bell 1998) (Barr Laboratories, Inc. was acquired by Teva Pharmaceutical Industries in 2008, and is this specific warfarin product is no longer available). Table 22.2 shows the USP specification, the brand at the time (DuPont) and Barr Laboratories, Inc. 5 mg content uniformity which demonstrates that the generic warfarin product was equivalent in content uniformity to that of the brand. The color and shape of the brand and generic warfarin 5 mg dosage form are shown in Figure 22.1.
- **They are bioequivalent to the comparative branded drug** in that they do not present a known or potential bioequivalence problem, and they meet an acceptable *in vitro* standard, or if they do present such a known or potential problem (such as solutions), they are shown to meet an appropriate

Table 22.2   USP content uniformity of brand and generic Warfarin 5 mg tablets.

USP specification	Coumadin® 5 mg tablet (DuPont)	Warfarin 5 mg tablet (Barr)
4.25–5.75 mg/tablet	4.63–5.38 mg/tablet	4.74–5.17 mg/tablet

(a) (b)

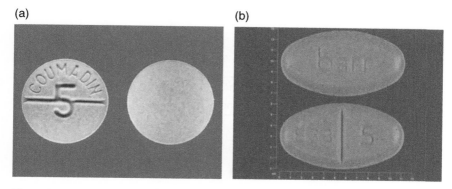

Figure 22.1 Dosage form comparison of the reference listed product Coumadin® 5 mg (left = A) to generic Warfarin (Barr Laboratories, Inc.) 5 mg tablet (right = B) (Courtesy of Jeff Harrison of Drugs.com) (Coumadin-5-2650 n.d.; Barr-833-5-373 n.d.).

bioequivalence standard. The definition of bioequivalence is the absence of a significant difference in the rate (maximum drug plasma concentration (C_{max}) and time to maximum drug plasma concentration (t_{max}) and extent (area under the plasma drug concentration-time curve (AUC)) to which the active ingredient or active moiety in pharmaceutical equivalents or pharmaceutical alternatives becomes available at the site of drug action when administered at the same molar dose under similar conditions in an appropriately designed study (Guidance for Industry 2001) – see Figures 22.2 and 22.3.

Figure 22.2 Intravenous (i.v.) and oral pharmacokinetic curves with C_{max}, t_{max}, and AUC.

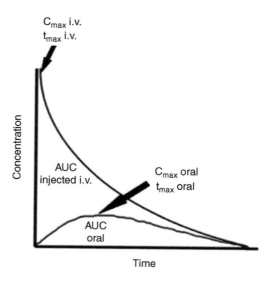

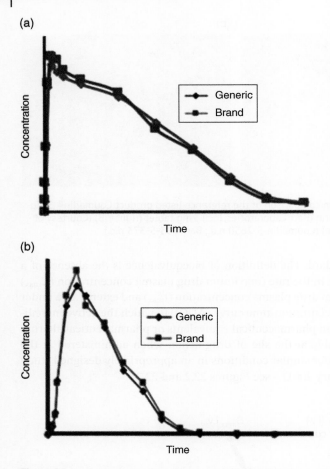

Figure 22.3 Intravenous (i.v. – A) and oral (B) bioequivalence curves between a generic and brand drug.

In order to determine the bioequivalence of a generic test drug to that of a reference brand drug, a randomized, crossover trial is usually conducted with both the generic and brand-name drug being assessed over time. For the generic drug to be deemed bioequivalent to the brand drug, the ratio of each pharmacokinetic (PK) characteristic (the rate (C_{max}) and extent (AUC)) of the generic drug to the reference drug is calculated. The ideal value of this ratio is 1:1, or just 1.00 (indicating a perfect match or perfect bioequivalence). However, since variability is inherent in human studies, the FDA bioequivalence requirement is that the 90% confidence interval of the PK ratio should lie between 0.80 and 1.25 (20). For the entire 90% confidence interval to meet this requirement, the mean PK value of the generic drug product should actually lay quite close to that of the reference branded drug control, making the variation between the generic drug product and the reference branded drug control usually very small, assuring therapeutic equivalence and interchangeability.

The Orange Book has two basic categories into which multisource drugs have been placed and are indicated by the first letter of the relevant therapeutic equivalence code. An "A" drug products that FDA considers to be therapeutically equivalent to other pharmaceutically equivalent products for which there are no known or suspected bioequivalence problems. These are designated AA, AN, AO, AP, or AT, depending on the dosage form or actual or potential bioequivalence problems that have been resolved with adequate in vivo and/or in vitro evidence supporting bioequivalence (and "AB" generic drug product). Drug products that FDA the considers not to be therapeutically equivalent to other pharmaceutically equivalent products drug products for which actual or potential bioequivalence problems have not been resolved by adequate evidence of bioequivalence are designated as "B" generic drug products.

Figure 22.4 illustrates the graphical bioequivalence data comparing the reference listed drug (DuPont's Coumadin®) to the test comparator generic drug (Barr Laboratories, Inc. Warfarin) at the 5 mg strength (Mengler and Bell 1998). As illustrated by the graph, the pharmacokinetic profiles were basically superimposable with a pharmacokinetic ratio close to 1 and the 90% confidence interval of 97–103% – the generic drug being the test drug and the brand drug being the control. A number of PK parameters are assessed in bioequivalent studies to assess how the rate and extent of the availability of the generic drug compares to the control. These parameters include at a minimum, the maximum serum

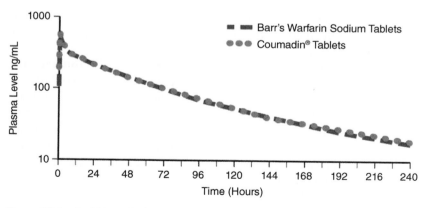

Figure 22.4 Oral bioequivalence curves between Coumadin® and Barr Pharmaceuticals, Inc. Warfarin 5 mg tablets.

concentration of a drug (C_{max}), time to the maximum serum concentration of a drug (t_{max}), half-life ($t_{1/2}$), elimination (k_e) and drug exposure.

Furthermore, clinical studies by Neutal and Smith (Neutel and Smith 1998) conducted a randomized crossover study to compare the efficacy by the international normalized ratio (INR) and tolerability of Barr warfarin sodium to the currently available Coumadin. Warfarin is a vitamin K antagonist used primarily to prevent venous thrombosis by inhibiting coagulation factors II, VII, IX, and X. The anticoagulation achieved by warfarin is monitored by a coagulation test known as the prothrombin time (PT). INR was introduced in an attempt to standardize the PT because different PT test reagents were not standardized. INR evaluation and testing is the method upon which patients taking warfarin are monitored to assure they have a consistent anticoagulation effect (or the therapeutic clinical effect) from their specific individualized dose of warfarin. The authors reported that after multiple crossovers between the brand and generic warfarin products, the INR for Coumadin® was 2.38 and for Barr's generic warfarin was 2.43, which was statistically and clinically insignificant. This data reinforces the FDA's concept of therapeutic equivalence that a generic drug shown bioequivalent to a reference listed brand product will have the same safety, efficacy, and clinical outcomes.

- **They are adequately labeled to that of the comparative branded drug.** The FDA requires generic drug labels to conform to the labeling of the brand product's RLD. Generally, the only differences in labeling between the brand and generic labels are the pharmacokinetics, manufacturer, and excipients if they differ. The generic label incorporates all of the safety and clinical information contained in the branded label. However, the holder of the NDA for the RLD will propose any necessary changes to the drug label when new information becomes available. Once the FDA approves the changes, the ANDA holders must update their labels with the corresponding modifications. But in cases where the RLD is withdrawn from the market, the NDA holder can no longer update the drug label; changes to the label may need to be made by the ANDA holders.

- **They are manufactured in compliance with Current Good Manufacturing Practice regulations (cGMPs).** The term "current" is included in cGMP to let pharmaceutical companies that they need to remain current with the latest regulations and technologies. cGMP are the FDA's formal regulations regarding the design, monitoring, control, and maintenance of pharmaceutical manufacturing processes and facilities. The cGMP regulations for drugs contain the minimum requirements for the methods, facilities, and controls used in manufacturing, processing, and packing of a drug product to assure that a product is safe for use, has the identity, strength, purity and quality and adheres to the procedures approved in the drug application (e.g. ANDA). The FDA inspectors determine whether the pharmaceutical manufacturer has the necessary facilities, equipment, and ability to produce the drug it intends to market. Inspections are conducted using the Code of Federal Regulations (CFR), and pharmaceutical facilities are usually inspected after any drug approval (pre-approval inspection) or generally every two years. Pharmaceutical inspections are conducted using the FDA's Title 21 of the CFR, which interprets the FFDCA and related statutes, including the PHSA. The pharmaceutical drug quality-related regulations appear in several parts of Title 21, including sections in parts 1-99, 200-299, 300-499, 600-799, and 800-1299.

Conclusions

The FDA classifies generic drugs as therapeutically equivalent to branded drug products if they are safe and effective, pharmaceutically equivalent, and bioequivalent to the branded reference drug product, adequately labeled and manufactured in compliance with cGMP regulations. The FDA follows a rigorous review process to make sure that, compared to the brand-name drug products, the proposed generic medications is pharmaceutically equivalent and contain the same active ingredient, strength, dosage form, route of administration, and is bioequivalent. The FDA assures that generic drugs perform the same way in the human body and have the same intended use as the name-brand medication. Health care professionals and consumers can be assured that FDA-approved generic drug products have met the same rigid standards as the branded reference listed drug. All generic drugs approved by FDA have the same high quality, identity, strength, purity, and stability as brand-name drugs. In addition, FDA inspects facilities to make certain the generic manufacturing, packaging, and testing laboratories comply with the same quality standards as those of brand-name drugs.

References

21 CFR § 314.3(b) (2022). CFR – Code of Federal Regulations Title 21, 21 CFR § 314.3(b), https://www.accessdata.fda.gov/scripts/cdrh/cfdocs/cfcfr/cfrsearch.cfm?fr=314.3 (6 January 2022).

21 Largest Generic Drug Companies (n.d.). https://finance.yahoo.com/news/21-largest-generic-drug-companies-185756541.html.

Approved Drug Products with Therapeutic Equivalence Evaluations (2020). 40th Edition, U.S. Department of Health and Human Services, Food and Drug Administration, Office of Medical Products and Tobacco, Center for Drug and Food Evaluation and Research Office of Generic Drugs, Office of Generic Policy.

Barr-833-5-373 (n.d.). https://www.drugs.com/imprints/barr-833-5-373.html (Courtesy of Jeff Harrison of Drug.com).

Boehm, G., Yao, L., Han, L., and Zheng, Q. (2013). Development of the generic drug industry in the US after the Hatch-Waxman Act of 1984. *Acta Pharmaceutica Sinica B* 3 (5): 297–311.

Coumadin-5-2650 (2022). https://www.drugs.com/imprints/coumadin-5-2650.html (Courtesy of Jeff Harrison of Drugs.com) 7 March 2022.

Drug Price Competition and Patent Term Restoration Act of 1984 (1984). Pub. L. 98-417, 98 Stat. 1585.

Food and Drug Administration (2009). Consumer Education: What You Should Know About Buying and Using Generic Drugs.

Generic (2021). Merriam-Webster.com *Dictionary*, Merriam-Webster, https://www.merriam-webster.com/dictionary/generic (accessed 1 January 2021).

Generic Drug Savings in the U.S (PDF) (2015). Washington, DC: Generic Pharmaceutical Association (GPhA). (accessed 16 June 2016).

Guidance for Industry (2001). Statistical Approaches to Establishing Bioequivalence. U.S. Department of Health and Human Services Food and Drug Administration Center for Drug Evaluation and Research (CDER). 3616fnl.PDF (fda.gov).

Matej Mikulic. Branded vs. generic U.S. drug prescriptions dispensed 2005–2018 (2019). Statista. (https://www.statista.com/statistics/205042/proportion-of-brand-to-generic-prescriptions-dispensed).

Mengler, C.J. and Bell, R.G. (1998). *In-Vivo Correlation of the Pharmacodynamic and Pharmacokinetic Behavior of Warfarin Sodium Tablets*. San Francisco, California: AAPS.

Neutel, J.M. and Smith, D.H.G. (1998). A randomized crossover study to compare the efficacy and tolerability of Barr warfarin sodium to the currently available Coumadin. *Cardiovascular Reviews and Reports* 19: 49–59.

Prasad, V. and Mailankody, S. (2017). Research and development spending to bring a single cancer drug to market and revenues after approval. *JAMA Internal Medicine* 177 (11): 1569–1575.

Sarpatwari A, Choudhry NK, Avorn J, Kesselheim AS (2015) Paying physicians to prescribe generic drugs and follow-on biologics in the United States. *PLoS Medicine* 12(3): e1001802. doi:https://doi.org/10.1371/journal.pmed.1001802.

State Policy Options To Reduce Prescription Drug Spending (2020). https://www.americanprogress.org/issues/healthcare/reports/2020/02/13/480415/state-policy-options-reduce-prescription-drug-spending (13 February 2020).

U.S. Food and Drug Administration (2009). Generic Drugs: Myths and Facts.

US Generic Drug Market (2019). Industry Trends, Share, Size, Growth, Opportunity and Forecast 2020-2025. IMARC Group. IMARC Services Private Limited. https://www.imarcgroup.com/us-generics-market#:~:text=The%20 US%20generic%20drug%20market,effects%2C%20and%20route%20of%20 administration.

Warfarin Sodium Tablet Monograph (2020). United States Pharmacopeia 43-National Formulary 38.

Chapter Self-Assessments: Check Your Knowledge

Questions:

- What is a generic drug?
- According to the USFDA, what does therapeutic equivalence mean?
- What is an authorized generic drug product?
- What is generic bioequivalence and how is it determined?
- What are current Good Manufacturing Practices?

Answers

- A generic drug is a product that compares to the branded drug (or reference listed drug product (RLD)) which is usually the first approved branded drug product) in dosage form, route of administration, strength, quality, safety, and performance characteristics. The generic drug must have the same intended use as listed in the drug label as the branded RLD and demonstrate that it is therapeutically equivalent to the RLD.
- The FDA classifies a generic drug as therapeutically equivalent those reference listed drug products (RLD (brand drug product)) that meet the following general criteria:
  - They are approved as safe and effective to the comparative branded drug
  - They are pharmaceutical equivalents
  - They are bioequivalent to the comparative branded drug
  - They are adequately labeled to that of the comparative branded drug
  - They are manufactured in compliance with Current Good Manufacturing Practice regulations (cGMPs).
- An authorized generic drug is an approved brand name drug that is marketed without the brand name on its label. It is the exact same drug product as the branded product and may be marketed by the brand name drug company or another company with the brand company's permission.

- In order to determine bioequivalence of a generic test drug to that of a reference brand drug, a randomized, crossover trial is usually conducted with both the generic and brand-name drug being assessed over time. For the generic drug to be deemed bioequivalent to the brand drug, the ratio of each pharmacokinetic characteristic (the rate (Cmax) and extent (AUC)) of the generic drug to the reference drug is calculated. The ideal value of this ratio is 1:1, or just 1.00 (indicating a perfect match or perfect bioequivalence). However, since variability is inherent in human studies, the FDA bioequivalence requirement is that the 90% confidence interval of the PK ratio should lie between 0.80 and 1.25 (20). For the entire 90% confidence interval to meet this requirement, the mean PK value of the generic drug product should actually lay quite close to that of the reference branded drug control, making the variation between the generic drug product and the reference branded drug control usually very small, assuring therapeutic equivalence and interchangeability.

- Current Good Manufacturing Practices (cGMP) are the FDA's formal regulations regarding the design, monitoring, control, and maintenance of pharmaceutical manufacturing processes and facilities. The cGMP regulations for drugs contain the minimum requirements for the methods, facilities, and controls used in manufacturing, processing, and packing of a drug product to assure that a product is safe for use, has the identity, strength, purity, and quality, and adheres to the procedures approved in the drug application.

Quiz

1   True or False. A majority of all filled prescriptions are generic drug products. (Answer True)

2   To be generically equivalent to a reference listed drug product, a generic manufacturer must show that the generic drug product is _____ and _____ to that of the reference drug product. Answer: Pharmaceutically and Therapeutically Equivalent

3   The Orange Book has two basic categories into which multisource drugs have been placed and are indicated by the first letter of the relevant therapeutic equivalence code. They are: (Answer b)
   A G (good) and B (bad)
   B A (therapeutically equivalent) and B (not therapeutically equivalent)
   C C (capsule) and T (tablet)
   D I (injection) and O (other dosage forms)

4 The concept of therapeutic equivalence applies only to drug products containing the _____ and does not encompass a comparison of different therapeutic agents used for the same condition. (Answer c)

  A Same color tablet

  B Same shape of tablet

  C Identical active ingredient(s)

  D All of the above

  E None of the above.

5 True or False: The FDA requires generic drug labels can have different labeling than that of the brand product's reference listed drug (RLD). (Answer False)

23

The Generic Drug Approval Process

Jim Ottinger, BSc

Executive Vice President of Regulatory and Quality, UroGen Pharma Ltd. New York, NY, USA

Introduction

More than 4 billion generic prescriptions were filled across the United States in 2018. When generic drugs are available, they are dispensed 97% of the time. They account for 90% of all prescriptions dispensed in the United States, while only accounting for 22% of total drug costs. Generic drugs saved the US healthcare system $292.6 billion in 2018, and estimates over the proceeding 10 years are that generic drugs saved nearly $2 trillion (The Case for Competition 2019). Generic drugs are a critical component of the US healthcare system.

The modern US generic drug industry began in 1984 with the passage of landmark legislation known as the Drug Price Competition and Patent Term Restoration Act, more colloquially called the Hatch-Waxman Act or Amendments, named after the legislators, that sponsored the Act, Senator Orrin Hatch and Congressman Henry Waxman. Hatch-Waxman was a compromise for the branded and generic drug industries, creating a new pathway for generic drugs while offering additional exclusivity for branded drugs. It amended the Federal Food, Drug, and Cosmetic Act under Section 505(j) to create a new abbreviated new drug application (ANDA) pathway for generic drugs. In return, the Act allowed the branded drug industry to claim five years of regulatory exclusivity for a new drug and three years of exclusivity for a new use of an established drug product. In addition, the patent term of a new drug product could be extended based on a formula, considering the product development and FDA review time.

The Hatch-Waxman Act was only the beginning of the generic drug pathway, and it has been amended several times; litigation surrounding individual products has established many legal precedents, and FDA has evolved generic drug policy. The objective of this chapter is to provide a high-level overview of the concepts behind generic drug development, submission, and approval.

Fundamentals of Drug Development, First Edition. Edited by Jeffrey S. Barrett.
© 2022 John Wiley & Sons, Inc. Published 2022 by John Wiley & Sons, Inc.
Companion website: www.wiley.com/go/Barrett/FundamentalsDrugDevelopment

ANDA Pathway

The ANDA pathway was created for drug products that are the "same as" the branded drug, meaning it has identical active ingredient(s), dosage form, strength, route of administration, and conditions of use as the branded product. The ANDA pathway was "abbreviated" in comparison to a new drug application (NDA) in that it allowed the FDA to rely on its previous findings of safety and effectiveness of the branded product, eliminating the need to repeat time-consuming and costly nonclinical and clinical studies. Therefore, approval of a generic drug product was based specifically on submission of comparable data on the pharmaceutical equivalence of the drug substance/drug product [chemistry, manufacturing, and controls (CMC)] and clinical bioequivalence (Food Drug and Cosmetic Act n.d.).

In certain cases, the ANDA pathway could also be used for a drug product not fully meeting the definition of "same as" if declared suitable by the FDA through a petition procedure. For example, a branded drug may be available in 100 mg or 200 mg strengths; however, if a generic drug manufacturer would like to offer a 150 mg strength, they would submit a suitability petition to have this declared as eligible for an ANDA even though the strength is not the same as the branded product (Code of Federal Regulations Title 21 n.d.). In cases where the drug product differs from the branded product, for example, a different salt of the active ingredient, and cannot be submitted as an ANDA, the generic firm may consider an alternative abbreviated pathway using the 505(b)(2) NDA approach (Guidance for Industry 2019).

Generic drug products are regulated by the FDA's Office of Generic Drugs (OGD), which is responsible for the review and approval of all ANDAs. The Office, which is located at the FDA's White Oak Campus in Maryland, consists of an immediate Office plus subordinate Divisions – Divisions of Bioequivalence Process Management, Therapeutic Performance, Orange Book Publication, and Regulatory Assessment.

The Orange Book

The identification, choice of, and comparison to the branded product is the beginning of the generic drug development process. The official source for the list of approved products is contained in the publication formally known as *Approved Drug Products with Therapeutic Equivalence Evaluations* but commonly referred to as the Orange Book, simply because the cover of the original publication was an unmistakably bright orange, apparently as it was released around Halloween over 40 years ago (Gingery 2020).

The Orange Book lists all approved drug products along with applicable patent and exclusivity information. The approved branded product selected by the generics company is known as the "Listed Drug," the "Reference Listed Drug," or the "RLD." The choice of the Listed Drug drives the development program, must be identified in the ANDA, and is used in the clinical bioequivalence study.

The Orange Book also lists any exclusivity and patent information that may be applicable to the Listed Drug, both critical data points to generic companies. The Hatch-Waxman Act established regulatory exclusivity for innovative new drugs, which drives the earliest timing of an ANDA submission. For a new drug containing a new chemical entity, no applicant can submit an ANDA for a period of 5 years following the branded drug's approval date, or 4 years following the approval date if the generic applicant certifies that the Listed Drug's patents are invalid or not infringed upon (Code of Federal Regulations Title 21 n.d.).

First-to-File

The ability to submit a generic drug application one year earlier by attempting to invalidate patents associated with the Listed Drug is a powerful incentive for the generic drug industry to adopt this strategy. Significant resources are dedicated to the objective of proving in court that existing patents listed in the Orange Book are invalid, unenforceable, or will not be infringed upon by the generic sponsor. These ANDAs contained certifications known as "Paragraph IV Certifications" which set off patent litigation between the generic and branded firms. Not only did this strategy allow submission one year earlier, but it had the additional incentive that the first successful generic applicant with a Paragraph IV Certification would be awarded its own 180 days of generic drug exclusivity. The 180 days exclusivity allowed the generic company to undercut the branded drug's price while maintaining robust margins, and this exclusivity is a critical component of the profitability of the generic drug industry. The introduction of multiple generic competitors following the end of the exclusivity period generally leads to rapid price declines and commodity pricing. The significance of the exclusivity led every generic company to focus on being first-to-file as the industry's "Holy Grail." This is a major point of differentiation between generic and branded drug development. On the branded side, relative to projections at the beginning of development, firms will routinely delay NDAs by weeks, months, and quarters due to slow enrollment, FDA advice, manufacturing, and other issues. For a generic firm with an ANDA eligible for first-to-file status, the delay of a single day can define its success or failure. For blockbuster branded drugs, generic firms will target the first-to-file date as their submission objective, those that submit successful applications will share the 180-day exclusivity, any firm that submitted even one day following the first-to-file date will be prohibited from launching their version and enter the market at commodity pricing.

The value of being first-to-file is still a coveted position, but it was even more so in the past. In the original legislation, exclusivity was awarded to the single company that submitted a successful Paragraph IV Certification. This led to a scene where representatives of generic drug companies would physically camp out and line up outside OGD offices to literally be the first company in line to file (the applications were paper-based at the time). To eliminate this circus atmosphere, the legislation was amended in 2003 so that exclusivity is now shared between any generic manufacturers that file a successful application on the same day following the 4-year innovative exclusivity period (Medicare Prescription Drug 2003).

Of course, not all approved branded drugs are protected with exclusivity or patents, and an ANDA may be submitted at any time for these products using a Paragraph 1 Certification (no patent in Orange Book), a Paragraph 2 Certification (Orange Book patent expiring) or a Paragraph 3 certification (ANDA can be approved after patent expiration). For smaller market size branded drugs with no exclusivity or patent protection, FDA also offers first-to-file exclusivity to encourage generic competition for products with limited use.

FDA Interactions

Another significant difference between innovative and generic drug product development is the mechanism to interact with the FDA during the development phase. On the branded side, meetings with FDA New Drug Review Divisions are common both at milestone time points (pre-IND, end of Phase 2, pre-NDA) and a variety of ad hoc meetings. Conversely, meetings with OGD for development advice have generally been non-existent, with only recent changes in approach for potential pre-ANDA meetings for complex generic products. Instead, generic firms can submit written correspondence to OGD seeking information on either generic drug development or post-approval submission requirements. These requests are known as "controlled correspondence." In the past, OGD had been slow to respond to these requests; however, more recent legislation described later in this chapter has improved timelines significantly. OGD has committed to review and respond to 90% of standard and complex controlled correspondence within 60 and 120 days, respectively, from the date of submission. Controlled correspondence is deemed to be "complex" if it involves the review of clinical content, bioequivalence protocols, or alternative bioequivalence approaches (Guidance for Industry 2020).

Submission and Content

ANDAs must comply with the requirements of the common technical document (CTD) format developed at the International Conference of Harmonisation and common to any drug application in the United States, Europe, and Japan. As of 5 May 2017, all ANDAs, and any submissions to an ANDA, must be submitted electronically in eCTD format through the FDA Gateway (Guidance for Industry 2019).

At a high level, the CTD is comprised of the following modules with specific highlights to follow:

- Module 1: Administrative Information and Prescribing Information
- Module 2: Summaries
- Module 3: Quality
- Module 4: Nonclinical
- Module 5: Clinical

Module 1 – Administrative Including Labeling

Current FDA guidelines should be consulted for a full review of the administrative components (letters, forms, certifications, etc.), which are included in Module 1.

Importantly, the labeling for the generic product is included in Module 1 of the application. The draft label and labeling for each strength and container are to be included. The prescribing information (PI), also referred to as the package insert, as well as any patient labeling (e.g. patient package insert, medication guide), is to be included as well. It is important to note that the PI for a generic product must be the same as the listed product (RLD), with certain exceptions allowed. An ANDA must contain a side-by-side comparison of the listed drug's labels vs. the generic drug labels, with all differences highlighted and annotated. The current Listed Drug's PI, any Patient Labeling, and one container label and one outer carton for each strength and package size must also be included.

Hatch-Waxman permits that a generic applicant may not pursue approval for all patents listed in the Orange Book for the Listed Drug. A branded company may continually expand the label for the innovative drug by staging new indications for use or pursuing pediatric exclusivity of the original indication. This evergreening of the drug product label would prohibit generic competition since the generic firm would risk infringement lawsuits from the innovative industry if the ANDA label included protected information. For example, an innovative company obtains approval for a drug for major depressive disorder (MDD) therefore securing 5 years of exclusivity. Four years later, they obtained approval for a new indication of a generalized anxiety disorder (GAD), adding 3 years of additional exclusivity to the brand. To avoid the delay of a generic entry to the originally approved indication, Hatch-Waxman permits ANDA applicants to certify if the applicant does not wish to pursue approval covered by all listed patents. In this example, once the exclusivity for MDD expires, the ANDA applicant can seek approval of a generic for the MDD indication. In the labeling, it will exclude or "carve out," the GAD indication. These carve-outs allow ANDA applicants to focus on the original patent listings and are a workaround to evergreening.

Once the ANDA is approved, it is the responsibility of the ANDA holder to update labeling to match any changes to the labeling of the listed product. In most circumstances, ANDA holders are not permitted to initiate independent updates to the labeling content, even in cases where the ANDA product is the only marketed version of a product due to the NDA withdrawal or market discontinuation of the branded NDA. In this case, only FDA can institute label changes.

Module 2

Module 2 will contain summaries of all the data presented in the ANDA, the Quality Overall Summary and Summary of Clinical Bioequivalence being the most important.

Module 3 - Quality (Chemistry, Manufacturing, and Controls)

While the ANDA process is abbreviated for nonclinical and clinical data, it is not abbreviated for the CMC data, which must demonstrate that the generic product is pharmaceutically equivalent to the listed drug. The CMC data requirements for generic drugs are essentially the same as for innovative drugs, with the possible exception of less stability data for the drug product in the original submission (6 vs. 12 months for NDAs). The FDA allows ANDA sponsors to amend the application with 12 months of stability data during the review cycle (Guidance for Industry 2014).

Module 4 - Nonclinical

An ANDA will typically not include data in Module 4, and the FDA will instead refer to nonclinical data from the Listed Drug. This module may include, however, any nonclinical reports to qualify impurity levels in the proposed specification.

Module 5 – Clinical

The bridge between establishing the safety and efficacy of the generic product to the Listed Drug is established by demonstrating the bioequivalence of the two products, typically evaluating the rate and extent of absorption in comparison to the Listed Drug in a clinical bioequivalence study, although other measures may also be used in certain circumstances. These data are all included in Module 5.

Bioequivalence is defined in the regulations as the absence of a significant difference in the rate and extent to which the active ingredient in pharmaceutical equivalents or pharmaceutical alternatives becomes available at the site of drug action when administered at the same dose under similar conditions in an appropriately-designed study (Code of Federal Regulation Title 21 n.d.). The rate of absorption is measured by $C_{max,}$ while the extent of absorption is measured by area under the curve (AUC). For highly variable drugs, the FDA generally recognizes that the absence of significant difference occurs if the rate and extent of absorption are within the range of 80–125% of that of the Listed Drug. For narrow therapeutic index drugs, the range is 90–111% (Guidance for Industry 2001; Yu and Bing 2014). Due to the criticality of these data to the ANDA pathway, the FDA routinely publishes product-specific guidance describing the Agency's current thinking on how to develop generic drugs. As of 2021, there are close to 2000 guidance documents that outline the type of study, study design, strengths, and population to be evaluated. Additional information on analytes, dissolution test methods, and sampling times will also be provided.

Many drug products cannot be assessed by a standard bioequivalence study, and there are many additional variables to evaluate for topical drugs, liposomal drugs, orally-inhaled drugs, nasal drugs, and others. Additional approaches such as in vitro/in vivo correlations and comparative pharmacodynamic approaches may be employed. Should no other options exist, a clinical endpoint study may be the only alternative.

Biowaivers for the requirement for in vivo bioavailability and bioequivalence studies may be obtained based on the Biopharmaceutics Classification System. Biowaivers may also be obtained for products with many strengths if the product uses a common formulation per strength. Information on biowaivers is included in the individual product-specific guidelines.

Bioequivalence must also be supported with robust dissolution profile comparisons between the generic drug and the listed drug.

Submission Fees and Approval Timelines

Prior to 2012, ANDAs were not required to pay a submission fee, commonly referred to as a user fee. Due to the growth in both the number of sponsors, the number of foreign facilities requiring inspection, and drugs eligible for the

ANDA pathway, the FDA review capabilities were overwhelmed. Long approval times exceeding 3 years were common, and in 2012, there were 2299 applications in the OGD backlog (Woodcock 2016). In order to speed the access of generic drugs, FDA and the generic drug industry collaborated on a user fee program in which the industry agreed to pay fees in exchange for FDA committing to certain performance goals. The proposal was closely modeled on the successful Prescription Drug User Fee Act (PFUDA), which was introduced for branded drugs in 1992. The Generic Drug User Fee Act (GDUFA) was passed by Congress as part of the Food and Drug Administration Safety and Innovation Act of 2012.

The agreement between the FDA and Industry was that GDUFA I was a five-year program eligible for re-negotiation on the basis of its ability to expedite the review and approval of ANDAs. The program met its goals and was re-negotiated with additional enhancements as GDUFA II in 2018.

The GDUFA program dramatically changed both the review paradigm and the generic drug industry itself. ANDAs submitted prior to GDUFA were subjected to three to four year review times, and that a "speed-to-filing" strategy was commonly used since applications could be amended with new data during the long review cycle. GDUFA ended that strategy with assignment of time penalties for the submission of ANDA amendments and the implementation of a "Refuse- to-Receive" designation, allowing FDA to not accept substandard applications for review. The changes required by GDUFA resulted in the submission of much higher quality original ANDAs.

GDUFA II expanded the program and addressed a broad range of issues including priority review for certain generics, guidance on complex generics, original and amended ANDA review timelines, and improved timelines for the review of controlled correspondence. In total, GDUFA II includes over 25 commitments.

For the 5 year range of the program (2018–2022), the FDA has committed to taking action on 90% of submitted ANDAs within 10 months, a remarkable achievement in view of the pre-GDUFA timelines. According to 2019 FDA metrics, the FDA is currently exceeding this goal at 97% of applications (Hahn 2019).

For its part, the industry pays application fees, program fees, and facility fees to fund the increases in staff at the FDA required to accelerate ANDA review and facility inspections. The fees change every year based on the projected FDA review and inspection workload. The fee for FY 2021 for an ANDA submission is $196 868. The program fees depend on the size of the generics company and range from $154 299 per year for small businesses to $1 542 993 for a large generics firm. Fees for facilities vary depending on the role and location; foreign locations are higher to fund the higher cost of FDA inspections (Federal Register 2020).

The GDUFA I and II programs, similar to their PDUFA counterparts for innovative drugs, have modernized the development, submission, and review standards for generic drugs. The accelerated availability of generic drugs plays an important role in containing US healthcare costs while maintaining FDA standards of quality, safety, and effectiveness.

References

Code of Federal Regulation Title 21 (2022). Section 320.1 Bioavailability and bioequivalence requirements. https://www.accessdata.fda.gov/scripts/cdrh/cfdocs/cfcfr/CFRSearch.cfm?CFRPart=320 (6 January 2022).

Code of Federal Regulations Title 21 (2022). Section 314.92 Drug Products for which abbreviated applications may be submitted. https://www.accessdata.fda.gov/scripts/cdrh/cfdocs/cfcfr/CFRSearch.cfm?CFRPart=320 (6 January 2022).

Code of Federal Regulations Title 21 (2022). Section 314.108 New drug product exclusivity. https://www.accessdata.fda.gov/scripts/cdrh/cfdocs/cfcfr/CFRSearch.cfm?CFRPart=320 (6 January 2022).

Federal Register (2020). Generic Drug User Fee Rates for Fiscal Year 2021.

Food Drug and Cosmetic Act (2018). Chapter V: Drugs and Devices, Section 505(j). https://www.fda.gov/regulatory-information/federal-food-drug-and-cosmetic-act-fdc-act/fdc-act-chapter-v-drugs-and-devices (28 March 2018).

Gingery, D. (2020). An Orange Book by any Other Name. . . Pink Sheet.

Guidance for Industry (2001). Statistical Approaches to Establishing Bioequivalence.

Guidance for Industry (2014). ANDAs: Stability Testing of Drug Substances and Products. Questions and Answers.

Guidance for Industry (2019). Determining Whether to Submit and ANDA or a 505(b)(2) Application.

Guidance for Industry (2019). ANDA Submissions – Content and Format.

Guidance for Industry (2020). Controlled Correspondence Related to Generic Drug Development.

Hahn, S.F.Y (2019). Performance Report to Congress for the Generic Drug User Fee Amendments.

Medicare Prescription Drug (2003). Improvement, and Modernization Act of 2003.

The Case for Competition (2019). Generic Drug & Biosimilars Access & Savings in the U.S. Report. Association for Accessible Medicines.

Woodcock, J. (2016). Implementation of the Generic Drug User Fee Amendments of 2012 (GDUFA) – House Testimony.

Yu, L. and Bing, V. (2014). FDA Bioequivalence Standards. Advances in the Pharmaceutical Sciences Series 13.

Chapter Self-Assessments: Check Your Knowledge

Questions:

- The development of a generic drug may take years, why is the ANDA regulatory pathway considered as abbreviated?
- Why is the Orange Book an important resource for a generic firm?
- Describe the benefits for a generic firm if their ANDA is considered "first-to-file"?
- Why did the Generic Drug User Fee Act do to modernize the generic industry?

Answers:

- The ANDA pathway is "abbreviated" as it eliminates the need for generic firms to repeat costly and time-consuming nonclinical and clinical studies.
- The Orange Book contains all the regulatory exclusivity and patent information for Branded products which are used to define the timelines and strategy for the development of new generic product entrants.
- An ANDA which is considered first-to-file is awarded 180 days of regulatory exclusivity, which allows for higher margin sales which is an important component of generic drug industry profitability.
- In exchange for dramatically reduced review timelines, the Generic Drug User Fee Act required that generic firms submit a high-quality ANDA to avoid a "refuse to receive" designation. The increased quality of ANDAs and rapid FDA review time has resulted in significantly faster introductions of generic products into the US Healthcare System.

Quiz:

1   In 2018, Generic drugs accounted for what percentage of prescriptions in the U.S.?

   A   10%
   B   25%
   C   50%
   D   90%

2   What legislation allowed for the submission of generic drugs via an abbreviated pathway?

   A   Generic Drug User Fee Act
   B   Drug Price Competition and Patent Term Restoration Act
   C   Prescription Drug User Fee Act
   D   Generics for Americans Act

3 A generic product must be the same as the branded drug, which one of the following is not a criterion used?

  A Active ingredients
  B Dosage form
  C Strength
  D Color

4 Why is a Paragraph IV certification a powerful tool for the generic drug industry?

  A It forces the generic drug firm to summarize its ANDA in 4 paragraphs.
  B It is a lawsuit against the branded drug product manufacturer.
  C It seeks to invalidate the branded drug product patents listed in the Orange Book.
  D It can accelerate the availability of generic drugs
  E c and d

5 A generic firm can seek FDA advice on development through which mechanism?

  A Requesting a preANDA meeting.
  B Through general correspondence with the FDA.
  C Using controlled correspondence
  D None of the above

24

Data Sharing and Collaboration
Jeffrey S. Barrett

Aridhia Digital Research Environment

The Pharmaceutical Industry's History of Collaboration

The roots of the pharmaceutical industry trace back to the apothecaries and pharmacies that offered traditional remedies as far back as the middle ages, offering a hit-and-miss range of treatments based on centuries of folk knowledge, as discussed in Chapter 1. The modern industry as we understand it today really has its origins in the second half of the nineteenth century. While the scientific revolution of the seventeenth century had spread ideas of rationalism and experimentation, and the industrial revolution had transformed the production of goods in the late eighteenth century, the marrying of the two concepts for the benefit of human health was a comparatively late development. An often-unrecognized complimentary occurrence was the academic-industrial collaborations happening currently as well as the intermingling of scientists in both settings prior to concerns regarding intellectual property (prior to many patent laws at that point as well). Pharmaceutical firms, first in Germany in the 1880s and more recently in the United States and England, established cooperative relationships with academic labs. The resulting exchange of research methods and findings drove a focus on dyes, immune antibodies, and other physiologically active agents that would react with disease-causing organisms. Postulated by Paul Ehrlich in 1906 following more than a decade of research, the concept that synthetic chemicals could selectively kill or immobilize parasites, bacteria, and other invasive disease-causing microbes would eventually drive a massive industrial research program that continues to the present. Table 24.1 describes several early examples of collaboration between the early pharmaceutical industry and academic and government stakeholders.

Fundamentals of Drug Development, First Edition. Edited by Jeffrey S. Barrett.
© 2022 John Wiley & Sons, Inc. Published 2022 by John Wiley & Sons, Inc.
Companion website: www.wiley.com/go/Barrett/FundamentalsDrugDevelopment

Table 24.1 Historical examples of early collaboration within the pharmaceutical industry.

Dates/era	Collaboration partners	Purpose
Early 1900s	Frederick Banting and colleagues at the University of collaborate with the scientists at Eli Lilly to purify the insulin extract.	Treatment of diabetes
1920–1940s	Government-supported international collaboration with the industry including Merck, Pfizer, and Squibb for mass producing penicillin during World War Two	Treatment for infections including pneumonia, gonorrhea, and rheumatic fever during wartime.
1930–1950s	Nobel Prize to Philip Hench (Mayo Clinic), Tadeus Reichstein (Basel University), and Edward Kendall (Mayo Clinic) for clinical application of corticosteroids; largescale synthesis of cortisone for use in clinical trials was developed in collaboration with a team at Merck.	Treatment of Rheumatoid arthritis
1950s	Selman Waksman (Rutgers University) won the Nobel Prize in 1952 for the discovery of streptomycin then persuaded George Merck to establish a production plant that provided the streptomycin used by Sir Geoffrey Marshall (Medical Research Council, United Kingdom).	Large scale production of streptomycin and first randomized clinical trial.

More recently, several pharmaceutical companies have taken the additional step of establishing research institutions with affiliations to academic centers of excellence in hopes of blending the academic culture of innovation with expertise rarely found outside of big pharma (Frearson and Wyatt 2010; Silber 2010). Prominent examples include the Genomics Institute of the Novartis Research Foundation (GNF) that is geographically located near strong academic sites including the Scripps Research Institute (TSRI), UCSD, and the Salk Institute for Biological Studies (Su et al. 2004). GNF was originally founded and directed by TSRI Professor Peter Schultz. Another example of the strong ties between highly innovative academic researchers and major pharmaceutical enterprises is the California Institute for Biomedical Research (Calibr) which was launched in 2012 via a partnership between Merck and Professor Schultz (Thayer 2017).

Direct partnerships between pharma/biotech's and academic centers are also becoming more common. Recently, Boehringer Ingelheim established a wide-ranging collaboration with the Harvard Medical School's ICCB-Longwood Screening Facility to initiate multiple RNAi screening programs surrounding research questions of mutual interest to both organizations. Leo Pharma has partnered with Professor Phil Baran (TSRI) to leverage that lab's expertise in natural product synthesis. Genentech and the UCSF School of Pharmacy have established research partnerships based upon mutual interest and complementary expertise within the general field of neurodegenerative disease. Indeed, collaborations between pharma and major research institutions are growing in popularity. High-profile pharma/academia collaborations include efforts at TSRI (Takeda, Merck, Pfizer, Janssen), Harvard (Ipsen, Pfizer, Roche, Sanofi), UCSF (GE Healthcare, Pfizer, Sanofi, Bayer), MIT (Novartis, Sanofi, Pfizer, Merck), The Broad Institute of Harvard and MIT (AstraZeneca, Roche, Novartis) and Vanderbilt (GlaxoSmithKline, Janssen, Bristol-Myers Squibb, AstraZeneca). A recently started syndicate of academics engaged in drug discovery [the Academic Drug Discovery Consortium (ADDC)] maintains up-to-date cataloging of major pharma/academia collaborations (http://www.addconsortium.org).

The goals of this chapter are to examine the collaboration landscape across the pharmaceutical industry examining the nature of these collaborations and the extent to which they are successful while also exploring current trends in the industry and healthcare in general. We will drill down specifically on data sharing as a key component of most collaboration efforts and assess current approaches as well as limitations and obstacles to sharing. Finally, we will explore the underlying technology around data sharing and project the future for both data sharing and collaboration as components of the industry's sustainability for the future.

Current Collaboration Landscape: Who, How and Why?

Its obvious from the historical examples (see Table 24.1) that the collaboration landscape for pharmaceutical research and development is diverse and dynamic. Depending on the phase of development collaboration partners are likely to change. Likewise, many of these relationships extend beyond a single compound in development and support a varied array of R&D and commercial platforms.

Many predict that in the future, no pharmaceutical company will be able to "profit alone." It will, rather, have to "profit together" by joining forces with a wide range of organizations, from academic institutions, hospitals, and technology providers to companies offering compliance programs, nutritional advice, stress management, physiotherapy, exercise facilities, health screening, and other such services. Figure 24.1 illustrates the likely impact of emerging trends on the pharmaceutical industry practices, particularly how it relates to future practices among the various stakeholders including the healthcare industries and technology partners. As the figure indicates many of the emerging trends are related to the sharing of outcome data and the future collaboration between the pharmaceutical and healthcare industries on assessing "value." As healthcare in general

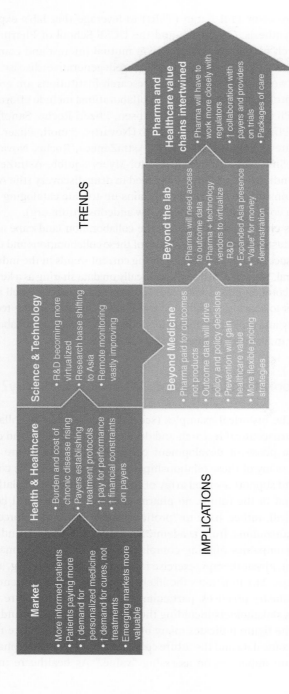

Figure 24.1 Emerging trends and their impact on collaboration in the pharmaceutical industry.

moves toward a more value-based economy, the pharmaceutical industry will likewise have to follow suite, which will no doubt affect their ability to recoup R&D investment as well as influence pricing strategies.

Data Sharing Considerations: Motivations and Incentives

The predominant benefit of data sharing and collaboration is accelerated scientific progress. Advances are clearly valuable to both the pharmaceutical industry and academic researchers, especially when translated into improved patient outcomes, reduced research costs, and decreased time in moving discoveries from the bench to the bedside. Despite the anticipated benefits, sharing research data must still be viewed as a work in progress [Barrett 2020]. There are a few obvious take-home messages that continue to resonate with the current state of affairs. Within the pharmaceutical industry, most real sharing is still occurring in the pre-competitive space or targeted in populations where the financial gains are modest or non-existent (e.g. pediatric oncology, rare diseases, and global health settings). Academic–industry collaborations are broad but difficult based on intellectual property (IP) considerations and other incentivization issues. Meaningful collaboration still requires mutual understanding, and sharing is still problematic for a variety of reasons.

Since 2014 the pharmaceutical industry has endorsed a commitment to share de-identified individual patient data upon request (PhRMA 2013). Two separate studies have confirmed that the extent to which this occurs within a reasonable time frame (two years) is 15% or less (Murugiah et al. 2016; Hopkins et al. 2018). Issues identified were highlighted by the lack of data sharing policies/processes and data sharing policy conditions that exclude access based on ongoing follow-up and regulatory activity. For the industry to sustain itself and embrace the innovation and collaboration necessary to thrive in a value-based healthcare system, it will have to learn to share in a manner it is unaccustomed to and addresses any IT and legal barriers in addition to adopting the requisite internal policies. Likewise, academic investigators will need to cope with internal IP concerns and embrace potential stewardship in an open manner. This will require a more transparent conversation with all relevant stakeholders where benefit: risk to sharing is objectively calibrated.

Technology

Since the mid-1980s, efforts have been made to institutionalize quantitative and qualitative data-sharing as a normative practice on the grounds of its scientific, financial, public policy, and pedagogical benefits. Obstacles have hampered widespread uptake of data-sharing, including technical, ethical, and cultural challenges, particularly around risks for research participants and opportunity costs for researchers. By "data sharing" we typically refer to the collection of practices, technologies, cultural elements, and legal frameworks that are relevant to transactions in any kind of information digitally, between different kinds of organizations. The mechanism to share data is complex, as the definition suggests, but strides in technology have indeed enabled the process of sharing data to happen in a much more efficient and secure manner than in the past. Still today, most of the data of interest are siloed and fragmented in different healthcare systems or public and private databases. This practice prevents the optimal usability of these sources and is against the desire for more intelligent healthcare inspired by big data. Security and privacy concerns and the lack of ensured authenticity trails of data bring even more obstacles to data sharing. Many privacy concerns such as HIPAA (Health Insurance Portability and Accountability Act) and GDPR (General Data Protection Regulation) have been codified and are legitimately focused on protecting the rights of both patients and volunteers and patients that participate in clinical trials. Any sharing strategy must likewise be able to meet these requirements while addressing functional requirements.

Requirements for data sharing are more commonly imposed by institutions, funding agencies, and publication venues in the medical and biological sciences than in the physical sciences. Requirements vary widely regarding whether data must be shared at all, with whom the data must be shared, and who must bear the expense of data sharing. Funding agencies such as the NIH and NSF tend to require greater sharing of data, but even these requirements tend to acknowledge the concerns of patient confidentiality, costs incurred in sharing data, and the legitimacy of the request. Private interests and public agencies with national security interests (defense and law enforcement) often discourage sharing of data and methods through non-disclosure agreements.

A variety of models for clinical trial data sharing have been proposed, planned, or implemented. The types of data that are shared differ across the models. Proposed models of data sharing have generally imposed some sort of restriction on the sharing of data that could directly or potentially identify trial participants, as well as data that reveal confidential information or trade secrets or might result in inaccurate analysis. Access to clinical trial data in current models ranges from essentially full access to de-identified data to fully restricted or no access.

In an open or public access model, data are made available, at a defined time, to any party who seeks them, for any purpose. For example, the EMA has announced that it will release, to any data requester that is a known entity to the agency, both summary and participant-level data (excluding, for example, personally identifiable data and information the EMA deems to be confidential) immediately after a regulatory decision about a new drug (Eichler et al. 2013; EMA 2013). In some models of data sharing, access is restricted to specific classes of user or for specific purposes. Requestors might need to demonstrate that they meet specified eligibility criteria. Some models require only the name and contact information of the requestor, while others require information about the proposed use of the requested data or how the data will be analyzed. Some models might also impose conditions relating to whether the data generators would receive credit in publications. In some cases, the actual data are not provided to the requestor. Instead, data holders might run specific data analyses for approved requestors and deliver to the requestors only the results of the requested analyses. In another model, recipients receive credentials to access and run queries on the data but are not able to download or obtain copies of the data. Data sharing can also take place indirectly, through a "trusted intermediary" or "honest broker," who either negotiates the conditions for data sharing (with the data provider retaining control over the data and its release) or takes full control of the data and brokers both the conditions for data release and the delivery to recipients.

From a technology standpoint, in addition to capacity, a data-sharing infrastructure needs to be capable of managing data access according to the strategies defined in a data-sharing agreement. As big data approaches become more widespread, newer technological solutions to data access may offer effective ways of achieving the benefits of sharing clinical trial data while mitigating its risks. These newer solutions are predicated on an approach to data query that differs from the traditional one with which most clinical trialists are familiar. In the traditional approach, data are brought to the query. That is, if a data requester wants to run a query, the requester obtains a copy of the data, installs the data on his/her own computer, and runs the query on the downloaded data. Because the data requester now holds a copy of the data, the original data holder has effectively lost control over access to the data.

Common data models include localized data stores (every data holder hosts its own data on its own server), one single centralized data store (all data are collected onto one central database), and a federated query model (databases are federated when independent geographically dispersed databases are networked in such a way that they can respond to queries as if all the data were in a single virtual database) (IOM Report 2015). Just because data are accessible does not mean they are usable. Data are usable only if an investigator can search and retrieve them, can make sense of them, and can analyze them within a single trial or combine them across multiple trials. Given the large volume of data anticipated from the sharing of clinical trial data, the data must be in a computable form amenable to automated methods of search, analysis, and visualization.

An exciting possible solution to data privacy and sharing concerns around patient-level data is the potential to use synthetic data for research purposes. Synthetic data can be broadly defined as any data applicable to a given situation that is not obtained by direct measurement. Data generated by a computer simulation is easily appreciated as synthetic data, but this would also include most applications of physical modeling, such as music synthesizers or flight simulators as well so it would be applicable to many industries.

In the medical and pharmaceutical context, specifically the ability to avoid concerns of privacy protection, the creation of synthetic data must be appreciated as an involved process of data anonymization – synthetic data is a subset of anonymized data (Mac Hanavajjhala et al. 2008). Synthetic data is used in a variety of fields as a filter for information that would otherwise compromise the confidentiality of aspects of the data. Privacy concerns that often prohibit the use of patient-level data such as HIPAA and GDPR then are rightly sensitive to the access of human information (i.e. name, home address, IP address, telephone number, social security number, credit card number, etc.) may be obviated with the use of synthetic data. While still an evolving field and approach, technology vendors continue to explore the market for this approach, and there is great hope that it will offer an acceptable solution assuming it can sufficiently recapitulate the populations of interest in sufficient detail. Likewise, the regulatory acceptance of such approaches will require additional qualification and validation, assuming this is perceived as an acceptable solution for real-world data/real-world evidence or placebo/control groups.

Collaboration Examples

Conducting the science and business of drug discovery and development through the difficulties funding deficits and regulatory challenges is now the unfortunate normal for industrial and academic/non-profit/government research labs. These challenges coincide at a time of extraordinary opportunities (Thomas and McKew 2014). The conventional and often convoluted road toward clinical approval is giving way to a more open approach where the complimenting strengths of pharma/biotechs and academic/government labs are being leveraged to hasten the translation of new discoveries into new drugs. Table 24.2 highlights some of the recent examples of diverse collaborations focused on various aspects of drug development. While many of them are focused on clinical trial activities, quite a few are also reliant on data and/or sample sharing and inform early stages of development as well.

Table 24.2 Current examples of data sharing and collaboration across diverse stakeholders involved in various aspects of pharmaceutical research and development.

Collaboration	Stakeholders	Goals
Children's Oncology Group (COG), https://www.childrensoncologygroup.org	PhRMA, NIH/NCI, experts in childhood cancer throughout the United States, Canada, and a number of international sites	• To understand the causes of cancer and find more effective treatments for the children • Improving the outcome for all children with cancer • More than 90% of children with cancer cared for at COG sites
FNIH Biomarker consortium, https://fnih.org/what-we-do/biomarkers-consortium	NIH, public and private institutions and partners, academic investigators	• Novel tissue imaging platforms to characterize tumor heterogeneity, tissue spatial connections/spatial heterogeneity. • A blood-based or remote sensing technology that can be used to replace solid tumor biopsies and to understand cancer of unknown origin or tumor heterogeneity to enable meaningful clinical intervention opportunities
National Cancer Institute's I-SPY 2 adaptive Phase 2 trial platform in adults with cancer, https://www.ispytrials.org/backgrounders/executive-summary-the-i-spy-2-trial	NIH/NCI and ten cancer centers across the US and FNIH Biomarker consortium	• The main objective of the I-SPY 1 TRIAL was to identify indicators of response to neoadjuvant chemotherapy that predict survival in women with high-risk (Stage II-III) breast cancer. • I-SPY 2 trial identifies women at the highest risk and introduces the most promising drugs in development that are individually targeted to the characteristics of each woman's tumor. The purpose of the study is to further advance our ability to practice personalized medicine.
RESPIRI-TB, part of the IMI AMR Accelerator Programme, https://www.imi.europa.eu/projects-results/project-factsheets/respiritb	EFPIA companies: Janssen; Research organizations, public bodies, non-profit groups; Small and medium-sized enterprises and mid-sized companies: Fund BV and Mitologics,	• To advance the development of new drug candidates that could be part of a new, more effective, shorter regimen to treat MDR-TB.
COVID-19 Prevention Trials Network (COVPN)	Four existing NIAID-funded clinical trials networks: the HIV Vaccine Trials Network (HVTN), based in Seattle; the HIV Prevention Trials Network (HPTN), based in Durham, NC; the Infectious Diseases Clinical Research Consortium (IDCRC), based in Atlanta; and the AIDS Clinical Trials Group	• To use real-world data providing actionable information about the prevalence of SARS-CoV-2 in specific populations highlighting individual risk factors for patients, helping to improve understanding of the disease, tailor public health interventions and strategies to mitigate risks for individuals and communities, and help stop the spread of SARS-CoV-2

The Case Against Data Sharing

Despite the anticipated benefits, sharing research data has yet to be widely adopted, as mentioned previously. When data is shared post-approval, it is often administrated with an honest-broker approach. Two of the more common examples include the Yale Open Data Access (YODA) project (YODA 2013), https://yoda.yale.edu, and the Duke Clinical Research Institute (DCRI, https://dcri.org/our-work/analytics-and-data-science/data-sharing) (IOM 2015) who serve in such a capacity for J&J and BMS respectively. While these represent a step in the right direction, sharing still based on low-risk, post-approval data only. Academic–industry collaborations are broad but difficult based on IP considerations and other incentivization issues. Meaningful collaboration still requires "skin in the game," and sharing is still difficult.

Some positive examples exist (list), but the sustainability of these is seemingly always in question. This is not an issue of technology (Boyd et al. 2009). Some of the bottlenecks include the value/overvalue of IP, a lack of resources and expertise, the lack of sharing history or culture of sharing, and the lack of trust for governance around sharing. Some potential solutions have been proposed, but these have mostly been implemented for academic collaborations (Boyd et al. 2007; Jarquín 2012), and industry has been slow to adopt more open data sharing within their organizations, preferring a more traditional Biostat/data management governance model enforced by SOPs focused on protecting clinical data.

Data Sharing for the Future

Recommendations for improving sharing within the context of drug development include necessary improvements to the manner and mechanism of internal Pharma sharing solutions. Specifically, these environments need to be more in line with a federated governance model (governance balanced between a central authority and constituent units) and based on IT solutions that permit more flexible sharing rules accommodating complex sharing with diverse internal partners, improved and broader data-sharing agreements reflecting the IP considerations of diverse external stakeholders and dynamic and sharing considerations that can change over time (e.g. new partners, change in partner relationships or revised agreements). The generation of an honest broker approach best practices and commercial solutions that reflect diverse stakeholders and accommodate global data privacy concerns would also help. Synthetic data may indeed provide an alternative to data sharing, but hopefully, this will represent a subset of an overall approach and not the full solution.

References

Barrett, J.S. (2020). Perspective on data-sharing requirements for the necessary evolution of drug development. *Journal of Clinical Pharmacology* 60 (6): 688–690. https://doi.org/10.1002/jcph.1607.

Boyd, A.D., Hosner, C., Hunscher, D.A. et al. (2007). An 'Honest Broker' mechanism to maintain privacy for patient care and academic medical research. *International Journal of Medical Informatics* 76 (5–6): 407–411.

Boyd, A.D., Saxman, P.R., Hunscher, D.A. et al. (2009). The University of Michigan Honest Broker: a web-based service for clinical and translational research and practice. *Journal of the American Medical Informatics Association* 16 (6): 784–791. https://doi.org/10.1197/jamia.M2985.

Eichler, H.G., Pétavy, F., Pignatti, F., and Rasi, G.G. (2013). Access to patient-level trial data: A boon to drug developers. *New England Journal of Medicine* 369 (17): 1577–1579.

EMA (European Medicines Agency) (2013). Publication and access to clinical-trial data. http://www.ema.europa.eu/docs/en_GB/document_library/Other/2013/06/WC500144730.pdf (accessed 1 December 2013).

Frearson, J. and Wyatt, P. (2010). Drug discovery in academia: the third way? *Expert Opinion Drug Discovery* 5 (10): 909–919. https://doi.org/10.1517/17460441.2010.506508. PMID: 20922062; PMCID: PMC2948567.

Hopkins, A.M., Rowland, A., and Sorich, M.J. (2018). Data sharing from pharmaceutical industry sponsored clinical studies: audit of data availability. *BMC Medicine* 16: 165.

Institute of Medicine. Board on Health Sciences Policy, Committee on Strategies for Responsible Sharing of Clinical Trial Data, Sharing Clinical Trial Data: Maximizing Benefits, Minimizing Risk. National Academies Press, 2015, ISBN: 0309316324, 9780309316323, pgs 153–155.

Jarquín, P.B. (2012). *Data Sharing: Creating Agreements In Support of Community-Academic Partnerships*. Colorado Clinical and Translational Sciences Institute & Rocky Mountain Prevention Research Center http://trailhead.institute/wp-content/uploads/2017/04/tips_for_creating_data_sharing_agreements_for_partnerships.pdf.

Mac Hanavajjhala, A., Kifer, D., Abowd, J. et al. (2008). Privacy: Theory meets Practice on the Map. 2008 IEEE 24th International Conference on Data Engineering. pp. 277–286.

Murugiah, K., Ritchie, J.D., Desai, N.R. et al. (2016). Availability of clinical trial data from industry-sponsored cardiovascular trials. *Journal of the American Heart Association* 5 (4): e003307.

Principles for responsible clinical trial data sharing. 2013. http://phrma-docs.phrma.org/sites/default/files/pdf/PhRMAPrinciplesForResponsibleClinicalTrialDataSharing.pdf (accessed February 2020).

Silber, B.M. (2010). Driving drug discovery: the fundamental role of academic labs. *Science Translational Medicine* 2: 30cm16.

Su, A.I., Wiltshire, T., Batalov, S. et al. (2004). A gene atlas of the mouse and human protein-encoding transcriptomes. *Proceedings of the National Academy of Sciences of the United States of America* 101 (16): 6062–6067. PMID: 15075390; PMC: PMC395923.

Thayer, A. (2017). HitGen works with Scripps and Calibr. *C&EN* 95 (18): 17.

Thomas, C.J. and McKew, J.C. (2014). Playing well with others! Initiating and sustaining successful collaborations between industry, academia and government. *Current Topics in Medicinal Chemistry* 14 (3): 291–293. https://doi.org/10.2174/1568026613666131127125351. PMID: 24283974; PMCID: PMC4337773.

Yale Open Data Access Report (2013). https://yoda.yale.edu.

Chapter Self-Assessments: Check Your Knowledge

Questions:
- Why is data sharing viewed as an essential component to the future of the pharmaceutical industry?
- Describe the current state of data sharing in the pharmaceutical industry?
- Describe data privacy concerns and how they impact data sharing approaches?
- What is synthetic data and how can it be part of the data sharing solution?

Answers:
- Many of the emerging trends in the pharmaceutical industry are related to the sharing of outcome data and the future collaboration between the pharmaceutical and healthcare industries on assessing "value." As healthcare in general moves toward a more value-based economy, the pharmaceutical industry will likewise have to follow suit, which will no doubt effect their ability to recoup R&D investment as well as influence pricing strategies.
- Within the pharmaceutical industry, most of real sharing is still occurring in the pre-competitive space or targeted in populations where the financial gains are modest or non-existent (e.g. pediatric oncology, rare diseases, and global health settings). Academic–industry collaborations are broad but difficult based on intellectual property (IP) considerations and other incentivization issues. Meaningful collaboration still requires mutual understanding, and sharing is still problematic for a variety of reasons.
- Many privacy concerns such as HIPAA (Health Insurance Portability and Accountability Act) and GDPR (General Data Protection Regulation) have been codified and are legitimately focused on protecting the rights of both patients and volunteers and patients that participate in clinical trials. Any sharing

strategy must likewise be able to meet these legal requirements while addressing functional requirements for any data sharing solution.

- Synthetic data is a subset of anonymized data; it is used in a variety of fields as a filter for information that would otherwise compromise the confidentiality of particular aspects of the data. Privacy concerns that often prohibit the use of patient-level data such as HIPAA and GDPR then are rightly sensitive to the access of human information (i.e. name, home address, IP address, telephone number, social security number, credit card number, etc.) may be obviated with the use of synthetic data. While still an evolving field and approach, technology vendors continue to explore the market for this approach, and there is great hope that it will offer an acceptable solution assuming it can sufficiently recapitulate the populations of interest in sufficient detail.

Quiz:

1   True or false. Direct partnerships between pharma/biotech's and academic centers are rare and becoming less likely based on intellectual property disputes.

2   Common data models include all but which of the following. Choose the best answer:
   A localized data stores (every data holder hosts its own data on its own server)
   B a single centralized data store (all data are collected onto one central database)
   C a federated query model (databases are federated when independent geographically dispersed databases are networked in such a way that they can respond to queries as if all the data were in a single virtual database)
   D All are correct

3   Obstacles have hampered widespread uptake of data-sharing, including _____, _____ and _____ challenges particularly around risks for research participants and opportunity costs for researchers. Choose the best answer:
   A technical, ethical, and cultural
   B religious, ethical, and cultural
   C personal, ethical, and technical
   D technical, cultural, and commercial
   E None of the above are correct

4   True or false. Just because data are accessible does not mean they are usable.

25

The Future of the Pharmaceutical Industry

Jeffrey S. Barrett

Aridhia Digital Research Environment

How We Got Here and Why We Cannot Sustain It

In the future of healthcare and specifically the pharmaceutical industry, the era of blockbuster drugs that treat large populations will likely wane. Many would say that this is already our reality. Instead, the pharmaceutical industry is on the fringe of an era where tailored therapies are developed to cure or prevent disease rather than treat symptoms (Batra et al. 2019). Twenty years from now, rather than picking up a prescription at the pharmacy, personalized therapies based on a diverse set of a patient's characteristics including their genomics, metabolome, microbiome, and other clinical information, might be manufactured or compounded just in time through additive manufacturing (the industrial production name for 3D printing, a computer-controlled process that creates three-dimensional objects by depositing materials, usually in layers). For many years, pharmaceutical companies decided what their products were worth and priced them accordingly. Today healthcare policymakers, payers, and patient groups are now playing an increasingly important role in the valuation process – and this trend will accelerate as healthcare expenditures everywhere continue to increase. The aging of the population, together with dietary changes and more sedentary lifestyles, is driving up the disease burden in both developed and developing countries. People's expectations are also rising as new therapies for treating serious illnesses like cancer reach the market. Global healthcare costs have risen commensurately; between 2000 and 2006, expenditure on healthcare as a percentage of gross domestic product (GDP) climbed in every country in the OECD (Convention on the Organization for Economic Co-operation and Development). Healthcare and R&D cost aside, and there is also a fundamental concern with the high attrition rates in later stages of development (Hutchinson and Kirk 2011;

Fundamentals of Drug Development, First Edition. Edited by Jeffrey S. Barrett.
© 2022 John Wiley & Sons, Inc. Published 2022 by John Wiley & Sons, Inc.
Companion website: www.wiley.com/go/Barrett/FundamentalsDrugDevelopment

Takebe et al. 2018) which give pharmaceutical and regulatory leaders pause with respect to the current paradigms (Zurdo 2013).

These are challenging times for pharma companies as economic, supply chain, and other forms of uncertainties abound. Even as many of these companies are focusing on therapies and vaccines for the recent pandemic (SARS-Cov-2 of 2020), leaders should think strategically about their investments – in terms of therapeutic area, digital technologies, and talent – in order to thrive in the future. Some companies have concluded that the pandemic has forced them to provide immediate attention to existing priorities (e.g. R&D, digital transformation, cyber). Despite this urgency, priorities should be selective and strategic; companies should prepare for risk but not let it hold them back.

The goal of this chapter is to review current trends in the pharmaceutical industry, particularly highlighting those that represent shifts in the industry regarding the manner in which they conduct research and development related to changes in healthcare economics. Both a historical context for these changes as well as likely changes in the industry in the short and long term are projected.

Relevant Trends

Based on emerging technology, we can be reasonably certain that digital transformation – enabled by radically interoperable data, AI, and open, secure platforms – will drive much of this change. It should be broadly appreciated that this digital transformation is broad-based with respect to the pharmaceutical industry spanning across the value chain, R&D, distribution, as well as for marketing and sales. Unlike today, many believe care will be organized around the consumer rather than around the institutions that drive our existing healthcare system (Milne and Kaitin 2010). Implicit in the expected digital transformation is the investment in data systems and services that support rapid and secure access to data and information.

Many consulting firms have conducted interviews of various stakeholders (e.g. Deloitte Center for Health Solutions interviewed a diverse group of thought leaders including futurists, venture capitalists, digital health leaders, and academics) to assess the current situation and gauge future directions. Five forces emerged that could alter the course of the biopharmaceutical sector. These forces represent both opportunities and threats to incumbents. They include prevention and early detection, custom treatments and personalized medicine, curative therapies, digital therapeutics, and precision intervention. In another assessment, more diverse respondents believe customized treatments, nonpharmacological intervention, and prevention and early detection will have the greatest effect in the life sciences industry in the next 10 years. While most are prepared to address nonpharmacological intervention and prevention and early detection, many respondents feel we are not prepared to tackle customized treatments (e.g. precision medicine approaches where each patient could receive earlier diagnoses, risk assessments, and optimal treatments) (Vogenberg et al. 2010).

The Case for Change

After more than a decade of cost-cutting, restructuring, transformations, and turnarounds, the pharmaceutical industry would appear to be suffering from change fatigue. Change has been at the forefront of pharmaceutical executives' minds for a long time, and pressures have been rising steadily. Since 2007, even before the economic crisis of 2008 and 2009 began, companies had been announcing cost-reduction measures, largely driven by the patent cliff. Following the financial crisis, a massive acceleration of the challenges in the external environment also occurred. Some had and continue to implement extensive layoffs and site closures. Many are undertaking transformations to reinvent their commercial model, restructure their R&D, streamline their manufacturing footprint, or all three. Among those doing so, more than half say their main goal is to reduce costs or improve productivity. Yet despite these efforts, the industry in aggregate has seen little improvement.

Many feel that the pharmaceutical industry is facing a productivity crisis. Despite extraordinary scientific achievements including completing the sequencing of the human genome, the rate at which the industry generates new products would seem to be shrinking. In 2002, the US Food and Drug Administration (FDA) approved only 17 new molecular entities (NMEs) for sale in the United States – a disappointing fraction of the 15-year high of 56 NMEs approved in 1996 and the lowest since 1983. For context, however, this decline occurred despite a doubling of research and development (R&D) spending by US-based pharmaceutical companies between 1995 and 2002. The same pattern is apparent in worldwide evaluation, where the annual number of new active substances approved in major markets fell by 50 percent during the 1990s while private-sector pharmaceutical R&D spending tripled. This situation prompted many headlines about the "dry," "weak," or "strangled" pipelines of many companies and claims that "Big Pharma's business model is bankrupt."

As there is an opportunity cost attached to R&D resources and investment, "bang for the buck" is a serious concern. Pharmaceutical R&D paid off well in previous decades, with statistical studies showing a historical correlation between the number of new drugs introduced and declines in mortality and other health indicators across a wide range of diseases. Nonetheless, progress has been disappointing in some areas: No new broad-spectrum antibiotics have been marketed in almost forty years, and many forms of cancer, as well as chronic diseases such as diabetes, Alzheimer's, Parkinson's, and schizophrenia, still lack effective well-tolerated treatments. Continuing growth in R&D spending represents an investment in overcoming these challenges, but this upward trajectory will be sustainable only if it can be paid for. As increased spending collides with intensifying pressure to contain healthcare costs, the factors driving the efficiency of the drug discovery and development process are being brought into sharp focus.

One Possible Future

In the pharmaceutical industry, innovation is often best measured by the number of new chemical entity (NCE) drugs approved by the FDA. Current efforts to encourage greater innovation have not succeeded to any great extent. Many are concerned that strategies have failed to address the issue of price and sustainability. One area that offers much promise is the promotion of public–private partnerships aimed at enabling the academic sector to generate more innovative high-risk ideas and then also do much of the work to "derisk" them (Wolinsky 2017). In this context, when the industry finally does engage in the process, they don't have to take on as much risk, they have a good idea about the patient population, they know the biomarker, they know that may be a prototype drug is already available and showing promise and can develop a streamline development plan with a high probability of technical success. The flaw in the approach, however, is that even when the lion's share of the drug development has already been done, the project can often seem to end up with a conventional, large Phase 3 trial model, and pay-back to the pharmaceutical companies is still based on the maximum the market will bear (see Figure 25.1 for the conceptualized "Valley of Death" scenario that is often used to describe the drug development cycle). A key component for future success would be to ensure new drugs can make it to market at a price that is more sustainable and better reflects the extent of a future public/philanthropic investment in their discovery and development. While many would like to see more drug development done in an academic setting, which, in theory would be more open to taking risks in search of high payoffs, and better at conducting "small, smart trials," cutting costs and development time, there are few current academic centers ready to take on this mantle. In addition to such smart trials, alternatives to large Phase 3 randomized control trials (RCT) would help as well. One possible solution could be more expanded use of real-world data (RWD) and real-world evidence (RWE) beyond the complimentary support for RCTs or safety surveillance.

One solution to the affordability problem could be offered by the increasing investment in drug development that is being financed through philanthropic

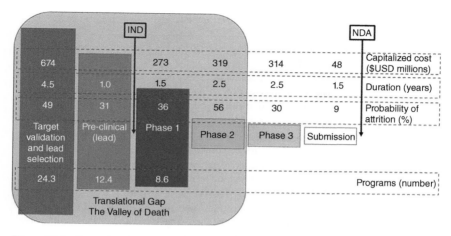

Figure 25.1 Drug-development cycle and the "valley of death" adapted from (Zurdo 2013).

foundations and charities. Philanthropy to support research or other noble causes had its roots in the early twentieth century when steel tycoon Andrew Carnegie created the Carnegie Foundation to "promote the advancement and diffusion of knowledge and understanding." But the idea of venture philanthropy – non-profits investing in for-profit companies for social good – started only in the late 1990s to support education and housing and was more recently taken up by disease-related charities. Over this time, many foundations have been formed to support certain disease populations, some of which affect a great number of patients (e.g. Michael J Fox Foundation for Parkinson's Disease patients) and others that support small numbers of patients that would fall in the category of rare diseases (e.g. Myasthenia Gravis Rare Disease Network (MGNet), Spinal Muscular Atrophy (SMA) foundation). Some of these efforts are tracked and coordinated through National Organization for Rare Disorders, Inc. (NORD), but many manage their philanthropic activities entirely on their own. Many of these have created small research infrastructures to solicit funding from members and direct this funding to areas of research focused on advancing cures either through academic research or direct funding to the private sector.

Between basic discovery research and late-stage development lies the critical step of proving the utility of a proposed drug. The funding gap that often occurs in this period has been referred to as the "valley of death." The risks are great and may be considered as not worth taking for products designed to treat rare and neglected diseases, which may ultimately yield a very limited return on investment.

To help fill this funding gap, U.S.-based foundations have increased their investments in discovery and development for new drugs specific to their

Table 25.1 Mechanisms to facilitate drug development.

Initiative category	Mechanism topic
Academic	• The academic research community needs to increase investments in technology that can improve target validation and drug safety.
Government	• Government research funding aimed at addressing health challenges needs to be more focused on forecast morbidity and the cost of care in the United States.
	• FDA needs to be adequately funded so it can partner with drug developers and direct the research being performed toward answering important regulatory questions.
Private sector	• Small Business Innovation Research and Small Business Technology Transfer regulations revisited and revised to allow for greater investment.
	• New incentives for high-risk investors can be created, perhaps through tax law.
	• Private disease foundations' provision of support to the academic community for discovery and to industry for development is beneficial and should be embraced.
	• Experienced investors need to be brought into the innovation process earlier.
	• The pharmaceutical industry and academia can work together to build a stronger US industry.

diseases of interest. In 2007, such groups invested approximately $75 million in biopharmaceutical companies, a 10-fold increase since 2000 (Gambrill 2007). In a recent IOM report (IOM 2009), Caskey put forth several suggestions for overcoming the impediments to new drug discovery and development. These have been summarized below in Table 25.1.

Another hopefully future revelation is the necessity of legal reform/evolution necessary to reinvigorate innovation in pharmaceutical research and development. As Keith Sawyer (Sawyer 2007) writes in his recent text on Group Genius, "To release the innovation potential of society we need to modify seven aspects of our legal system to create a closer match to the natural behavior of the collaborative web." While his thesis is focused on internet-based technologies and industries, his suggestions for future innovation are broadly applicable across industries.

1) Reduce copyright terms
2) Reward small sparks
3) Legalize modding
4) Free the employees

5) Mandatory licensing
6) Pool patents
7) Encourage industry-wide standards

The importance of patent protection obviously increases with the size of the investment needed to achieve innovation in any field. If innovation can be purchased cheaply, patent protection is relatively unimportant. This was premise behind much of the "innovation through acquisition" phase of the 1980s and 1990s in the pharmaceutical industry. Regarding patents, as the investment cost escalates, however, patent protection becomes far more important. In the pharmaceutical industry, where the cost of an NCE drug has escalated so dramatically (Wouters et al. 2020), the assurance of strong patent protection has become increasingly crucial to the future of the industry. For pharmaceutical products, therefore, the seventeen-year patent term has become a legislative figment. In reality, a drug patent has a much shorter effective life. As a result, incentives to invest in pharmaceutical R&D have been substantially reduced. The erosion of effective patent life for pharmaceuticals began about 20 years ago. It coincides with the erosion in pharmaceutical innovation, as measured by the yearly FDA approval of NCE drugs. While reduced pharmaceutical innovation is often viewed as the result of this effective loss of patent life, an alternative perspective is that this should incentivize pharmaceutical sponsors to plan better and to consider pooling patents with collaborators when possible (item 6 earlier).

Modding in general terms refers to the act of modifying anything, be it hardware, software, or anything else to perform a function not originally conceived or intended by the designer or achieve a pre-defined new specification. In the pharmaceutical context, modding similarly refers to building upon other patented products to create a modified product or service to achieve a different or new outcome (e.g. indication, advantage, etc.). Obviously, modding may sometimes infringe the legal rights of the copyright owner. Some nations have laws prohibiting modding and accuse modders of attempting to overcome copy protection schemes. In the United States, the Digital Millennium Copyright Act (DMCA) has set up stiff penalties for mods that violate the rights of intellectual property owners, but much of this is focused on software pirating. The issue of pharmaceuticals is more complex. Drugmakers contend that profits fund the research that produces breakthrough treatments and that expiring patents hamper their ability to develop new drugs. Longer-lasting patents, they say, would protect the profits that they need to keep innovative products moving through the pipeline. Critics question that assumption and contend that there is no proof of a link between patent life and innovation. In their view, drug companies focus on developing the most marketable drugs instead of the most urgently needed medications. Extending patents would serve mainly to boost drug companies' profits, not to encourage the innovation needed to address the world's unmet medical needs. Likewise, mods being generally prohibited serves to limit innovation by reducing the creative input to come solely from the original sponsor.

On a positive note, the scientific foundation on which pharma rests is improving exponentially, thanks to massive increases in processing power, advances in genetics and genomics, and new data management tools. For the last half-century, computers have been doubling in performance and capacity every 18 months. This revolution has transformed biomedical research. In 2001, it cost US$95 million to read an entire human genome. Today, two leading manufacturers are developing machines that can do so for as little as $1000 – in a matter of hours. Inexpensive gene sequencing will let doctors diagnose and treat patients based on information about their individual genomes. And, by 2020, genetic testing will be part of mainstream medical practice in some countries. Technological developments have also paved the way for electronic medical record (EMR) systems that capture vast quantities of outcomes data. Numerous healthcare providers in the mature and growth markets alike are building the necessary infrastructure. Meanwhile, with sophisticated data sharing, processing, and mining techniques, scientists can easily collaborate and make better sense of what they see. In effect, two changes are taking place concurrently. Our technologies for collecting biological data are improving by many orders of magnitude. Our technologies for synthesizing and analyzing that data are also becoming much cheaper and more efficient. Together, these advances will help pharma break through some of the barriers that have previously held it back.

References

Batra, N., Betts, D., and Davis, S. (2019). Forces of change: the future of health, Deloitte Insights.

Gambrill, S. (2007). Venture philanthropy on the rise. *The CenterWatch Monthly* 14 (8): 6–14.

Hutchinson, L. and Kirk, R. (2011). High drug attrition rates – where are we going wrong? *Nature Reviews Clinical Oncology* 8: 189–190. https://doi.org/10.1038/nrclinonc.2011.34.

Institute of Medicine (US) Forum on Drug Discovery, Development, and Translation (2009). *Breakthrough Business Models: Drug Development for Rare and Neglected Diseases and Individualized Therapies: Workshop Summary*, 2. Washington (DC): National Academies Press (US) Current Model for Financing Drug Development: From Concept Through Approval. https://www.ncbi.nlm.nih.gov/books/NBK50972.

Milne, C.P. and Kaitin, K.I. (2010). Impact of the new US health-care-reform legislation on the pharmaceutical industry: who are the real winners? *Clinical Pharmacology and Therapeutics* 88 (5): 589–592. https://doi.org/10.1038/clpt.2010.167. PMID: 20959844; PMCID: PMC3017719.

Sawyer, K. (2007). *Group Genius: The Creative Power of Collaboration*. Basic Books.

Takebe, T., Imai, R., and Ono, S. (2018). The current status of drug discovery and development as originated in United States Academia: the influence of industrial and academic collaboration on drug discovery and development. *Clinical and Translational Science* 11 (6): 597–606. https://doi.org/10.1111/cts.12577. Epub 2018 Jul 30. PMID: 29940695; PMCID: PMC6226120.

Vogenberg, F.R., Isaacson Barash, C., and Pursel, M. (2010). Personalized medicine: part 1: evolution and development into theranostics. *Pharmacy and Therapeutics* 35 (10): 560–576. PMID: 21037908; PMCID: PMC2957753.

Wolinsky, H. (2017). Charities and the lure of capitalism: Philanthropies dedicated to finding cures for rare diseases explore new models for funding and cooperation to accelerate research and drug development. *EMBO Reports* 18 (4): 519–522. https://doi.org/10.15252/embr.201744065. Epub 2017 Mar 8. PMID: 28274952; PMCID: PMC5376960.

Wouters, O.J., McKee, M., and Luyten, J. (2020). Estimated research and development investment needed to bring a new medicine to market, 2009–2018. *JAMA* 323 (9): 844–853. https://doi.org/10.1001/jama.2020.1166.

Zurdo, J. (2013). Developability assessment as an early de-risking tool for biopharmaceutical development. *Pharmaceutical Bioprocessing* 1 (1): 29–50.

Chapter Self-Assessments: Check Your Knowledge

Questions:

- Explain what is meant by "the era of blockbuster drugs" and discuss why it is not sustainable?
- What does change fatigue refer to in the context of the current situation for the pharmaceutical industry?
- Explain the opportunity cost attached to R&D resources and investment and the concerns with the "bang for the buck" view of the current situation?
- How can public-private partnerships benefit the productivity of the pharmaceutical industry in the future?

Answers

- A blockbuster drug is an extremely popular drug that generates annual sales of at least $1 billion for the company that sells it. Examples of blockbuster drugs include Vioxx, Lipitor, and Zoloft. More than half of the revenue of major pharmaceutical companies and above one-third of the total pharmaceutical revenues came from the sales of these blockbuster drugs. In general, blockbuster drugs are used for common ailments, such as diabetes, cholesterol, high blood pressure, and cancer, many of which require extended periods of treatment without a cure. Because these drugs come with a patent, the pharmaceutical company is the only company allowed to sell it for a specified period of time, often many years. In essence, the pharma company has a monopoly on this drug and can charge any price. When the patent expires, many companies flood the market with generic versions of the drug at a significantly reduced price than the original, wiping away the monopoly and creating a competitive market. This significantly cuts into the original drug's sales, negatively impacting the pharma company that created it. Questions concerning the fate of these blockbuster drugs and the sustainability of the model are beginning to surface as they are approaching their patent expiration dates, and as they are expected to face significant competition from generic versions.

- After more than a decade of cost-cutting, restructuring, transformations, and turnarounds, the pharmaceutical industry would appear to be suffering from **change fatigue**. Change has been at the forefront of pharmaceutical executives' minds for a long time, and pressures have been rising steadily. Since 2007, even before the economic crisis of 2008–2009 began, companies had been announcing cost-reduction measures, largely driven by the patent cliff. Following the financial crisis, a massive acceleration of the challenges in the external environment also occurred. Some had and continue to implement extensive layoffs and site closures. By and large, these measures have had only limited success, and change fatigue refers to the lack of confidence in continuing to explore these measures to spur innovation.

- Pharmaceutical R&D has paid off well in the past, with studies showing a historical correlation between the number of new drugs introduced and declines in mortality and other health indicators across a wide range of diseases. Nonetheless, progress has been disappointing in some areas: No new broad-spectrum antibiotics have been marketed in almost forty years, and many forms of cancer as well as chronic diseases such as diabetes, Alzheimer's, Parkinson's, and schizophrenia still lack effective, well-tolerated treatments. Continuing growth in R&D spending represents investment in overcoming these challenges, but this upward trajectory will be sustainable only if it can be paid for.

- One area that offers much promise is the promotion of public–private partnerships aimed at enabling the academic sector to generate more innovative high-risk ideas and then also do much of the work to "derisk" them. In this context, when the industry finally does engage in the process, they don't have to take on as much risk, they have a good idea about the patient population, they know the biomarker, they know that may be a prototype drug is already available and showing promise and can develop a streamline development plan with a high probability of technical success.

Quiz:

1 True or false. With respect to trends, the aging of the population, together with dietary changes and more sedentary lifestyles, is driving up the disease burden in both developed and developing countries.

2 Complete the sentence with the best answer. Based on emerging technology, we can be reasonably certain that _____ – enabled by radically interoperable data, AI, and open, secure platforms—will drive much of this change. Choose the best answer:
   A global health
   B real-world evidence
   C precision medicine
   D digital transformation

3 In 2002 the US Food and Drug Administration (FDA) approved only seventeen new molecular entities (NMEs) for sale in the United States – a disappointing fraction of the fifteen-year high of fifty-six NMEs approved in 1996 and the lowest since 1983. For context, however, this decline occurred despite a _____ of research and development (R&D) spending by US-based pharmaceutical companies between 1995 and 2002. Choose the best answer:
   A tripling
   B doubling
   C quadrupling
   D reduction

4 Over time many foundations have been formed to support certain disease populations, some of which affect a great number of patients and others which support small numbers of patients that would fall in the category of rare diseases. Which **is not** an example of a foundation that has supported the pharmaceutical industry directed to a certain patient population?
   A Michael J Fox Foundation for Parkinson's Disease patients
   B Myasthenia Gravis Rare Disease Network (MGNet),
   C Spinal Muscular Atrophy (SMA) foundation
   D Kayne West Foundation (KWF)
   E American Society of Cataract & Refractive Surgery (ASCRS) Foundation

Glossary

A Abbreviated New Drug Application (ANDA): An application to the FDA containing data for the review and potential approval for a generic drug product.

Adverse Drug Reaction (ADR): An injury caused by taking medication. ADRs may occur following a single dose or prolonged administration of a drug or result from the combination of two or more drugs. An ADR is a special type of AE in which a causative relationship can be shown.

Adverse Effect (AE): An adverse effect is an undesired harmful effect resulting from a medication or other intervention such as surgery. An adverse effect may be termed a "side effect" when judged to be secondary to a main or therapeutic effect.

B Bioanalysis: A sub-discipline of analytical chemistry covering the quantitative measurement of xenobiotics (drugs and their metabolites, and biological molecules in unnatural locations or concentrations) and biotics (macromolecules, proteins, DNA, large molecule drugs, metabolites) in biological systems.

Biomarker: A biological molecule found in blood, other body fluids, or tissues that is a sign of a normal or abnormal process or of a condition or disease. A biomarker may be used to see how well the body responds to a treatment for a disease or condition. Also called molecular marker and signature molecule.

Biosimilar: A biosimilar (also known as follow-on biologic or subsequent entry biologic) is a biologic medical product that is almost an identical copy of an original product that is manufactured by a different company. Biosimilars are officially approved versions of original "innovator" products and can be manufactured when the original product's patent expires.

Fundamentals of Drug Development, First Edition. Edited by Jeffrey S. Barrett.
© 2022 John Wiley & Sons, Inc. Published 2022 by John Wiley & Sons, Inc.
Companion website: www.wiley.com/go/Barrett/FundamentalsDrugDevelopment

C Capital expenditure plan: A plan that includes the people and procedures a business relies on to evaluate long-term needs and assess long-term business requirements. The plan describes the long-term planning needs and business growth objectives, helping the business to prioritize and plan for capital asset purchases

Case report form (CRF): A printed, optical, or electronic document designed to collect the data that is described in the protocol for each trial subject.

Chemistry, Manufacturing and Controls (CMC): A functional group that supports the manufacture of a pharmaceutical or biologic product, all relevant specific manufacturing processes, product characteristics, and product testing to ensure that the product is safe, effective, and consistent between batches.

Chemistry, Manufacturing and Controls (CMC): The collection of pharmaceutical data to assure the quality of both the drug substance and drug product which are submitted for review by the FDA in an IND or NDA.

Clinical Operations: A functional group (team of individuals) that ensures proper planning, conduct, patient safety, and data quality while fostering good communication between study sites and sponsors.

Clinical Pharmacology: The study of drugs in humans. It is underpinned by the basic science of pharmacology, with added focus on the application of pharmacological principles and methods in the real world. In the laboratory setting, they study biomarkers, pharmacokinetics, drug metabolism, and genetics.

Code of Federal Regulations (CFR): The codification of the general and permanent rules and regulations (sometimes called administrative law) published in the Federal Register by the executive departments and agencies of the federal government of the United States.

Confidentiality Information Memorandum (CIM): A document drafted by a Mergers & acquisitions (M&A) advisory firm or investment banker used in a sell-side engagement to market a business to prospective buyers.

Contract Manufacturing Organization (CMO): A contract manufacturing organization (CMO), sometimes called contract development and manufacturing organization (CDMO), is a company that serves other companies in the pharmaceutical industry on a contract basis to provide comprehensive services from drug development through drug manufacturing.

Controlled Correspondence: A mechanism for generic firms to seek information from the FDA on either generic drug development or post-approval submission requirements.

D Drug monograph: A publication that specifies for a drug (or class of related drugs) the kinds and amounts of ingredients it may contain, the conditions and limitations for which it may be offered, directions for use, warnings, and other information that its labeling must contain.

DSMB (Drug Safety Monitoring Board): An independent group of experts who monitor patient safety and treatment efficacy data while a clinical trial is ongoing.

E Effectiveness: The extent to which a drug achieves its intended effect in the usual clinical setting

Efficacy: The maximum response achievable from a pharmaceutical drug in research settings and to the capacity for sufficient therapeutic effect or beneficial change in clinical settings.

European Medicines Agency (EMA): An agency of the European Union (EU) in charge of the evaluation and supervision of medicinal products. Prior to 2004, it was known as the European Agency for the Evaluation of Medicinal Products or European Medicines Evaluation Agency (EMEA).

Excipients: An inactive substance that serves as the vehicle or medium for a drug or other active substance as part of a drug product formulation. Common excipients include things like coloring agents, preservatives, binders, and fillers.

Expiry date: The length of time for which an item remains usable, fit for consumption, or saleable. In the pharmaceutical context, shelf life is the period of time, from the date of manufacture, that a drug product is expected to remain within its approved product specification while stored under defined conditions. The expiry date, also known as the shelf life, is typically expressed in units of months, i.e. 24 months, 36 months, to a maximum of 60 months.

F FDA: An agency within the US Department of Health responsible for protecting public health by assuring the safety, effectiveness, quality, and security of human and veterinary drugs, vaccines, and other biological products, and medical devices. The FDA is also responsible for the safety and security of the food supply, cosmetics, dietary supplements, tobacco, and products that give off radiation.

Food effect trial: Bioavailability study conducted for new drugs and drug products during the IND period to assess the effects of food on the rate and extent of absorption of a drug when the drug product is administered shortly after a meal (fed conditions), as compared to administration under fasting conditions.

G Generic Drug User Fee Act (GDUFA): A law designed to speed access for safe and effective generic drugs to the US healthcare system. GDUFA enables FDA to assess industry user fees to bring greater predictability and timeliness to the review of generic drug applications.

Generic: A generic drug product is one that is comparable to an innovator drug product in dosage form, strength, route of administration, quality, performance characteristics, and intended use.

GLP: Specifically refers to a quality system of management controls for research laboratories and organizations to try to ensure the uniformity, consistency, reliability, reproducibility, quality, and integrity of chemical (including pharmaceuticals) nonclinical tests.

GMP: A system for ensuring that products are consistently produced, tested, and adequately controlled according to quality standards appropriate to their intended use.

GXP: A general abbreviation for the "good practice" quality guidelines and regulations. The "x" stands for the various fields, including the pharmaceutical and food industries, for example, good agricultural practice or GAP. GxP guidelines were established in the United States by the Food and Drug Administration (FDA). They aim to ensure that businesses working in regulated industries manufacture products that are safe and fit for use, meeting strict quality standards throughout the entire process of production.

H HEOR, health economics and outcomes research: The field of study that examines the need to select therapeutic "interventions" from multiple treatment options, including biopharmaceuticals, medical devices, and healthcare services. As the benefits and costs of these interventions can range dramatically, and the benefits can be economic, clinical, both, or HEOR may include hard to measure costs or benefits the patient experiences directly. HEOR can help healthcare decision-makers – including clinicians, governments, payers, health ministries, patients, and more – to adequately compare and choose among the available options.

HHS, Health and Human Service: The US Government's principal agency for protecting the health of all Americans and providing essential human services, especially for those who are least able to help themselves. HHS accomplishes its mission through programs and initiatives that cover a wide spectrum of activities, serving and protecting Americans at every stage of life, from conception.

I Informed consent: Permission granted in the knowledge of the possible consequences, typically that which is given by a patient to a doctor for treatment with full knowledge of the possible risks and benefits. From a drug development perspective, informed consent refers to the process by which a patient learns about and understands the purpose, benefits, and potential risks of medical or surgical intervention, including clinical trials, and then agrees to receive the treatment or participate in the trial.

Intellectual Property (IP): A work or invention that is the result of creativity, such as a manuscript or a design, to which one has rights and for which one may apply for a patent, copyright, trademark, etc.

International Council on Harmonization (ICH): An initiative that brings together regulatory authorities and pharmaceutical industry to discuss scientific and technical aspects of pharmaceutical product development and registration. The mission of the ICH is to promote public health by achieving greater harmonization through the development of technical guidelines and requirements for pharmaceutical product registration.

Investigational New Drug (IND): An application to the US FDA by which a pharmaceutical company or investigator obtains clearance to conduct human clinical trials on an investigational new drug. Regulations are described primarily in 21 CFR § 312. Similar procedures are followed in the European Union, Japan, and Canada.

Investigators Brochure (IB): A comprehensive document that summarizes the body of information about an investigational product or study drug authored by the study sponsor and provided to clinical sites and investigators prior to study conduct. The IB is typically included in the IND with the study protocol and provided to regulatory authorities for review as well.

J Juvenile: A child or young person who is not yet old enough to be regarded as an adult. In a drug development context, a juvenile is someone less than 18 years of age. In drug regulation, pediatric patients are defined as children younger than age 17 with specific age ranges for the categories of neonate, infant, child, and adolescent.

K KPD Model: Kinetic-pharmacodynamic (KPD) models are used to predict the time course and magnitude of drug effects in the absence of pharmacokinetic (PK) data. In this approach, a virtual compartment representing the biophase in which the concentration is in equilibrium with the observed effect is used to extract the PK component from the pharmacodynamic data alone.

L Life cycle management (LCM): The process of management of a good as it moves through the typical stages of its product life: development and introduction, growth, maturity/stability, and decline. In the pharmaceutical development context, it also refers to the strategies employed to maintain and extend a product's life to maximize the time over which the product generates adequate income to justify manufacturing costs and recoup R&D investments.

M Metabolite: A metabolite is the intermediate end-product of metabolism. The term metabolite is usually restricted to small molecules. Metabolites from chemical compounds, whether inherent or pharmaceutical, are formed as part of the natural biochemical process of degrading and eliminating the compounds.

N New Drug Application (NDA): The NDA application is the mechanism through which drug sponsors formally propose that the FDA approve a new pharmaceutical for sale and marketing in the U.S. The data gathered during the animal studies and human clinical trials of an Investigational New Drug (IND) become part of the NDA.

Non-disclosure agreement (NDA): A legally binding contract that establishes a confidential relationship. The party or parties signing the agreement agree that sensitive information they may obtain will not be made available to any others.

O Off-target effects: On-target refers to exaggerated and adverse pharmacologic effects at the target of interest in the test system. Off-target refers to adverse effects as a result of modulation of other targets; these may be related biologically or totally unrelated to the target of interest.

P Patent Cliff: The term refers to the phenomenon of patent expiration dates and an abrupt drop in sales that follows for a group of products capturing a high percentage of a market.

Patent: A government authority or license conferring a right or title for a set period, especially the sole right to exclude others from making, using, or selling an invention. A patent is a form of intellectual property.

Pharmacodynamics (PD): the branch of pharmacology concerned with the effects of drugs and the mechanism of their action. It includes the study of the biochemical, physiologic, and molecular effects of drugs on the body and involves receptor binding (including receptor sensitivity), post-receptor effects, and chemical interactions.

Pharmacokinetics (PK): The branch of pharmacology concerned with the movement of drugs within the body. It includes the study and quantification of the bodily absorption, distribution, metabolism, and excretion of drugs.

Pharmacometrics: The branch of science concerned with mathematical models of biology, pharmacology, disease, and physiology used to describe and quantify interactions between xenobiotics and patients, including beneficial effects and side effects resultant from such interfaces.

Phase 1: The phase of development that represents the initial introduction of an experimental drug or therapy to humans. This phase is the first step in the clinical research process involved in testing new or experimental drugs.

Phase 2: The second phase of clinical trials or studies for an experimental new drug, in which the focus of the drug is on its effectiveness. Phase 2 trials typically involve hundreds of patients who have the disease or condition that the drug candidate seeks to treat.

Phase 3: The phase of development conducted to confirm and expand on safety and effectiveness results from Phase 1 and 2 trials, to compare the drug to standard therapies for the disease or condition being studied, and to evaluate the overall risks and benefits of the drug. Regulatory requirements typically insist that two well-controlled, Phase 3 trials concluding pre-established (and agreed upon) metrics for safety and efficacy be conducted prior to submission of an NDA.

Phase 4: Testing in humans that occurs after a drug (or other treatment) has already been approved by the Food and Drug Administration (FDA) and is being marketed for sale. Occasionally, the FDA (or equivalent regulatory authority) approves a drug for general use but requires the manufacturer to continue to monitor its effects; during this phase, the drug may be tried on slightly different patient populations than those studied in earlier trials.

PK/PD: A technique that combines the two classical pharmacologic disciplines of pharmacokinetics and pharmacodynamics. Most often, PK/PD refers to modeling such data.

PMDA: The Pharmaceuticals and Medical Devices Agency (PMDA) is the government organization in Japan in charge of reviewing drugs and medical devices, overseeing post-market safety, and providing relief for adverse health effects. The organization is the counterpart to the US-based Food and Drug Administration (FDA).

Population PK/PD: The study of variability in drug concentrations and response within a patient population receiving clinically relevant doses of a drug of interest. The methodologic approach behind population PK/PD is often nonlinear, mixed-effect modeling.

Prescribing Information: A document that summarizes the safe and effective use of a drug. The label is intended to be both informative and accurate, non-promotional, false, or misleading, and without implied claims or suggestions for use if evidence of safety or effectiveness is lacking. It is based whenever possible on data derived from human experience. The details of the product label contents and requirements are codified by the US Government in the Code of Federal Regulations under "General Requirements for Prescription Drug Labeling," (21 CFR 201.56). The Prescribing information is also called a package insert, professional labeling, direction circular, package circular and is often referred to as "labeling" or "the label."

Proof-of-concept (POC): Evidence, typically derived from an experiment or pilot project, which demonstrates that a design concept, business proposal, etc., is feasible. In the pharmaceutical context, Proof-of-concept trials provide initial evidence for target use in a specific population, the most appropriate dosing strategy, and duration of treatment. A significant goal in designing an informative and efficient POC study is to ensure that the study is safe and sufficiently sensitive to detect a preliminary efficacy signal (ie, a potentially valuable therapy). Proof-of-concept studies help avoid resources wasted on targets/molecules that are not likely to succeed.

Proof-of-Mechanism (POM): Refers to Early Clinical Drug Development in Phase 1, often performed in healthy volunteers but also patients. Studies attempt to demonstrate adequate drug exposure at the target site of action, showing that the drug interacts with the intended molecular receptor or enzyme and also showing that the drug affects cell biology in the desired manner and direction.

Proof-of-Principal (POP): Proof of Principle (POP) is a realization of a particular process to prove its feasibility. It is a method of supplying evidence to support whether a product or service has the potential to be successful in certain applications. In the pharmaceutical context, Proof of Principle studies are an early stage of clinical drug development when a compound has shown potential in animal models and early safety testing. This step often links between Phase-1 and dose-ranging Phase-2 studies.

Q Quality Assurance (QA): The maintenance of a desired level of quality in a service or product, especially by means of attention to every stage of the process of delivery or production.

Quality of Life (QOL): QOL is defined as the degree to which an individual is healthy, comfortable, and able to participate in or enjoy life events. Within the arena of healthcare, quality of life is viewed as multidimensional, encompassing emotional, physical, material, and social well-being. Quality of life is an important consideration in medical care. Some medical treatments can seriously impair quality of life without providing appreciable benefit, whereas others greatly enhance the quality of life.

R Real World Data (RWD): Data derived from a number of sources that are associated with outcomes in a heterogeneous patient population in real-world settings, such as patient surveys, clinical trials, and observational cohort studies. Real-world data refer to observational data as opposed to data gathered in an experimental setting such as a randomized controlled trial (RCT). They are derived from electronic health records (EHRs), claims and billing activities, product and disease registries, etc.

Real World Evidence (RWE): The clinical evidence regarding the usage and potential benefits or risks of a medical product derived from analysis of RWD. RWE can be generated by different study designs or analyses, including but not limited to randomized trials, including large simple trials, pragmatic trials, and observational studies (prospective and/or retrospective).

Return on Investment (ROI): A performance measure used to evaluate the efficiency of an investment or compare the efficiency of several different investments. ROI tries to directly measure the amount of return on a particular investment, relative to the investment's cost. To calculate ROI, the benefit (or return) of an investment is divided by the cost of the investment. The result is expressed as a percentage or a ratio.

S SAP: Refers to Systems, Applications, and Products in Data Processing; is a German multinational software corporation that makes enterprise software to manage business operations and customer relations.

Source: A place, person, or thing from which something comes or can be obtained. In the pharmaceutical context, source typically refers to the place or organization specifically where, raw materials or supplies of any kind are purchased for incorporation into various aspects of drug development including the product itself.

Special populations: Groups of people with needs that require special consideration and attention in a drug development setting. Patients can belong to more than one special population, but the distinguishing feature is that they represent a departure from mainstream patients.

Stability: Pharmaceutical product stability may be defined as the capability of a particular formulation to remain within its physical, chemical, microbiological, therapeutic, and toxicological specifications while in a specific container closure system. There are many factors that can affect the stability of a pharmaceutical product. These include the stability of the active drug(s), interactions between active and inactive ingredients, the dosage form, manufacturing process, the container system, and the environment for shipping, handling, and storage.

Standard of care (SOC): Also referred to as best practice, SOC is viewed as a guideline for the appropriate treatment of a condition, as established by formal or informal consensus among experts on that condition. Basically, the standard of care for the treatment of a disease is whatever most physicians agree is the best way to treat that disease.

Statistical Analysis Plan (SAP): A document containing a detailed elaboration of the principal features of the analysis described in a clinical trial protocol, and which includes procedures for statistical analysis of the primary and secondary variables and other data.

Sunshine Act: In the pharmaceutical context, this refers to the Physician Payments Sunshine Act, not to be confused with Government in the Sunshine Act. The Sunshine Act requires manufacturers of drugs, medical devices, biological and medical supplies covered by the three federal healthcare programs Medicare, Medicaid, and State Children's Health Insurance Program (SCHIP) to collect and track all financial relationships with physicians and teaching hospitals and to report these data to the Centers for Medicare and Medicaid Services (CMS). The goal of the law is to increase the transparency of financial relationships between healthcare providers and pharmaceutical manufacturers and to uncover potential conflicts of interest.

Supply Chain: Supply chain is a set of players, processes, information, and resources that transfers raw materials and components to finished products or services and delivers them to the customers. It includes suppliers, intermediaries, third-party service providers, and customers. It also includes all of the logistics activities, manufacturing operations and activities with and across marketing, sales, product design, finance, and information technology.

Surrogate Marker: A surrogate marker or endpoint has been defined as "a biomarker intended to substitute for a clinical endpoint," the latter being "a characteristic or variable that reflects how a patient feels, functions, or survives."

T Test Product: Typically refers to the new drug formulation developed in comparison to a reference product that represents the current standard or comparator in the context of a bioequivalence trial.

V Vaccine: a suspension of attenuated or killed microorganisms (viruses, bacteria, or rickettsia), administered for prevention, amelioration, or treatment of infectious diseases.

Vector: In medicine, a carrier of disease or of medication. For example, in malaria, a mosquito is the vector that carries and transfers the infectious agent. In molecular biology, a vector may be a virus or a plasmid that carries a piece of foreign DNA to a host cell.

Virus: a disease-causing agent that is too tiny to be seen by the ordinary microscope, that may be a living organism or maybe a special kind of protein molecule, and that can only multiply when inside the cell of an organism.

W WHO: A part of the United Nations that deals with major health issues around the world. The World Health Organization (WHO) sets standards for disease control, healthcare, and medicines; conducts education and research programs; and publishes scientific papers and reports.

World Bank: The World Bank is an international organization dedicated to providing financing, advice, and research to developing nations to aid their economic advancement. The bank predominantly acts as an organization that attempts to fight poverty by offering developmental assistance to middle- and low-income countries.

Self-Assessment Quiz Answers

Chapter 1:

1) False
2) a
3) a
4) d

Chapter 2:

1) b
2) True
3) f
4) a

Chapter 3:

1) a
2) True
3) c
4) d

Chapter 4:

1) a
2) False
3) True
4) c

Chapter 5:

1) b
2) True

Fundamentals of Drug Development, First Edition. Edited by Jeffrey S. Barrett.
© 2022 John Wiley & Sons, Inc. Published 2022 by John Wiley & Sons, Inc.
Companion website: www.wiley.com/go/Barrett/FundamentalsDrugDevelopment

3) c
4) d

Chapter 6:

1) False, 1.46 million
2) b
3) f
4) b

Chapter 7:

1) False, not true for oncology and rare diseases
2) a
3) True
4) e

Chapter 8:

1) d
2) d
3) a
4) True

Chapter 9:

1) d
2) True
3) c
4) b

Chapter 10:

1) c
2) False. Age requirements during Phase 1 are typically bounded between 18 years at the lower end (hence excluding pediatrics) with an upper bound often at 55–65 years. Also, women who could possibly become pregnant are typically excluded, and BMI is typically capped at 25–30 kg/m^2 (hence obese subjects are excluded). This is not the case for Phase 3 populations which must reflect the demographics of the intended use of the drug.
3) a
4) d

Chapter 11:

1) False
2) d

3) f
4) a
5) True

Chapter 12:

1) True
2) a
3) d
4) a

Chapter 13:

1) b
2) a
3) True
4) False

Chapter 14:

1) a
2) b
3) e
4) False – Phase 1 trials, not Phase 3

Chapter 15:

1) True
2) d
3) d
4) a

Chapter 16:

1) c
2) True
3) d
4) e

Chapter 17:

1) b
2) a
3) c
4) a

Chapter 18:

1) b
2) a
3) f
4) c

Chapter 19:

1) False
2) False
3) d
4) a

Chapter 20:

1) True
2) d
3) True
4) a

Chapter 21:

1) True
2) c
3) b
4) a

Chapter 22:

1) True
2) Pharmaceutically and Therapeutically Equivalent
3) b
4) c
5) False

Chapter 23:

1) d
2) b
3) d
4) e
5) c

Chapter 24:

1) False
2) d
3) a
4) True

Chapter 25:

1) True
2) d
3) b
4) d

Index

Fundamentals of Drug Development, First Edition. Edited by Jeffrey S. Barrett.
© 2022 John Wiley & Sons, Inc. Published 2022 by John Wiley & Sons, Inc.
Companion website: www.wiley.com/go/Barrett/FundamentalsDrugDevelopment

Printed and bound by CPI Group (UK) Ltd, Croydon, CR0 4YY

16/04/2025

14658415-0003